# Advancements of Forward Osmosis in Food Processing Industries

This book provides a comprehensive exploration of forward osmosis (FO) technology and its transformative impact on the food processing industry including fundamental principles, its current applications, and the emerging trends that promise to revolutionize food processing.

With detailed case studies, insights from industry experts, and a forward-looking perspective, it guides readers on how to leverage FO technology to improve efficiency and innovation in food processing, leading to better-quality products and reduced operational costs.

**Key features:**

- Highlights the potential applications of FO and the fundamentals of FO modelling to make them industrially viable for large-scale applications
- Provides new insight into innovative technologies for product development and process improvement
- Discusses detailed cost–benefit analysis (CBA) of FO process, along with comparative analysis with competing technologies
- Explains the investment scenarios, return on investment (ROI), global market trends, regulatory considerations, and safety standards
- Includes real-world examples and case studies for practical integration of the FO process in liquid food and beverage processing industries

This book is aimed at industrial professionals, academicians, and graduate students in food processing, membrane technology, and design/optimization.

# Advancements of Forward Osmosis in Food Processing Industries

Senthilmurugan Subbiah, Ananya Khasnabis
and Sharad K. Gupta

CRC Press
Taylor & Francis Group
Boca Raton London New York

CRC Press is an imprint of the
Taylor & Francis Group, an **informa** business

Designed cover image: Senthilmurugan Subbiah, Ananya Khasnabis and Sharad K. Gupta

First edition published 2026
by CRC Press
2385 NW Executive Center Drive, Suite 320, Boca Raton FL 33431

and by CRC Press
4 Park Square, Milton Park, Abingdon, Oxon, OX14 4RN

*CRC Press is an imprint of Taylor & Francis Group, LLC*



*Library of Congress Cataloging-in-Publication Data*
[Insert LoC Data here when available]

ISBN: 978-1-041-10526-8 (hbk)
ISBN: 978-1-041-11015-6 (pbk)
ISBN: 978-1-003-65787-3 (ebk)

DOI: 10.1201/9781003657873

Typeset in Times
by Apex CoVantage, LLC

# Contents

# About the Authors

**Senthilmurugan Subbiah** is a distinguished professor in the Department of Chemical Engineering at the Indian Institute of Technology Guwahati. He is highly regarded for his expertise in process systems engineering, with research interests spanning membrane technology, process modelling, simulation, optimization, and control. Prof. Subbiah's work is noted for its focus on the development of sustainable and innovative solutions for complex chemical processes, particularly in the areas of water treatment, desalination, and greywater management. Throughout his career, Prof. Subbiah has published extensively in leading international journals and presented his findings at numerous conferences worldwide. His research emphasizes both theoretical and practical aspects, bridging the gap between academic innovation and industrial application. He has successfully led several funded research projects and has collaborated with industry partners to advance the implementation of novel process technologies. In addition to his research contributions, Prof. Subbiah is dedicated to teaching and mentoring, playing a pivotal role in guiding graduate and undergraduate students in their academic and research pursuits. His commitment to education and research excellence has earned him recognition and respect in the academic community. Prof. Subbiah continues to contribute to the field of chemical engineering, driving advancements in process systems and sustainable engineering practices.

**Ananya Khasnabis** is a dedicated researcher and engineer specializing in membrane technology, greywater treatment, and desalination. With a strong academic foundation and practical experience in chemical and food technology, Dr. Khasnabis has made significant contributions to the development of sustainable and innovative solutions in the fields of food processing and wastewater treatment. Her research interests have expanded to include advanced membrane processes, such as forward osmosis, with a particular focus on their applications in food processing. With her research work, she continues to explore innovative applications of forward osmosis in food processing, aiming to contribute to the advancement of this field with practical, scalable solutions.

**Sharad K. Gupta** is a distinguished faculty member at the Indian Institute of Technology Delhi. He is renowned for his expertise in Chemical Engineering, particularly in the areas of Process Systems Engineering, Process Optimization, and Control. Prof. Gupta has an extensive academic background with significant contributions to research and development in process design and simulation, optimization techniques, and advanced process control strategies. He has published numerous papers in high-impact journals, collaborated on various research projects, and mentored several Ph.D. and master's students. Prof. Gupta is also actively involved in industry collaborations, aiming to bridge the gap between academic research and practical industrial applications. His work is highly regarded in the field, making him a respected figure in chemical engineering.

# Preface

Amid growing demand for sustainable, energy-efficient, and high-quality food processing technologies, the Forward Osmosis (FO) has emerged as a promising membrane-based separation process, offering unique advantages such as low energy consumption and preservation of product quality, particularly in the concentration of thermosensitive food and dairy products.

This book, *Advancements of Forward Osmosis in Food Processing Industries*, offers a comprehensive and interdisciplinary examination of FO technology. It captures FO's scientific evolution, technical innovations, and industrial applications, highlighting its relevance to modern food engineering challenges. Highlighting extensive research and practical insights, this volume aims to serve as both an academic reference and a practical guide for scientists, engineers, policymakers, and industry practitioners.

One of the distinctive features of this book is its holistic overview of mathematical modelling in FO processes. Mathematical models are crucial for understanding membrane transport dynamics, predicting performance, and optimizing process configurations. Accurately simulating key phenomena, such as water flux, solute diffusion, fouling, and concentration polarization (CP), enables the rational design and scale-up of FO systems. This also plays a pivotal role in improving FO technologies' economic feasibility and commercial viability.

This book systematically explores FO technology, emphasizing its applications in food processing, technological advancement, practical implementations, and future prospects. The content structure includes:

Chapter 2 "Forward Osmosis Theory": In this chapter, the theoretical aspects of the FO processes are thoroughly examined, and essential components of the FO systems such as draw solution (DS), membrane characteristics, and membrane modules are introduced. Additionally, operational parameters, influencing factors, and strategies to overcome common challenges such as CP and membrane fouling are discussed.

Chapter 3 "Forward Osmosis System Modelling, Design, and Optimization": This chapter introduces the modelling of the FO process, providing a comprehensive guide to developing and validating mathematical models. It details the fundamental equations governing mass and solute transfer, parametric studies, and advanced optimization strategies. Practical guidelines for designing FO systems specifically tailored for liquid food concentration applications are also presented.

Chapter 4 "Application of Forward Osmosis in Liquid Food and Dairy Processing Industries": In this chapter, diverse applications of FO in food and dairy industries are explored in detail. This chapter also covers the emerging applications such as FO-assisted crystallization.

Chapter 5 "Economic Analysis and Market Trend": This chapter provides in-depth economic analysis comparing FO technology with other competitive technologies like reverse osmosis (RO) and evaporation. It covers CBAs, investment scenarios, and return-on-investment calculations and discusses global market trends, regulatory frameworks, and safety standards relevant to FO.

Chapter 6 "Emerging Trends and Future Prospects in Forward Osmosis for Liquid Food Processing": This chapter addresses recent innovations and future prospects of FO, including advancement in membrane technology, integration with other technologies, and the use of artificial intelligence (AI). Potential applications, market trends, life-cycle assessments, and regulatory considerations are also analysed.

Chapter 7 "Conclusion": The concluding chapter synthesizes key findings, highlights major advancements, and provides insights into the future trajectory of FO technologies in food processing. It summarizes the overall benefits, remaining challenges, and strategic directions for future research and commercial adoption.

Ultimately, this book aspires to advance the adoption of FO on a commercial scale in the food and beverage processing industries. By aligning rigorous scientific principles with practical insights and real-world case studies, it underscores the potential of FO not just as an emerging technology but as a scalable, economic, and environmentally responsible solution.

# Acknowledgement

As authors, we collectively extend our heartfelt gratitude to all those who, in various ways, contributed to the completion of this book. The journey of writing has been enriched by countless sources of knowledge, inspiration, and encouragement that guided our efforts throughout.

We are thankful for the collective wisdom, resources, and ideas that have shaped this work. Every reference, discussion, and influence has played a significant role in bringing clarity and depth to the thoughts presented here.

Above all, we acknowledge the readers, whose curiosity and engagement are the true motivation behind this book.

# 1 Introduction to Forward Osmosis Process

## 1.1 BACKGROUND

Concentration processes are crucial in extending shelf life, reducing transportation costs, and improving product stability in the global food processing industry. Whether in liquid food or dairy, removing water facilitates better packaging and logistics and enables manufacturers to adjust product composition for specific end-use applications.

Despite its industrial significance, concentration remains one of the most energy-intensive stages in food processing. Thermal-based techniques dominate commercial concentration systems, yet as consumer preferences shift towards minimally processed foods and environmental regulations become more stringent, there is growing pressure on the industry to adopt alternatives that are both sustainable and capable of preserving the delicate flavours and nutritional qualities of foods.

For years, traditional methods, particularly thermal evaporation (including atmospheric and vacuum evaporation), have been vital in enhancing shelf life and optimizing logistics. These methods work by removing water through heat application, thus increasing the total solids content of a product. Atmospheric evaporation typically occurs in open pans or tubular evaporators, where the feed is heated until water vaporizes. Although effective, this method is energy-intensive and often leads to thermal degradation of heat-sensitive nutrients, flavours, and colours.

To address these challenges, vacuum evaporation offers an improved alternative. In this process, evaporation occurs under reduced pressure, which lowers the boiling point of water, enabling concentration at lower temperatures. This approach saves energy and helps preserve the organoleptic and nutritional qualities of food. However, conventional evaporation techniques still face difficulties such as high energy consumption, scaling, and fouling of equipment, particularly with viscous or thermally sensitive fluids.

While thermal systems remain prevalent, membrane-based concentration methods like RO have been introduced as partial alternatives, although they are not traditionally classified as conventional methods. Nonetheless, evaporation and vacuum evaporation continue to dominate the industry due to their high throughput, technological maturity, and versatility for a wide range of liquid foods, including fruit juices, dairy concentrates, and sugar syrups.

An exciting development in food processing is FO, an emerging separation process gaining significant attention for its low energy consumption and gentle operating conditions, making it ideal for sensitive materials such as food and dairy products. Unlike pressure-driven processes like RO, FO utilizes the natural osmotic pressure difference between a feed solution (FS) and a DS to drive water transport through a

DOI: 10.1201/9781003657873-1

semi-permeable membrane. This occurs without requiring high hydraulic pressure, which helps preserve the integrity of heat-sensitive compounds and minimizes thermal degradation risks.

Before delving into its application for food and dairy concentration, understanding the FO process's fundamentals is important. FO is driven by the osmotic gradient created between the FS (e.g. fruit or vegetable juice) and the DS. Water moves from the low osmotic pressure region (FS) to the high osmotic pressure region (DS), passing through the semi-permeable membrane and leaving behind concentrated solutes in the feed. This makes FO a highly attractive method for concentrating liquids while avoiding the extreme heat conditions associated with traditional thermal processes that often compromise product quality.

The energy efficiency and low-pressure operation of the FO process make it particularly suited for the food and dairy industries, where preserving high product quality is paramount. FO offers a major advantage over traditional methods like evaporation for concentrating food and dairy products because it operates at lower temperatures. This helps retain the product's nutritional value, flavours, and colours, minimizing heat damage—especially for heat-sensitive items like juices, milk, and yoghurt.

The mathematical principles of FO, including water flux and reverse solute flux (RSF) equations, can be adapted for designing and optimizing these processes for food applications. By controlling factors like the osmotic pressure difference and membrane properties (e.g. permeability and selectivity), desired concentrations can be achieved without compromising product quality. Additionally, FO can be integrated with other techniques, such as membrane distillation (MD) or RO, to further enhance its efficiency in large-scale food and dairy processing.

Adopting FO for food and dairy applications requires a solid understanding of these core principles and the ability to adjust operational parameters, such as DS composition and flow configurations, to meet specific concentration goals. As FO technology advances, its integration into food and dairy processing represents a promising avenue for achieving sustainable, high-quality production methods.

## 1.2 FORWARD OSMOSIS—OVERVIEW AND HISTORICAL EVOLUTION

Osmosis is a natural phenomenon that occurs when solvent molecules move from an area of lower solute concentration to an area of higher solute concentration through a semi-permeable membrane. This process is driven by the difference in osmotic pressure between the two solutions. Osmosis plays a vital role in numerous biological and industrial processes, particularly in the fields of water purification and food preservation.

For instance, in the pickling process, osmosis occurs because of the difference in solute concentration between the food (less solute concentration) and the brine or vinegar (more solute concentration). This imbalance causes water to leave the food and allows the brine or vinegar to enter. As water exits, the food's cells shrink, reducing water activity and making it harder for microbes to survive. At the same time, the brine infuses its flavour into the food. Overall, osmosis not only preserves the food but also enhances its taste and texture by replacing water with a flavourful, protective solution.

FO is an osmotically driven membrane separation process that leverages the natural osmotic pressure difference between two solutions to induce water transport across a semi-permeable membrane. Although FO is based on the principle of osmosis, the term "FO process" has been coined to distinguish it from the RO process. Both RO and FO processes utilize semi-permeable membranes to separate water from aqueous solutions effectively, although their driving forces are different. While the driving force in the RO process is created by hydraulic pressure, concentration or osmotic pressure gradient is the driving force for the FO process. In the FO process, two solutions of different osmotic pressures are separated by a semi-permeable membrane. The membrane comprises two main components: a support layer (SL) and a separation layer. The SL confers mechanical strength, while the separation layer ensures that only the solvent can pass through. In contrast to pressure-driven processes, FO does not rely on external hydraulic pressure. In the FO process, due to the osmotic pressure difference ($\Delta\pi$), the solvent diffuses from the low-concentration FS to the high-concentration DS through a semi-permeable membrane. The movement of solvent results in simultaneous concentration and dilution of FS and DS, respectively (Figure 1.1).

In the eighteenth century, osmosis was primarily an experimental curiosity in natural philosophy and was first described by Jean-Antoine Nollet, but its engineered use—particularly in food and water treatment—is relatively recent. In the 1960s, the critical development occurred with Loeb and Sourirajan's pioneering work on asymmetric cellulose acetate (CA) membranes. In the initial stage of FO research, RO membranes were predominantly employed as the primary membrane materials. However, the thickness of their SLs induced severe CP, leading to a significant decline in the separation efficiency of the FO process. By the 1990s, cellulose triacetate (CTA) membranes were developed explicitly for FO applications.

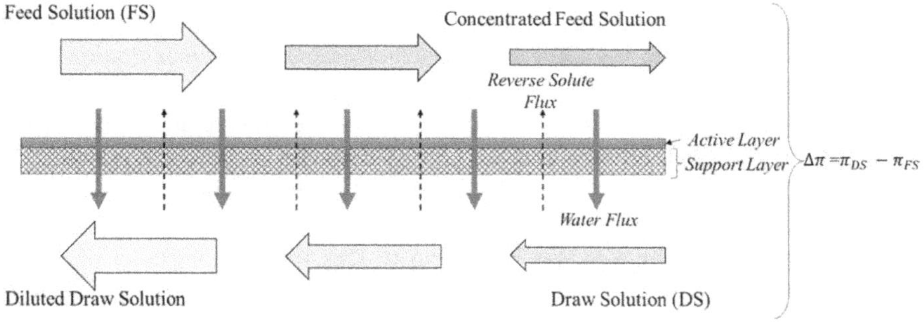


**FIGURE 1.1** Schematic representation of forward osmosis process with membrane active layer facing feed solution (AL-FS) (Note: osmotic pressure gradient, $\Delta\pi = \pi_{DS} - \pi_{FS}$, where $\pi_{DS}$ and $\pi_{FS}$ represent osmotic pressure of draw and feed solutions, respectively)

Nevertheless, these membranes exhibited limitations such as low mechanical strength and insufficient porosity, compromising their overall performance. Later, in the late 1990s and early 2000s, a renewed scientific interest arose due to advancements in membrane development, particularly thin-film composite (TFC) membranes with enhanced permeability, mechanical strength, and selectivity. Subsequent advancements introduced various FO membranes, such as aquaporin (AQP)-based biomimetic membranes, unsupported FO membranes, and self-assembled membranes (1). Despite their innovative designs, these membranes still faced challenges such as poor mechanical strength, complex fabrication processes, and high costs. A significant breakthrough occurred in the 2010s, when the TFC membrane was created to enhance FO performance. The fabrication processes involved interfacial polymerization (IP) to form a polyamide (PA) active layer (AL) on top of an SL via a cross-linking process between the aqueous phase of m-phenylenediamine (MPD) and the organic phase of trimesoyl chloride (TMC)/*n*-hexane. This process demonstrated a superior water flux. Upon this innovation, Oasys Water developed the first spiral-wound TFC-PA-based FO membrane element. To further optimize FO membranes, particularly in terms of enhancing water flux, improving selectivity, and mitigating ICP, current research efforts were predominantly focused on the modification of TFC-FO membranes.

The terminology "Forward Osmosis" was introduced in 1976 by Kessler and Moody for the extraction of potable water from saline water (2). However, it only became common after 2006. Prior to that, researchers used to describe the process as "Direct Osmosis" or "Direct Osmosis Concentration" (3). Interestingly, the application of direct osmosis (or FO) for the production of concentrated foodstuffs is not a recent invention. A group of Eastern European farmers used this technique to prepare a range of juice concentrates by simply immersing a bag made of semi-permeable material into a concentrated sodium chloride solution. Due to the osmotic pressure gradient, water was forced to move from the juice to the osmotic brine. Utilizing the technique, a highly concentrated juice was obtained, but the process was very slow. Based on the traditional method, the pioneering work for the concentration of fruit (grape) juice by osmotic concentration was reported by Popper et al. (1966) (4) using a first-generation RO membrane (CA) in both flat and tubular configurations. Initially referred to as "dialysis" by Popper et al. (1966), Beaudry and Lampi (1990) renamed the process as "direct osmosis". The authors demonstrated juice concentration using a very thin (25–100μm) semi-permeable membrane.

Direct osmosis operates under low-pressure and low-temperature conditions, making it a viable alternative to conventional evaporation for the concentration of heat-sensitive products such as raspberry juice (5). Using a tubular membrane module, improved performance in the concentration of tomato juice was observed when employing thinner membranes and low-viscosity DSs like sodium chloride, compared to more viscous media such as glucose or sucrose solutions (6). In the early 2000s, the term "direct osmosis" was renamed to "forward osmosis" to enhance the clarity of the process in scientific terms and standardize terminology across the related technologies.

Over time, numerous researchers emphasized the advantages of the FO process over traditional thermal and pressure-driven membrane technologies (7). In the

2000s, breakthroughs in membrane technology, such as improvements in material durability and efficiency, contributed to the more effective FO process using commercial FO membranes (8,9). By the late 2010s, hybrid membrane systems were explored for efficient draw solute regeneration, contributing to sustainable practices, with a commercially available biomimetic membrane with low structural material and superior membrane performance. Several studies demonstrated the feasibility of concentrated liquid foods using commercially available AQP inside a hollow-fibre (HF) membrane. Around the 2020s, researchers also started exploring the feasibility of utilizing food preservatives (such as potassium sorbate, sodium metabisulphite, sodium sulphite, sodium benzoate, calcium propionate, sodium diacetate, and edetate disodium) (10,11). Current ongoing research is focused on enhancing membrane performance, fouling resistance, improving scalability, and optimizing energy efficiency in large-scale food processing. The evolution of the FO process in the liquid and food processing industry is demonstrated in Figure 1.2.

Furthermore, compared to conventional separation technologies, FO offers several potential advantages such as low energy requirements, mild operating conditions, and low fouling propensity. Furthermore, due to its simple operating design, FO can be seamlessly integrated with other processes (such as RO, MD, and crystallization), enabling high-efficiency hybrid solutions for challenging separations. However, despite these advantages, the FO process has not yet been fully commercialized due to several challenges such as low membrane flux, CP, RSF, and high energy cost for the draw solute regeneration process (12).

Considering that FO is an osmotically driven membrane process and owing to its lack of hydraulic pressure, it is commonly believed that FO utilizes less energy. However, studies suggest that FO uses the same amount of energy (13). Additionally, the effect of fouling on the FO process is typically less severe than pressure-driven membrane processes. However, the low fouling behaviour is partly associated with

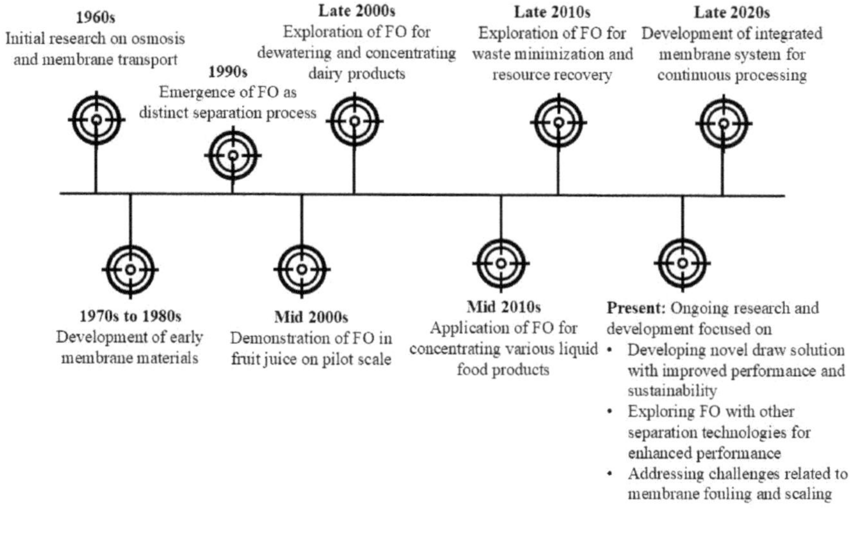


**FIGURE 1.2** Evolution of forward osmosis applications in food processing and membrane technology

the low flux obtained in the FO process. The effect of fouling on FO performance is typically less severe as the fouling layer is less dense, and therefore, the fouling is more reversible in nature. As a result of its lower fouling tendency, FO is capable of recovering clean water without extensive pretreatments.

However, it is important to realize that, unlike most water purification processes, the final product of the FO process is not like clean water but a diluted DS. Thus, in most cases, the FO process needs to be integrated with another separation step to recover the diluted DS and produce clean water. Considering this, FO can be regarded as an energy-efficient process if the DS does not need to be regenerated, such as dewatering using desalination brine or concentrated fertilizer as DS. Reportedly, for desalination, the FO–RO unit consumes more electricity than a stand-alone RO unit (14).

## 1.3 THEORETICAL BACKGROUND

Osmosis is one of the most fundamental and widely observed natural phenomena, influencing processes ranging from biological cell function to industrial separation technologies. It is defined as the spontaneous movement of solvent through a semi-permeable membrane from a region of lower solute concentration to a region of higher solute concentration.

The osmotic movement of water is fundamentally driven by the difference in chemical potential (μ) of the solvent across a semi-permeable membrane. Due to the presence of solute, the chemical potential of water in a solution is lower than that of pure water. The chemical potential of water in a solution can be expressed as:

$$\mu = \mu^0 + RT \ \ln \ a_w \tag{1.1}$$

where the term $\mu^0$ represents the chemical potential of pure water, R represents the universal gas constant (8.314 J $mol^{-1}$ $K^{-1}$), T represents the absolute temperature (K), and $a_w$ represents the water activity.

Osmotic pressure arises as the pressure required to equalize the chemical potential across a membrane, preventing the net movement of water. Thus, osmotic pressure (π) is defined as the pressure that must be applied to the solution side to stop osmosis.

In an ideal dilute solution, osmotic pressure (π) can be approximated using van't Hoff equation, derived by analogy to the ideal gas law:

$$\pi = iMRT \tag{1.2}$$

where the terms i, M, R, and T represent the van't Hoff factor, molarity of solute (mol $L^{-1}$), universal gas constant, and temperature (K) of the solution, respectively.

This expression suggests a linear relationship between osmotic pressure and solute concentration, assuming the solute behaves ideally and does not interact significantly with the solvent or other solutes. While van't Hoff law provides a useful starting point, real solution often deviates from the ideality due to solute–solute interactions (at higher concentration), solute–solvent interactions, and non-linearities in the activity–concentration relationship. Thus, the modified equation for a real solution may be rewritten as:

$$\pi = \phi iMRT \tag{1.3}$$

where $\phi$ accounts for nonideal behaviour.

At high concentration, the differences between Equations (1.2) and (1.3) become significant, necessitating empirical or semi-empirical corrections.

From a thermodynamic perspective, osmosis is a process that increases entropy by mixing solvents and solutes across a barrier. The Gibbs free energy change ($\Delta G$) for water transport can be expressed as:

$$\Delta G = -nRT \ln\left(\frac{a_{w,DS}}{a_{w,FS}}\right) \tag{1.4}$$

where $a_{w,DS}$ and $a_{w,FS}$ represent the water activities of the DS and FS, respectively.

The spontaneous negative $\Delta G$ drives water from the feed to DS until equilibrium is reached. This explains why FO does not require external mechanical energy input apart from fluid circulation and minimal hydraulic head.

The mechanism of FO is fundamentally different from that of conventional pressure-driven membrane processes. While RO depends on the application of external hydraulic pressure greater than that of osmotic pressure, FO harnesses the natural osmotic pressure gradient between two solutions to drive solvent (water) without mechanical compression. To optimize FO system design, understanding the specific transport processes including water movement, solute behaviour, and CP is vital.

In the FO process, the sole driving force is the osmotic pressure difference across the membrane, and it may be described as:

$$\Delta\pi = \left(\pi_{DS} - \pi_{FS}\right) \tag{1.5}$$

where the terms $\pi_{DS}$ and $\pi_{FS}$ represent the osmotic pressure of DS and FS, respectively.

High water permeability and high solute rejection are key performance requirements for FO membranes. In the given text, water permeability is expressed as Lp, defined on the basis of volumetric water flux. Although water permeability is commonly denoted as either Lp or A in the literature, Lp is used throughout this study for consistency. The water transport across the FO membrane is described as:

$$J_w = \mathrm{Lp}\left(\Delta\pi - \Delta P\right) \tag{1.6}$$

where the terms Lp, $\Delta\pi$, and $\Delta P$ represent water permeability coefficient, osmotic pressure gradient, and applied hydraulic pressure, respectively. For the FO process,

$\Delta P = 0$; for RO, $\Delta P > \Delta\pi$; and for pressure-retarded osmosis (PRO), $\Delta\pi > \Delta P$. The flux direction of the permeating water in FO, PRO, and RO is illustrated in Figure 1.3 (a). The FO point, PRO zone, and RO zone, along with the flux reversal point, are illustrated in Figure 1.3 (b) (15).

Alongside the water movement, a phenomenon called RSF, where solutes from the DS diffuse to FS. The RSF can be described as:

$$J_s = B(C_{DS} - C_{FS}) \tag{1.7}$$

where the terms $B$, $C_{DS}$, and $C_{FS}$ represent the solute permeability coefficient and the solute concentration in the DS and FS, respectively.

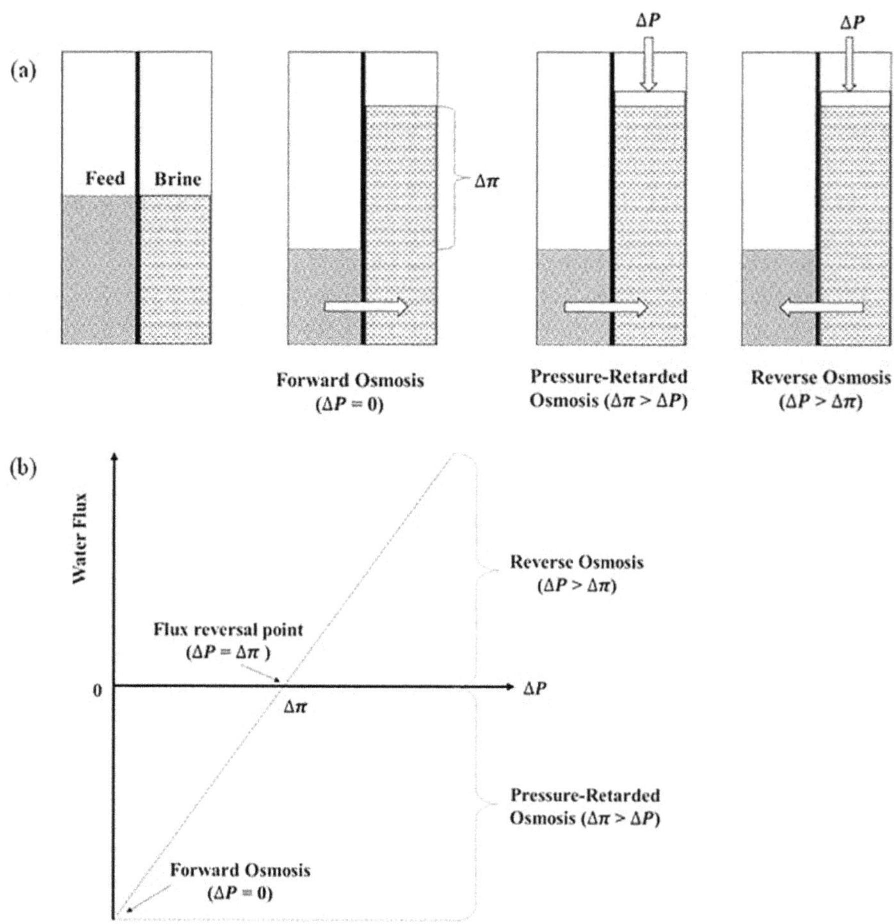


**FIGURE 1.3** (a) Solute flow in FO, PRO, and RO processes; and (b) direction and magnitude of water flux as a function of applied pressure in FO, PRO, and RO

High RSF is undesirable because it reduces the effective osmotic driving force, contaminates the feed product, and wastes draw solutes. The movement of water across the membrane involves convective transport through the AL, diffusion through the porous SL, and entry into the DS bulk phase.

The membranes' AL determines water selectivity, while the SL influences how effectively water can flow and how much internal concentration polarization (ICP) occurs.

At the microscopic level, water moves through tiny free-volume elements within the polymer matrix of the AL. The shape, size, and chemistry of these free-volume elements directly influence membrane performance. The net driving force of the AL is determined by the osmotic pressure gradient and internal SL structure.

Although FO is often described as a low-energy process, certain forms of energy are still required for pumping energy for FS/DS and regeneration of DS (if required). However, the overall energy consumption is dramatically lower than thermal evaporation or RO at high osmotic pressures, especially for concentrating high-viscosity liquid foods.

## 1.4 FORWARD OSMOSIS IN FOOD AND DAIRY PROCESSING

The food and dairy processing industries represent some of the most sensitive sectors, where product quality, nutritional value, and energy efficiency must be carefully balanced. The concentration of liquid food is one of the major unit operations in food processing industries. A vast amount of liquid food is industrially concentrated to reduce storage, handling, packaging, and transportation costs. Unlike traditional concentration methods, such as thermal evaporation and pressure-driven membrane processes (such as RO), FO utilizes osmotic gradients rather than mechanical or thermal energy. This unique operational feature allows for effective concentration of sensitive food products, preserving essential nutrients, flavours, and organoleptic properties that are otherwise degraded during conventional thermal methods. FO addresses these drawbacks by concentrating juices, beverages, and food extracts under ambient temperatures, which helps maintain nutritional integrity and taste profiles. Recent research demonstrated FO's efficiency in concentrating various liquid foods without compromising flavour, colour, and bioactive compounds.

The food industry offers one of the most diverse and promising landscapes for the application of the FO technology. One of the primary applications is the concentration of fruitjuices. Conventional concentration techniques often lead to considerable losses in the flavour and nutritional quality of juices. FO technology enables water removal under mild conditions, preserving delicate flavour, vitamins, and polyphenols. Multiple studies have reported successful concentration of orange, apple, pomegranate, and grape juices using FO membranes, achieving twofold to fivefold concentration factors while retaining over 90% of original aroma compounds.

Beyond juices, FO has demonstrated remarkable potential in the concentration of plant extract for use in functional foods and nutraceutical products. Plant extracts derived from green tea and coffee are highly sensitive to heat, making FO's low-temperature operation highly advantageous. Compared to traditional methods, the concentration of green tea extract by FO resulted in significantly higher retention of catechins.

Another emerging area for FO application is in the development of functional beverages and sports drinks. These products often require the concentration of bioactive compounds or electrolytes while maintaining low-sugar content. FO processes can be finely tuned to achieve selective concentration, thereby allowing the formulation of healthier beverage options with retained functional benefits.

In dairy processing, whey and milk are traditionally concentrated using energy-intensive evaporation and drying processes, whereas FO provides a gentle alternative that effectively concentrates dairy products to high dry matter content, essential for cheese, yoghurt, concentrated milk, and milk powder production. Pilot-scale studies demonstrated the feasibility of concentrating skim milk from an initial concentration of 8–21% dry matter with minimal energy usage compared to traditional evaporators or pressure-driven membrane systems (16).

FO also plays a pivotal role in the concentration and recovery of whey proteins. As a by-product of cheese manufacturing, whey is rich in biologically active proteins, yet its economic value is often under-exploited due to the challenges of gentle concentration. FO facilitates the concentration of whey protein isolates (WPIs) and whey protein concentrates (WPCs) while minimizing protein denaturation, aggregation, and functional loss.

Additionally, FO has been successfully employed for whey concentration, a crucial process for the recovery of valuable proteins and lactose. Lactose recovery represents another domain where FO exhibits strong advantages. Concentrating lactose-rich via FO enhances the crystallization process downstream, leading to higher yields and purities. Compared to conventional methods, FO systems achieve a high concentration factor with several studies demonstrating a concentration of whey up to 35% dry solids, thus significantly reducing downstream drying energy requirements and improving process economics.

Furthermore, the gentle operating conditions of FO make it particularly suitable for concentrating bioactive food ingredients such as proteins, polyphenols, and nutraceuticals. Unlike conventional methods, FO can concentrate these sensitive compounds without denaturation or loss of activity, thereby enabling the production of high-quality functional ingredients. Practical applications include protein recovery from tuna cooking juices and polyphenol extraction from olive mill wastewater (OMWW), both of which demonstrate FO's promise as an environmentally sustainable and economically attractive technology for value-added ingredient recovery. Beyond simple concentration, FO can also be coupled with crystallization and related processes. Such membrane-assisted crystallization strategies not only enable the selective recovery of high-purity compounds but also provide additional operational and economic benefits.

In conclusion, FO is poised to become an integral part of the future food processing industry landscape. Its gentle, energy-efficient, and sustainable nature addresses the food processing industries' evolving needs, positioning FO as an essential technology in the quest for enhanced product quality, sustainability, and economic viability.

## REFERENCES

1. Jamadi-Torab J, Malekzadeh A, Rahbari-Sisakht M, Ismail AF. State-of-the-Art in forward osmosis membrane process: Materials, fabrication, and applications. Chem Eng Process Process Intensif. 2025 May;110385.

2. Kessler JO, Moody CD. Drinking water from sea water by forward osmosis. Desalination. 1976 Jun;18(3):297–306.
3. Chang ZH, Sum JY, Lau WJ, Ang WL, Teow YH, Ooi BS, et al. Current state-of-the-art of non-reverse osmosis-like forward osmosis technology. J Memb Sci. 2024 Nov;711:123209.
4. Popper K, Camirand WM, Nury F, Stanley WL. Dialyzer concentrates beverages. J Food Eng. 1966;38(1966):102–4.
5. Wrolstad RE, McDaniel MR, Durst RW, Micheals N, Lampi KA, Beaudry EG. Composition and sensory characterization of red raspberry juice concentrated by direct-osmosis or evaporation. J Food Sci. 1993 May 26;58(3):633–7.
6. Petrotos KB, Quantick P, Petropakis H. A study of the direct osmotic concentration of tomato juice in tubular membrane—Module configuration. I. The effect of certain basic process parameters on the process performance. J Memb Sci. 1998;150(1):99–110.
7. Jiao B, Cassano A, Drioli E. Recent advances on membrane processes for the concentration of fruit juices: A review. J Food Eng. 2004;63(3):303–24.
8. Ravindra Babu B, Rastogi N, Raghavarao K. Effect of process parameters on transmembrane flux during direct osmosis. J Memb Sci. 2006;280:185–194 p.
9. Garcia-Castello EM, McCutcheon JR, Elimelech M. Performance evaluation of sucrose concentration using forward osmosis. J Memb Sci. 2009 Aug;338(1–2):61–6.
10. An X, Hu Y, Wang N, Zhou Z, Liu Z. Continuous juice concentration by integrating forward osmosis with membrane distillation using potassium sorbate preservative as a draw solute. J Memb Sci [Internet]. 2019;573(December 2018):192–9. Available from: https://doi.org/10.1016/j.memsci.2018.12.010
11. Zhang K, An X, Bai Y, Shen C, Jiang Y, Hu Y. Exploration of food preservatives as draw solutes in the forward osmosis process for juice concentration. J Memb Sci. 2021 Oct 1;635:119495.
12. Ibrar I, Altaee A, Zhou JL, Naji O, Khanafer D. Challenges and potentials of forward osmosis process in the treatment of wastewater. Crit Rev Environ Sci Technol. 2020 Jul 2;50(13):1339–83.
13. Park C, Kim Y, Kim D. Design of forward osmosis desalination configurations: Energy and energy perspectives. Appl Sci. 2025;15(16):9168. https://doi.org/10.3390/app15169168
14. McGovern RK, Lienhard VJH. On the potential of forward osmosis to energetically outperform reverse osmosis desalination. J Memb Sci. 2014 Nov;469:245–50.
15. Elewa MM. Emerging and conventional water desalination technologies powered by renewable energy and energy storage systems toward zero liquid discharge. Separations. 2024 Oct 11;11(10):291.
16. Chen GQ, Artemi A, Lee J, Gras SL, Kentish SE. A pilot scale study on the concentration of milk and whey by forward osmosis. Sep Purif Technol [Internet]. 2019;215(October 2018):652–9. Available from: https://doi.org/10.1016/j.seppur.2019.01.050

# 2 Forward Osmosis Theory

## 2.1 DRAW SOLUTION

The availability of a suitable DS/solute is crucial for advancing the FO process. In the FO process, since DS is the main source of the driving process, the performance of the FO process greatly depends on the selection of a suitable DS. As an osmotically driven membrane process, the draw solute should be able to significantly reduce the water chemical potential and consequently generate a high osmotic pressure in the FO process. On the other hand, the draw solute is also expected to be easily regenerated for reuse and to produce purified water. In addition to generating high osmotic pressure and being easily regenerated, an ideal DS must satisfy the following criteria: low RSF, low viscosity, low fouling propensity, chemical stability, compatibility with the membrane, low cost, and non-toxic (Figure 2.1).

Draw solutes can be broadly categorized as non-responsive and responsive draw solutes (*Figure 2.2*) (1). The draw solutes that do not have a significant change in their water affinity in response to stimuli (such as temperature, pH, electromagnetic field, or light) are referred to as non-responsive draw solutes, whereas responsive draw solutes, upon exposure to stimuli, undergo substantial changes in water affinity, thereby allowing ease of draw solute regeneration while maintaining sufficiently high drawing ability.

### 2.1.1 Non-responsive Draw Solute

#### 2.1.1.1 Inorganic Salts

The feasibility of inorganic salts such as monovalent ($NaCl$, $KCl$, $KBr$, $KNO_3$, $NaHCO_3$, and $NH_4Cl$) and divalent salts ($MgSO_4$, $CuSO_4$, $MgCl_2$, and $CaCl_2$) has been widely investigated as DS due to their low cost, low molecular weight, high solubility, and strong osmotic pressure. Noticeably, due to their larger hydrated radius and greater electrostatic repulsion, divalent/multivalent salts exhibit lower RSF than monovalent salts. However, due to higher diffusivity, the monovalent salts display higher water flux. Aiding its high water solubility, high osmotic pressure, and easy regeneration using the RO process, $NaCl$ is frequently used as DS for various FO processes. Divalent salts such as $MgCl_2$, though, exhibit lower water flux. However, due to the large molecular weight of magnesium ions, the $MgCl_2$ exhibited lower RSF. The feasibility of $MgSO_4$ was also proposed for desalination due to its high solubility, availability, and potential regeneration by a nanofiltration (NF) membrane to reduce regeneration cost. Unfortunately, NF regeneration of $MgSO_4$ produces brackish water that requires additional treatment for removal, thereby increasing the number of filtration stages of the FO–NF system.


DOI: 10.1201/9781003657873-2

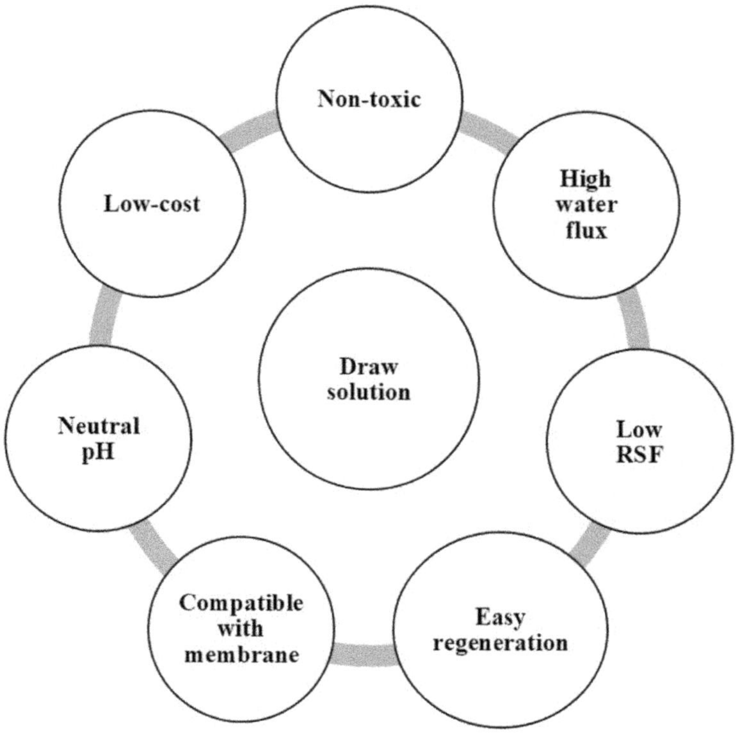


**FIGURE 2.1** Essential characteristics of an ideal draw solute

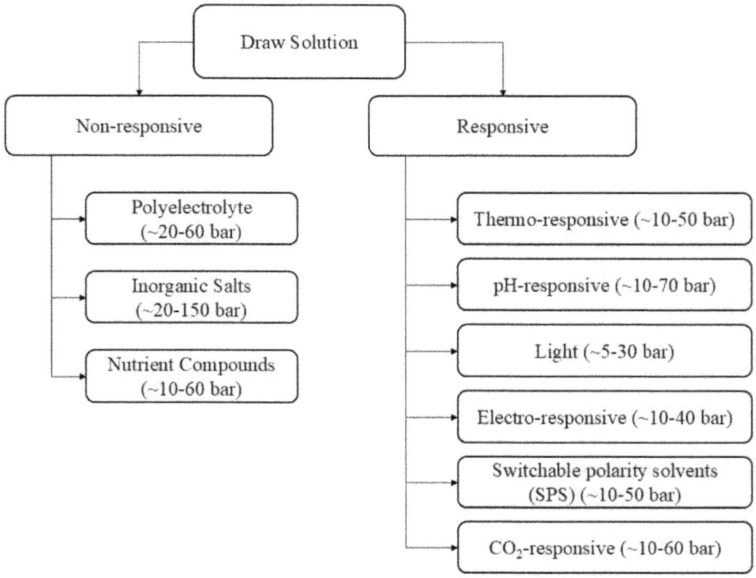


**FIGURE 2.2** Classification of draw solution

To overcome the limitations posed by single-component inorganic salts, recent investigations have explored the potential of multicomponent inorganic salt mixtures, such as combinations of NaCl, KCl, $MgCl_2$, and $Na_2SO_4$, as draw solutes (2). Compared to single-component DSs, these multi-salt mixtures offer several distinct advantages. Primarily, the synergistic interactions among multiple salts enhance overall osmotic pressure at relatively lower total concentrations, thereby increasing the water flux and efficiency of the FO process. Additionally, these mixtures effectively reduce RSF, as differences in ionic charges, hydrated radii, and diffusivities among component ions help balance solute transport, limiting unwanted diffusion across the membrane. Moreover, multicomponent salt mixtures notably mitigate scaling potential. Individual salts such as $Ca^{2+}$ and $Mg^{2+}$ often contribute to significant scaling and fouling issues when used alone; however, when combined with other salts, the likelihood of precipitation and scaling is reduced, thereby extending membrane life and enhancing operational stability. Furthermore, salt mixtures often possess improved overall solubility, allowing higher osmotic pressures to be achieved without encountering solubility limitations or crystallization.

The multicomponent approach also enables greater operational flexibility and the potential for customizing solutions for specific applications, such as nutrient delivery or wastewater reclamation, by optimizing the composition to achieve desired osmotic performance and minimize environmental impact. Economically, the use of mixtures composed of readily available and cost-effective salts could reduce overall DS costs, making FO processes more sustainable and economically viable for larger-scale implementation.

#### 2.1.1.2 Polymer and Organic Molecules

Apart from simple inorganic salts, many water-soluble polymeric/organic compounds were also investigated as non-responsive draw solutes. Linear poly(sodium acrylate) (PSA) was explored as DS for the FO process for seawater desalination. Compared to simple inorganic salts with lower molecular weights, these polyelectrolytes exhibited significantly reduced RSF, attributed to their larger molecular weights. The potential of organic salts (molecular weight 100–1000 g $mol^{-1}$) as DS for the FO process was also investigated, such as hexavalent phosphazene salts, ethylenediamine tetra-acetic-based salts, zwitterions, hydroacid complexes, and ionic liquids (ILs). The study revealed that the DS has lower viscosities and generates higher water flux than their polymeric counterparts, while generating lower RSF than inorganic salts (3).

In recent years, surfactants and their micellar DSs have gained the attention of many researchers (4,5). Surfactants offering moderately high osmotic pressure at concentrations above CMC are responsible for the stable water flux. Micellar solutes such as Triton X-100 (TX-100), cetylpyridinium chloride (CPC), and sodium dodecyl sulphate (SDS) can produce consistent flux and less RSF than inorganic salts while attaining a high level of regeneration through ultrafiltration (UF) (4). The combination of TX-100 with EDTA sodium salts as a DS significantly reduces RSF while sustaining a water flux of 4.6 L $m^{-2}$ $h^{-1}$ against a 35 g $L^{-1}$ NaCl FS (6). Reportedly, the interaction between the hydrophobic tails of surfactants and the hydrophobic membrane created a layer of surfactants on the membrane surface, restricting the reverse movement of the draw solute. Furthermore, a novel DS, composed of NaCl combined with Triton X-114 (TX-114), was developed to achieve enhanced water flux and significantly reduced RSF (7). It was observed that the mixture of 3 M *NaCl*/0.6 mM TX114 can significantly improve the water flux in the FO process (21.26 L $m^{-2}$ $h^{-1}$) with reduced specific RSF (SRSF = 1.82 g $L^{-1}$) and can be efficiently regenerated using the MD process. Another low-energy method for recovery of micelles from diluted DS is the Kraft point method. In the Kraft point approach, at temperatures below the Kraft temperature, micelles cannot develop, and therefore, an excess of insoluble monomers produced by the initial micelle causes them to crystalize, followed by paper filtration (4).

### 2.1.2 Responsive Draw Solutes

As mentioned in the previous section, contrary to the goal of the FO process, the regeneration of inorganic/organic DS was energy-intensive and uneconomical. To overcome this, researchers came across the application of stimuli-responsive materials that can change their solubility with the simulation of magnetism, pH, solar, temperature, or electricity. The application of responsive materials as DS has been recently explored for desalination, heavy metal rejection, and wastewater treatment.

#### 2.1.2.1 Thermo-Responsive Draw Solute

In thermos-responsive DS, temperature was the key factor for the phase transition of the DS. Broadly, the critical solution temperature can be categorized as low critical solution temperature (LCST) and upper critical solution temperature (UCST). For DS with LCST, the thermo-responsive DS and solvent should be separated into two phases by elevating the temperature above LCST. Similarly, for DS with UCST, the solution can be separated into two phases by reducing the temperature below UCST. Table 2.1 summarizes the FO performance of the thermally responsive draw solute, and *Figure 2.3* provides a schematic overview of the FO process using thermo-responsive hydrogel as the draw solute.

Reportedly, there are two mechanisms for the phase transformation of thermo-responsive DS. Firstly, the thermosensitive functional materials (such as poly(*N*-isopropylacrylamide) "PNIPAM", tributyl-*n*-tetradecyl-phosphonium-based ILs "$P_{444}$", and tributyl-*n*-tetradecyl-ammonium-based ILs "$N_{444}$") were shrunken or stretched due to their change in hydrophilicity with temperature variations. Secondly, long-chain polymers such as ammonium iodide salts, poly (*N*-isopropylacrylamide-co-acrylic acid), and glycerol-oligo (ethylene oxide)$_m$-block-(butylene oxide)$_n$ oligomers could achieve shrinking or stretching by themselves with temperature variation. Furthermore, according to the material type of the thermo-responsive DS, it is classified as gel, IL, and polymer.

- Gel: As DS, gels were investigated due to their unique network structure that facilitates the attraction of water molecules. The attraction of water molecules to the gel network is driven by osmotic pressure, leading to stretching of chains between network nodes. The gel can be further categorized as a

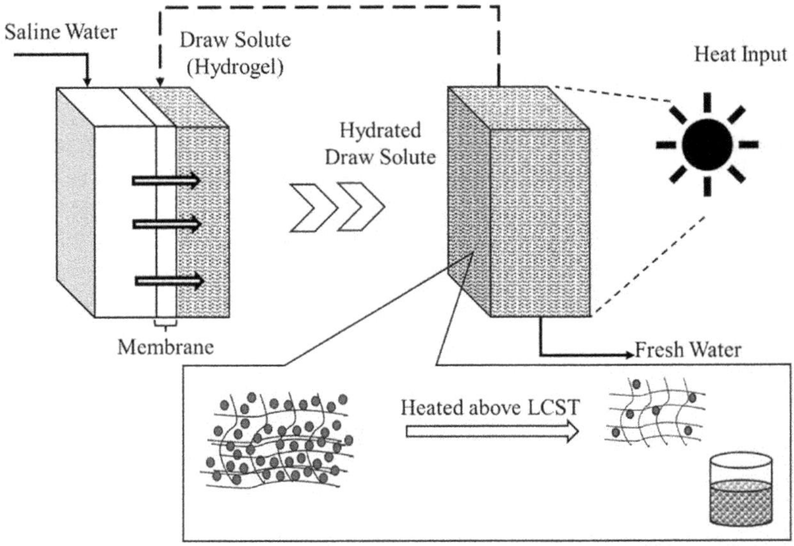


**FIGURE 2.3** Schematic representation of the forward osmosis process using thermo-responsive draw solute

hydrogel (large volume and wider contact surface) and a microgel (smaller in particle volume and larger in specific surface area).

Hydrogels are three-dimensional networks of polymeric chains that are cross-linked by either physical or chemical bonds. They are able to trap a large volume of water attracted by a high concentration of hydrophilic groups. For thermos-responsive hydrogel, two modification methods can be incorporated to improve water-drawing capacity. Firstly, increasing the number of hydrophilic groups in the hydrogel enhances water entrapment in the gel network. This can be achieved by incorporating hydrophilic materials like polyvinyl alcohol (PVA), acrylic acid (AA), and polyglutamic acid (γ-PGA), which contains carboxyl (-COOH), amino ($-NH_2$), or hydroxyl (-OH) groups.

Secondly, the pores of hydrogels provide enough channels for water molecules to move and be stored. During synthesis, additives such as polyethylene glycol (PEG) was often used to improve the pore structure of hydrogels. PEGs do not participate in the hydrogel polymerization reaction, but they are incorporated in the hydrogel network to create a dense porous structure that enhances the hydrophilic properties of the hydrogel. However, an excess of pore structure would result in a high cross-linking density, which may consequently reduce the hydrogel's elasticity, thereby hindering the hydrogel's shrinking and water recovery capacity at the thermosensitive temperature.

- Ionic Liquid: ILs are salts with low melting points that can become liquid at low temperatures, usually below 100 °C at atmospheric pressure (8). For

**TABLE 2.1**
**FO performance of thermally responsive draw solute**

| Category | Membrane Type | Draw Solution | Feed Solution | FO Mode | Water Flux, L m$^{-2}$ h$^{-1}$ | RSF, g m$^{-2}$ h$^{-1}$ | Ref. |
|---|---|---|---|---|---|---|---|
| Ionic liquids | TFC | SADMII (0.8 M) | DI water | AL-DS | 63.60 | 8.22×10$^{-3}$ | (9) |
| | TFC | Zn-Bet-Tf2N (0.2 M) | DI water | AL-DS | 14.90 | - | (10) |
| | TFC | P4444TFA (70% wt.) | DI water | AL-DS | 10.00 | 0.5 | (11) |
| Hydrogels/ microgels | TFC | NIPAM-AA-SSA (0.3 g) | DI water | AL-FS | 2.36 | - | (12) |
| | CTA | PNIPAM/γ-PGA/ PEG (0.2 g) | 2000 mg L$^{-1}$ NaCl | AL-DS | 0.41 | - | (13) |
| | TFC | GO-modified NPHs (0.2 g) | DI water | AL-FS | 4.30 | - | (14) |
| | CTA | NIPAM-AA microgels (0.1 g) | 2000 mg L$^{-1}$ NaCl | AL-DS | 23.80 | - | (15) |
| Polymers | TFC | PSSS-PNIPAM (33.3 wt.%) | 0.6 M NaCl | AL-DS | 3.50 | 2.0 | (16) |
| | TFC | PSSP5 (0.46M) | DI water | AL-DS | 16.28 | 1.27×10$^{-3}$ | (17) |
| | CTA | PEB 61 wt.% | water | AL-DS | 4.49 ± 0.10 | 0.79 ± 0.21 | (18) |
| | CTA | EB 52 wt.% | water | AL-DS | 4.14 ± 0.25 | 1.83 ± 0.30 | |

*RSF: reverse solute flux; TFC: thin-film composite; CTA: cellulose triacetate; DI water: deionized water; GO: graphene oxide; SADMII: 1-sodium acetate-2,3-dimethyl imidazolium iodide; Zn-Bet-Tf2N: multinuclear zinc complex designed from betainium bis(trifluoromethyl-sulfonyl)imide; $P_{4444}$TFA: tetrabutylphosphonium as the cation with trifluoroacetate; NIPAM-AA-SSA: N-isopropylacrylamide-acrylic acid-sewage sludge ash; PNIPAM/γ-PGA/PEG: N-isopropylacrylamide/ionic polyglutamic acid/polyethylene glycol; NPHs: nanocomposite polymer hydrogels; NIPAM-AA: N-isopropylacrylamide-co-acrylic acid; PSSS-PNIPAM: poly (sodiumstyrene-4-sulfonate-co-n-isopropylacrylamide); PSSP: poly(tetrabutylphosphonium styrenesulfonate); PEB: pentaerythritol-oligo(ethylene oxide)m-block-(butylene oxide)n oligomer; EB: oligo(butylene oxide)n-block-oligo(ethylene oxide)m-block-oligo(butylene oxide)n oligomer*

ILs to be considered potential draw solutes, they must exhibit high water solubility, a high degree of ionization, non-flammability, and good thermal and chemical stability (9). For example, carboxylate-functionalized imidazolium ILs (CFIMILs) as draw solute exhibited negligible RSF with superior FO performance.

- Polymer: Thermosensitive polymers are a promising class of DS because of their strong ability to draw water and ease of regeneration. After drawing, the thermosensitive polymer DS can be regenerated by heating. This causes their long chain structures to shrink and intertwine at temperatures higher than the LCST. As a result, the water molecules separated from the hydrophilic groups on the polymer's long chain structure, making DS regeneration easier. Then, the thermosensitive polymer DS was typically separated from the solvent using methods such as MD, UF, and autonomous precipitation. After separation, the DS was utilized in the subsequent FO cycle when the temperature was below the LCST.

PNIPAM is a typical thermo-responsive polymer that requires hydrophilicity improvement to be used as a DS. To achieve this, PNIPAM is polymerized with AA, sodium acrylate (SA), and sodium-4-styrenesulfonate (SSS). Reportedly, PNIPAM-AA and PNIPAM-SSS had an average flux of 2.95 L $m^{-2}$ $h^{-1}$ and 4.0 L $m^{-2}$ $h^{-1}$, respectively. Due to the high molecular weight and negative charge, the RSF was observed to be negligible (8).

### 2.1.2.2 Magnetism-Responsive Draw Solute

The potential of water-soluble magnetic nanoparticles (MnPs) as DS was investigated by numerous researchers. The studies revealed that the magnetic-sensitive MnPs can be efficiently recycled using the magnetic field and scarcely generate any RSF (19). MnPs such as $Fe_3O_4$ are capable of generating osmotic pressure and can be efficiently regenerated using a magnetic field (Figure 2.4).

The greatest challenge for MnP application was the strong aggregation of particles, hindering the reuse of DS after recovery. Ultrasonication was usually served to disperse the MnPs in DS. However, ultrasonication significantly impacts the MnPs' magnetism and responsive rate (20). To overcome these, the modified MnPs were widely favoured for the reduction of aggregation, elevation of FO performance, and enhancement of recycle sustainability.

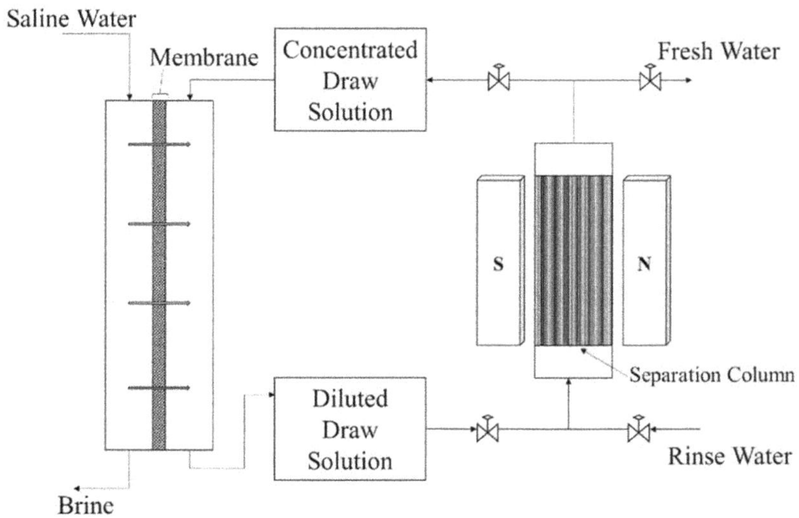


**FIGURE 2.4** FO desalination using magnetic nanoparticles as draw solute (17)

##### *2.1.2.2.1 Enhancing Magnetic Nanoparticles for Improved Stability and Performance in Forward Osmosis Application*

The efforts to modify MnPs have focused on reducing aggregation and extending their operational lifespan. The key strategies to achieve this involve electrostatic repulsion and controlling particle size. The choice of hydrophilic grafting materials significantly impacts the electrostatic repulsion properties of MnPs, resulting in variations of water flux and recyclability when used as DS in the FO system. For instance, for deep eutectic solvent (DES) grafting, MnPs referred to as poly(DES)@ Fe3O4 (21). The modification method employed ring-opening polymerization on a magnetic core, followed by introducing choline chloride (a hydrogen bond acceptor) to a glycerol surface. Using branched poly(DES)@ $Fe_3O_4$ as DS, the water flux was 17.9 L $m^{-2}$ $h^{-1}$, 12.3 L $m^{-2}$ $h^{-1}$, and 8.5 L $m^{-2}$ $h^{-1}$ against DI water, 0.2 M NaCl, and 0.3 M NaCl as FSs, respectively. The DES-MnPs demonstrated excellent stability and recyclability, maintaining performance across three magnetic isolation cycles.

Silicon dioxide ($SiO_2$) has gained attention as an interlayer for MnP modifications due to its strong interaction with $Fe_3O_4$. For instance, MnPs coated with SiO2 and sodium alginate sulphate (SAS), labelled as $Fe_3O_4$@$SiO_2$-SAS, exhibited average water fluxes of 12.8 L $m^{-2}$ $h^{-1}$, 12.4 L $m^{-2}$ $h^{-1}$, and 12.0 L $m^{-2}$ $h^{-1}$ during the first three operational cycles (21). Similarly, grafting poly[2-(dimethylamino)ethyl methacrylate] onto SiO2-coated MNPs yielded a water flux of 3.4 LMH with consistent performance across multiple recycling processes. SiO2 coatings not only facilitated the incorporation of additional hydrophilic groups but also increased particle size, enhancing stability (22).

##### *2.1.2.2.2 Effect of Coating Layer Thickness on Forward Osmosis Process*

The coating layer thickness of MnP significantly impacts their performance, particularly in terms of their dispersion, water extraction capability, and responsiveness in FO systems. Reportedly, according to the degree of agglomeration, the diameter of uncoated MnP varied from 3 nm to 357 nm. The low aggregation corresponds to the smaller particle diameter, resulting in both enhanced dispersion and modification potential. In the FO system, the MnPs with a smaller diameter typically display higher osmotic pressure and water extraction capacity. Additionally, increasing the concentration of particles and modifying with proper functional groups can further promote the FO performance.

The magnetic responsiveness of MnPs is a critical factor for the efficiency of magnetism-responsive DSs. However, modification of MnPs often reduces their magnetic responsiveness due to the insulating effect of thick coating layers. For instance, the MnPs grafted with 77% sodium polyacrylate (diameter 2.7 nm) exhibited a 70% reduction in maximum saturation magnetism (23). The low magnetism results in a decrease in the DS responsive efficiency. Reportedly, the previous studies have proved that grafting agglomeration materials or groups on MnPs may decrease the water flux and increase stability for recycling (24,25). Thus, achieving an optimal balance among magnetic sensitivity, water extraction performance, and particle dispersibility remains a crucial goal in MnP modification.

If the modified coating layer is not thick, particle agglomeration can occur under external magnetic fields. For example, the hydrophilic coatings, such as succinate-functionalized hyperbranched polyglycerol (SHPG) and hyperbranched polyglycerol carboxylate (HPGC), resulted in a particle diameter of 5.9±1.4 nm (26–28). However, their recovery using magnetic separation often led to aggregation, necessitating UF for assistance.

The moderate coating layer on MnPs with an appropriate particle diameter was found to be crucial for DS response and water flux. The maximum diameter of pectin-coated MnPs "P-MnPs" (520±28 nm) was significantly higher than that of uncoated MnPs (127±5 nm), and the maximum saturation susceptibility of pectin-coated MnPs was about 10% of the uncoated one (29). Despite these drawbacks, P-MnPs maintained recoverability using a simple magnet and exhibited high osmotic pressure, achieving a water flux of 26.6 LMH at 0.5 wt%. However, larger particle sizes generally resulted in lower dispersion and reduced osmotic pressure compared to smaller particles. However, not all coating materials effectively balance anti-agglomeration and water extraction performance. For instance, PEG dicarboxylic acid (PEG-(COOH)2)-coated MNPs with a SiO2 interlayer achieved a maximum water flux of only 4 LMH, with a response rate of 63.37% compared to uncoated MNPs (30).

Thus, the thickness of the coating layer plays a pivotal role in the performance of the modified MnPs. The thin coating fails to prevent agglomeration, leading to reduced responsiveness, while the thick coating hinders the magnetic properties, affecting the response rate and dispersion. The choice of coating materials also influences the agglomeration and water extraction capabilities, underscoring the need for careful optimization to achieve high-performing MnPs in FO applications.

##### *2.1.2.2.3 Integration of Magnetic and Thermo-Responsive Draw Solutes for Enhanced Forward Osmosis Process*

The development of magnetic-responsive draw solutes has led to an advancement in the FO process by integrating additional responsive properties to enhance water flux and overall responsiveness. PNIPAM, a thermos-responsive polymer with an LCST of 32 °C, was grafted onto MnPs (31). At temperatures above LCST, PNIPAM undergoes a phase transition, curling into a compact form and enabling efficient solid–liquid separation on a macroscopic scale. Combining magnetic sensitivity with the thermal responsiveness of PNIPAM improved the draw solute's overall responsiveness. Studies showed a 100% recovery rate under an applied magnetic field and temperatures above the LCST (32). When modified with PNIPAM and PSA, the resulting MNPs achieved a water flux of 11.66 LMH using DI water as the FS, significantly outperforming PSA-MNPs, which reached 3.8 L $m^{-2}$ $h^{-1}$ (33). Further enhancement was achieved by incorporating the polyelectrolyte co-polymer poly(sodium styrene-4-sulfonate) (PSSS) with PNIPAM-grafted MNPs. This formulation produced PSSS-PNIPAM@MNPs, a hybrid material with a water flux of 14.90 LMH using DI water as FS. The optimized composition of this co-polymer (15 wt.% PSSS and 85 wt.% PNIPAM, labelled as MNP5–15SN) exhibited exceptional thermal sensitivity, forming visible precipitates above its LCST (~32°C), which were easily separable from the DS. The precipitate could redissolve below the LCST, enabling reuse over multiple cycles with only a minimal flux reduction, retaining 86.6% of its initial flux after five cycles (34). In addition to thermo-responsive materials, $CO_2$-responsive draw solutes such as poly[2-(dimethylamino)ethyl methacrylate] (PDMA)-modified MnPs were developed. These solutes precipitate under a nitrogen-enriched environment ($CO_2$-free), allowing separation with an external magnetic field. Reconstitution for reuse involved exposing the aggregates to $CO_2$, which protonated the PDMA chains, increasing their solubility. The PDMA-MnP system demonstrated stable performance across multiple cycles with negligible flux decline, further underlining the potential of hybrid-responsive DS for sustainable FO applications (32). Recent studies demonstrate that grafting materials significantly influence membrane performance. Branched poly(DES)@$Fe_3O_4$ achieved a high water flux of 17.9 L $m^{-2}$ $h^{-1}$, whereas $Fe_3O_4$@$SiO_2$ modified with SAS exhibited flux rates up to 12.8 L $m^{-2}$ $h^{-1}$. Citrate-modified membranes (Cit-MnP1) presented variable flux performance ranging from 6.2 to 17.6 L $m^{-2}$ $h^{-1}$, while PSA enhanced flux from 1.72 to 5.32 L $m^{-2}$ $h^{-1}$ upon modification with manganese particles (MnPs). Notably, tri-sodium citrate membranes (Cit-MnPs) achieved exceptionally high flux levels (34.77 L $m^{-2}$ $h^{-1}$). Hydrophilic siloxane agents yielded moderate flux (~2.13 L $m^{-2}$ $h^{-1}$), and ferrohydrogel cross-linked membranes ($Fe_3O_4$@$SiO_2$@MPS) performed moderately at 10.2 L $m^{-2}$ $h^{-1}$. Increasing iron oxide content in SA and *N*-isopropylacrylamide membranes significantly enhanced their responsive rate, reaching up to 94.5% at 16 wt.%. IL-based membranes (C8-$FeCl_4$) displayed moderate flux (10.4 L $m^{-2}$ $h^{-1}$), indicating suitability for specialized membrane applications. For food-related uses, membranes employing SAS ($Fe_3O_4$@$SiO_2$-SAS) and tri-sodium citrate (Cit-MnPs) are especially promising due to their high flux, biocompatibility, and food-grade acceptance. Among these, the tri-sodium citrate membranes (Cit-MnPs), with their superior flux performance, are particularly advantageous, given citrate's recognized safety and widespread approval for food processing. Table 2.2 summarizes the grafting materials and FO performance of magnetism-responsive DS.

#### 2.1.2.3 $CO_2$-Responsive Draw Solute

Recently, a series of $CO_2$-switable polymers, such as poly(*N*,*N*-dimethylallylamine) (PDMAAm), 1-cyclohexylpiperidine (CHP), poly [2-(dimethylamino)ethyl methacrylate] (PDMA), and 2-(diethylamino)ethyl methacrylate (DEAEMA), was often explored for the synthesis of $CO_2$-responsive DS. Reportedly, the initial purpose was to design a polymer that could show $CO_2$ switchability with high osmotic pressure under $CO_2$ and low osmotic pressure under air. For the response of DS, $CO_2$ is necessarily removed to reduce the solubility and dissolve DS. Either $N_2$-purging or heating is used to discharge $CO_2$ from DS.

In another study, $CO_2$ and a thermally dual-responsive polymer, poly [2-(dimethylamino)ethyl methacrylate] (PDMAEMA), are used as the draw solute for FO desalination (35). Reportedly, the given polymer can reversibly switch between the protonated and deprotonated states in aqueous solution by purging with $CO_2$ or an inert gas. Once protonated, the polymer becomes a polyelectrolyte and is able to generate sufficiently high osmotic pressure for FO desalination of seawater.

**TABLE 2.2**
**Overview of the forward osmosis process for magnetic-responsive draw solute**

| Grafting Material | Classifications | Feed Solution | Water Flux, $L\ m^{-2}\ h^{-1}$ | Responsive Rate, % | Ref. |
|---|---|---|---|---|---|
| Branched poly(deep eutectic solvent) DES | 0.5 g $L^{-1}$ branched poly (DES)@$Fe_3O_4$ | Deionized (DI) water | 12.0 | 100 | (21) |
| | 1.5 g $L^{-1}$ branched poly (DES)@$Fe_3O_4$ | | | | |
| | 2.5 g $L^{-1}$ branched poly (DES)@$Fe_3O_4$ | | | | |
| | 3.5 g $L^{-1}$ branched poly (DES)@$Fe_3O_4$ | | 17.9 | | |
| Sodium alginate sulphate (SAS) | 0.02 g $mL^{-1}$ Fe3O4@$SiO_2$-SAS | DI water | 3.50 | 100 | (36) |
| | 0.04 g $mL^{-1}$ $Fe_3O_4$@$SiO_2$-SAS | | 6.00 | | |
| | 0.06 g $mL^{-1}$ $Fe_3O_4$@$SiO_2$-SAS | | 12.8 | | |
| Citrate (Cit) | Cit-MnP1 | Pure water | 6.2 | 100 | (23) |
| | Cit-MnP1 | | 17.6 | | |
| | Cit-MnP1 | | 13.7 | | |
| Polysodium acrylate (PSA) | PSA | DI water | 1.72 | 100 | (22) |
| | PSA-MnPs | | 5.32 | | |
| Hydrophilic siloxane agents | SiCOOH-MN | DI water | 1.81 | 100 | (22) |
| | SiPEG-MN | 500 ppm dissolution of methylene blue | 2.13 | 100 | |
| | | 1000 ppm dissolution of methylene blue | 1.41 | | |
| | | 500 ppm dissolution of methylene blue | | | |
| Tri-sodium citrate | 2 g $L^{-1}$ Cit-MnPs-1 | DI water | 34.77 (3 h) | | (36) |
| | 2 g $L^{-1}$ Cit-MnPs-2 | | 28.41 (3 h) | | |

| | | | | | |
|---|---|---|---|---|---|
| Chitosan and dehydroascorbic acid (DDA) | 0.01 g $L^{-1}$ DHAA-coated $Fe_3O_4$ | DI water | - | 100 | (37) |
| 3-(Trimethoxysilyl)propyl methacrylate (MPS) | Methylene bisacrylamide cross-linked with $Fe_3O_4$@$SiO_2$ incorporated ferrohydrogel | DI water | 5.72 | - | (24) |
| | $Fe_3O_4$@$SiO_2$@MPS cross-linked ferrohydrogel | | 10.2 | - | |
| Sodium acrylate and *N*-isopropylacrylamide | 0 wt.% iron (III) oxide | 2000 mg $L^{-1}$ NaCl | - | - | (38) |
| | 2 wt.% iron (III) oxide | | - | < 10 | |
| | 4 wt.% iron (III) oxide | | - | 10–20 | |
| | 8 wt.% iron (III) oxide | | - | 30 | |
| | 16 wt.% iron (III) oxide | | - | 66 | |
| | 0.1 g 16 wt.% iron (III) oxide | | - | 33.45 | |
| | 0.4 g 16 wt.% iron (III) oxide | | - | 94.5 | |
| 1-Alkyl-3-methylimidazolium tetrachloroferrates | C8-$FeCl_4$ | DI water | 10.4 | - | (39) |

On the other hand, the deprotonated draw solute is thermally responsive and precipitates at temperatures above its LCST at around 40 °C, enabling the ease of draw solute recovery (Figure 2.5). The key advantage of this dual-responsive system lies in its reversible transformation between a highly ionized state, optimal for FO performance, and a thermally induced precipitated state, which simplifies recovery. Table 2.3 summarizes the FO performance for $CO_2$-responsive DS.

#### 2.1.2.4 Solar Energy-Responsive Draw Solute

The solar-responsive draw solute can be regenerated by harnessing solar energy (42). The solar energy-sensitive characteristics are contributed by the high solar irradiation-sensitive materials that are either synthesized with the draw solute or coated on its surface.

The potassium-functionalized carbon nanofibres (K/CNF) coated with triethylene glycol (TEG), referred to as TED-K/CNF, demonstrated significant osmotic pressure and responsiveness to solar energy. The DS was reconcentrated by evaporating the water, aided by the high solar sensitivity of K/CNF under simulated solar flux. Depending upon the amount of TEC coverage, the water flux and photothermal response rate of TEG-K/CNF draw solute were influenced. For instance, at 10 vol.% TEG, the given draw solute achieved a water flux of 7.5 L $m^{-2}$ $h^{-1}$ and a photothermal efficiency of 80%. Furthermore, by increasing the TEG to 20 vol.%, the draw solute achieved a water flux of 13.3 L $m^{-2}$ $h^{-1}$ with a slightly reduced photothermal

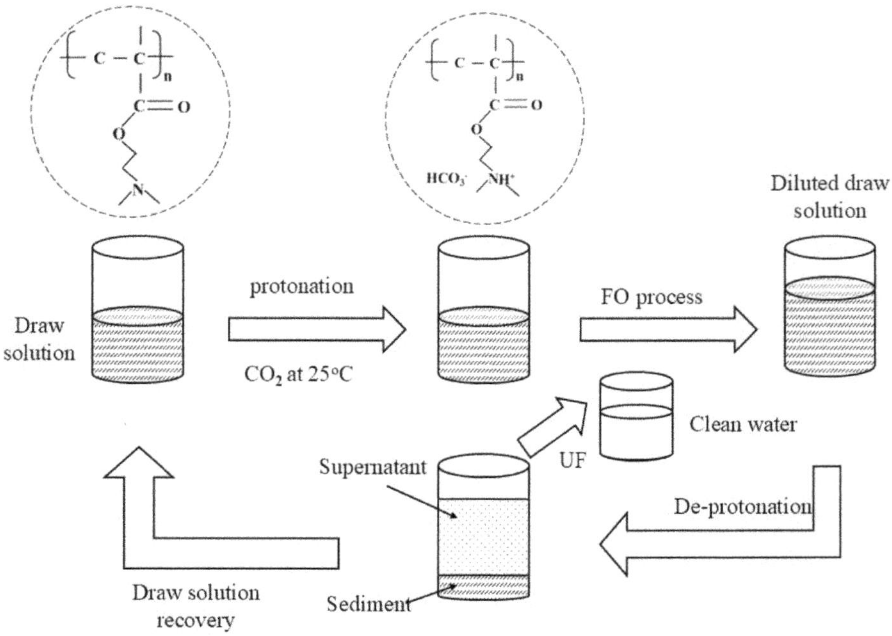


**FIGURE 2.5** Schematic diagram of the dual-responsive draw solute for forward osmosis desalination

**TABLE 2.3**
**Overview of FO performance using $CO_2$-responsive draw solute**

| $CO_2$-Responsive Material | Feed Solution | Water Flux ($Lm^{-2}h^{-1}$) | Responsive Rate | Ref. |
|---|---|---|---|---|
| 30 wt.% g-PDMAAm | 1.75% NaCl | - | 99.9% | (40) |
| | 3.50% NaCl | - | | |
| 50 wt.% g-PDMAAm | 1.75% NaCl | 44.00 | - | |
| | 3.50% NaCl | - | | |
| CHP | DI water | 38.0 ± 4.0 | - | (41) |
| | 1.3% w/v of urea | 65.0 ± 4.0 | | |
| | 1.0% w/v NaCl | 69.0 ± 2.0 | | |
| l-PDMAAm | 0.2 wt.% NaCl | 89 | ~98 wt.% | |

*PDMAAm: poly(N,N-dimethylallyl amine); CHP: 1-cyclohexylpiperidine; l-PDMAAm: linear poly(N, N-dimethylallylamine)*

efficiency of 77%. Reportedly, after five cycles of the FO process, the water flux remained steady at 12.9 L $m^{-2}$ $h^{-1}$, and the product water was found to be comparable with the potable water standard, free of TEG and K/CNF (43).

Reduced graphene oxide (rGO)-composite hydrogels were synthesized for use as a draw solute in desalination processes (44). Reportedly, the composite hydrogel was synthesized by incorporating 0.3 wt.% to 3 wt.% rGO into PSA and poly(sodium acrylate)-poly(*N*-isopropylacrylamide) (PSA-NIPAM). Compared to pure hydrogel, the composite hydrogels (PSA-1.2 wt.% rGO and PSA-NIPAM-1.2 wt.% rGO) exhibited increased water flux and swelling capacity. Additionally, pure water from the composite hydrogel was harvested using solar energy. The rGO within the hydrogel absorbed solar radiation and converted it into thermal energy, increasing the temperature within the hydrogel network. This thermal effect caused the hydrogel to shrink, releasing water molecules effectively.

### 2.1.2.5 pH-Responsive Draw Solute

Based on the pH of DS, the pH-responsive draw solute alters its dissolved state in a solvent. The pH-sensitive characteristics of pH-responsive draw solute are contributed from their negative ions/groups. For example, EDTA speciation of ethylene diamine tetra acetic acid disodium salt (EDTA-2Na) (45), sulphonate ions of 1,4-bis(3-propane-sulphonate sodium)-piperazinediethanesulfonic acid disodium-sulphate and (1–(3-aminopropyl)-imidazole) propane-sulfonate (APIS) (46–47), and molybdenum acid ions of $(NH_4)_6Mo_7O_{24}$ and $Na_6Mo_7O_{24}$ (48) have been reported in literature.

The structure, ionization state, or hydrodynamic diameter of these negative ions/groups undergoes changes when they encounter hydrogen ions. The highly protonated forms of EDTA, such as $H_3[EDTA]^-$, $H_4[EDTA]$, and $H_5[EDTA]^+$, were immiscible in water and precipitated at pH below 3.0, enabling the separation and recovery of EDTA derivatives. Similarly, the protonation of sulfonate ions neutralizes their charge, causing them to precipitate as hydrogen sulphate at pH 2.0 (46,47). Molybdenum acid ions, such as $HMo_7O_{24}^{5-}$ and $H_2Mo_7O_{24}^{4-}$, display distinct pH-responsive behaviour (52). Table 2.4 summarizes the FO performance and responsive rate of pH-responsive draw solute, since the pH-responsive draw solutes can drag water molecules through the hydrogen bonds and form a polymeric network. As a result, at the same concentration, the pH-responsive draw solute exhibited higher osmotic pressure and water flux than the NaCl aqueous solution (46).

Due to the presence of oxygen atoms and nitrogen atoms, the molybdenum acid ions and sulfonate ions formed coordination bonds connecting water molecules. For instance, the main trunk of $(NH_4)_6Mo_7O_{24}$ and $Na_6Mo_7O_{24}$ consisted of coordination bonds between multiple molybdenum ions and oxygen atoms. These oxygen atoms will form hydrogen bonds with water molecules. Similarly, the oxygen and nitrogen atoms of sulfonate ions could also form hydrogen bonds with the water molecules.

In a laboratory-scale FO process, the energy demand for pumping fluids is approximately 0.25 kWh $m^{-3}$, which is notably lower compared to other methods, such as MD, RO, and NF. However, assessing the energy needed to recover responsive DS is crucial for evaluating the feasibility and commercial viability of using these solutions in FO applications.

In summary, the responsive draw solute represents a transformative approach in advancing FO technology. The versatility of responsive draw solutes stems from their ability to phase-separate from water under external stimuli such as temperature, magnetism, $CO_2$, pH, or solar energy. Thermo-responsive draw solute utilizes thermosensitive materials like PNIPAM, $P_{4444}$, and $N_{4444}$ to undergo structural changes in response to temperature shifts. Magnetism-responsive draw solute leverages magnetic fields for regeneration, with improved dispersibility and hydrophilicity achieved through hydrophilic modifications. Similarly, $CO_2$-responsive draw solutes utilize reversible chemical interactions with $CO_2$ to achieve phase separation, whereas pH-responsive draw solutes undergo structural or ionization changes under acidic conditions. In contrast, solar-responsive draw solutes harness light energy to drive water recovery with minimal external energy input. These responsive draw solutes offer impressive performance metrics, with water flux ranging from 0.06 L $m^{-2}$ $h^{-1}$ to 69.00 L $m^{-2}$ $h^{-1}$ and high stability across multiple regeneration cycles. The low RSF and cost efficiency further enhance their suitability for industrial applications. With promising applications in wastewater treatment, pharmaceuticals, energy reduction, and economic assessment, the responsive draw solute presents a compelling opportunity for FO technology. Furthermore, their adaptability, energy efficiency, and cost-effectiveness positioned them as pivotal elements in the future development of the FO process. Table 2.4 summarizes the FO performance of the pH-responsive draw solute.

## 2.2 FORWARD OSMOSIS MEMBRANE

The FO membrane is the heart of the FO application, and the development of an ideal membrane is an active area of research for both academia and many industrial companies. The manufacture and modification of FO membranes have gained immense importance over the past few years.

**TABLE 2.4**
**Overview of forward osmosis performance of pH-responsive draw solute**

| pH-Responsive DS | Classification | Feed Solution | Water Flux (L m$^{-2}$ h$^{-1}$) | pH | RSF (g m$^{-2}$ h$^{-1}$) | Ref. |
|---|---|---|---|---|---|---|
| Poly-oxometalates | $(NH_4)_6Mo_7O_{24}$ | DW | 16.4 | < 2.0 | 0.0 | (47) |
| | $Na_6Mo_7O_{24}$ | DW | 14.2 | | | |
| | $(NH_4)_6Mo_7O_{24}$ | 2000mg/L | 10.0 | | | |
| | $Na_6Mo_7O_{24}$ | glutathione | 9.6 | | | |
| Polyacrylic acid sodium salts | 5 wt.% 2000 PAA-Na | DW | 8.44±0.21 | 4.35± 0.22 | 0.018± 0.002 | (48) |
| | 20 wt.% 2000 PAA-Na | | 17.26±0.54 | 7.78± 0.41 | 0.11± 0.004 | |
| | 20 wt.% 5000 PAA-Na | | - | 3.30± 0.43 | 0.098± 0.003 | |
| | 5 wt.% 50000 PAA-Na | | 4.22 | 3.86± 0.17 | 0.001–0.004 | |
| | 20 wt.% 50000 PAA-Na | | 9.1 | 5.46± 0.36 | - | |
| (1–(3-aminopropyl)-imidazole) propane-sulfonate (APIS) | 1 M APIS | DW | 20.0–25.0 | 2.0 | ~0.0 | (45) |
| EDTA-2Na | 0.25 M EDTA-2Na | DW | 4.0 | 2.0 | 0.2±0.0 | (44) |

Similar to RO membrane, FO membrane normally consists of a dense AL (or selective layer) of 0.1–1.0 μm thickness, supported by a porous thick SL (or substrate layer) of 100–200 μm thickness. While dense AL is responsible for achieving high water permeability and solute rejection, a material with high mechanical strength should be selected for the SL to stabilize the membrane structure under high-pressure operations. However, in the FO process, since there is low or no hydraulic pressure involved, there is no stringent demand on mechanical strength. Thus, in the FO process, the major concern lies in the design strategy to maximize water flux and minimize solute reverse flux.

### 2.2.1 Design Criteria of Forward Osmosis Membrane

The FO membrane design hinges on balancing efficiency and longevity through several key criteria. Engineers must optimize the SL's structure to minimize ICP, fine-tune surface chemistry for water transport and fouling resistance, and select materials that withstand chemical and thermal stresses encountered in real-world applications. Mechanical robustness is essential for module integration and large-scale deployment, but it must not come at the expense of water flux. These performance trade-offs demand a holistic approach—each design choice influences permeability, selectivity, and durability. With these principles in mind, the story of FO membrane fabrication unfolds in the next section, revealing the methods and innovations that bring these criteria to life.

*Transport Properties:* A fundamental design goal for FO membranes is to maximize water transport while minimizing solute leakage. The water permeability coefficient Lp (often expressed in L $m^{-2}$ $h^{-1}$) quantifies the membrane's capacity to transmit water, and higher Lp-values directly translate to higher water flux under a given driving force. Conversely, the solute permeability B (L $m^{-2}$ $h^{-1}$) measures the tendency of DS to diffuse across the membrane; a low B-value (high solute retention) is crucial to prevent RSF and preserve the osmotic driving gradient. An ideal FO membrane thus exhibits a high Lp and low B—yielding high water flux with minimal solute crossover—as noted in FO literature. These intrinsic transport parameters can be combined as a selectivity ratio (B/Lp), providing a quantitative basis to compare membrane performance. In practice, TFC PA FO membranes tend to achieve significantly higher Lp-values than CA membranes, enabling greater water flux, whereas CA membranes (though lower in permeability) offer superior chemical robustness. Table 2.5 underscores this trade-off: Most TFC-FO membranes display higher water permeability and a lower normalized salt permeability than CA-based FO membranes. This implies TFC membranes pass water more readily while limiting solute diffusion, owing to their ultrathin selective layers and highly porous supports. Engineering of the AL involves balancing thickness and cross-linking density—a thinner AL raises Lp but can increase B, so achieving high flux without sacrificing rejection requires careful control of polymer structure and chemistry (e.g. incorporating highly selective PA skins or AQP-like channels). Ultimately, optimizing Lp and B in tandem is paramount: FO membranes should facilitate rapid water uptake under osmotic pressure yet maintain high solute rejection to minimize RSF and ensure high osmotic efficiency.

*Support Layer Structure and Internal Concentration Polarization (ICP):* Unlike RO, FO operates with little-to-no hydraulic pressure, which relaxes the need for extreme mechanical reinforcement in the SL (49). This allows FO membrane supports to be engineered for mass transfer efficiency rather than pressure tolerance. The porous SL plays a critical role in mitigating ICP—the build-up of diluted DS or concentrated FS within the membrane that can drastically reduce driving force. Key structural attributes of the support include thickness, tortuosity, and porosity, which are often captured by the structural parameter S (with units of length). The S-value is defined by the relationship $S = t \cdot \tau / \varepsilon$ (where t is support thickness, $\tau$ is tortuosity, and $\varepsilon$ is porosity); a lower S indicates a thinner, more porous, and less tortuous support matrix that offers less resistance to diffusive transport. Low S-values are highly desirable in FO membranes because they minimize ICP and thus sustain higher water flux (50). In practical terms, this means designing the SL to be as thin as possible while maintaining sufficient porosity and wettability to allow draw solutes to diffuse quickly out of the membrane. For example, modern TFC-FO membranes often use ultra-porous polymer substrates or nanofibre composite supports to reduce diffusion path lengths. However, there are engineering trade-offs: An extremely thin or open support can be mechanically fragile and may require a backing layer for handling, which in turn increases the effective S-value. Indeed, many high-performance TFC FO membranes still employ a polyester fabric backing for module integration, and this added thickness can exacerbate ICP or hinder water penetration into the support. The SLs of some commercial TFC-FO membranes remain relatively thick and thus require further innovation in support structure to drive down S (51). Overall, the design criterion is to minimize ICP through an optimized SL: one that is thin, highly porous, and hydrophilic (fully wetted), thereby facilitating swift internal diffusion of solutes. Reducing the structural parameter improves the membrane's net osmotic driving force and efficiency, which is critical for achieving higher sustained flux in FO operation.

*Surface Chemistry*: FO membrane surfaces (the AL in contact with FS or DS) must be engineered to promote water transport and resist fouling. Hydrophilicity is a key surface property: a hydrophilic membrane surface and support readily absorb water, ensuring that pores remain filled and preventing de-wetting or air entrapment that would worsen ICP. Hydrophilic surfaces also form a hydration layer that can repel hydrophobic foulants, thus mitigating fouling tendencies. Accordingly, advanced FO membranes often incorporate hydrophilic polymers or surface coatings to lower the water contact angle and improve wetting. FO membranes are usually negatively charged and hydrophilic in nature to enhance solute rejection and water affinity (52). The negative surface charge (commonly found in PA ALs) helps repel anionic solutes, contributing to higher salt rejection, and can also hinder the adhesion of similarly charged colloidal foulants via electrostatic repulsion. Surface roughness is another critical factor: A smoother membrane surface is less prone to foulant deposition, whereas high roughness (e.g. the "ridge-and-valley" morphology of some PA layers) provides niches for particles, biofilms, and organic macromolecules to attach. Thus, an ideal FO membrane exhibits a smooth, hydrophilic AL with minimal charge-neutral areas for foulants to adsorb. This design criterion improves fouling resistance, maintaining stable flux over time and facilitating easier cleaning. In practice, there is a balance between creating an ultrathin selective layer for performance and controlling its morphology: IP conditions are tuned to yield a dense yet smooth PA skin without excessive roughness or defect formation. Researchers have found that incorporating certain additives or surface modifiers during membrane fabrication can yield smoother, more hydrophilic surfaces that significantly reduce fouling with negligible loss in permeability. For example, hydrophilic monomers or coatings (e.g. zwitterionic polymers and PEG) on the AL have been shown to decrease organic fouling by creating a protective hydration layer. Anti-fouling design extends to biofouling control as well: FO membranes may be functionalized with antibacterial agents (such as silver nanoparticles or antimicrobial polymers) to prevent microbial growth on the surface. TFC-FO membranes, while high-performing, can suffer from biofouling over prolonged operation—their surfaces can become "instantly polluted" by bacteria and thus often require biocide dosing or special storage conditions. The design criterion here is to achieve robust fouling resistance through surface chemical properties: a combination of hydrophilicity, appropriate surface charge, smoothness, and possibly biofilm-resistant functionalities. These features ensure that the membrane maintains high performance in real feed streams, with slower flux decline and easier cleaning intervals. In summary, an optimal FO membrane surface actively inhibits foulant adhesion and exhibits stable performance in the face of contaminants, all while permitting efficient water transport.

*Chemical and Thermal Stability*: FO membranes must withstand the chemical environment of both FS and DS, as well as any cleaning regimens, over long operational periods. A key criterion is chemical stability of the membrane materials, particularly resistance to hydrolysis, oxidation, or other degradation pathways. Many FO processes involve feed waters or draws that contain chlorine or other oxidizing agents (e.g. when treating wastewater or using NaOCl for disinfection). PA TFC membranes, although high in performance, are notoriously sensitive to chlorine and can suffer from chain scission (chlorine attack on amide bonds), resulting in rapid performance loss. In contrast, CA membranes exhibit much higher tolerance to chlorine and wider pH stability, which historically made them suitable for applications where oxidant exposure is likely. CA-based FO membranes demonstrate "excellent structural stability and broad chemical compatibility" across varied conditions. This advantage in chemical durability, however, comes at the expense of lower water permeability and flux, as previously mentioned. Thus, there is a performance trade-off between TFC and CA membranes: The former excel in flux and selectivity but require feed streams free of certain chemicals (or dechlorination steps), while the latter offer robustness to harsh chemistries with moderate performance. An ideal FO membrane would combine the best of both—high flux and rejection along with resistance to chemical attack. Researchers are exploring chlorine-resistant PA chemistries and surface coatings (e.g. sacrificial protective layers) to improve TFC membrane durability without sacrificing performance. In addition to chemical stability, thermal stability is a consideration if FO systems operate at elevated temperatures (for instance, when utilizing low-grade waste heat or in FO–MD hybrid systems). The membrane should maintain its structural integrity and performance at the operating temperature range, typically up to 40–50 °C for polymeric membranes (53). Thermal stability is closely tied to the membrane material: CTA can undergo hydrolysis at high temperature or extreme pH, while certain PA TFCs might soften if temperatures approach their glass transition. Therefore, designing FO membranes with polymers that remain stable (chemically inert and mechanically strong) under the expected chemical and thermal conditions is essential for longevity. In practice, FO membranes are often operated under gentle conditions (neutral pH, ambient temperature, and no chlorine) to maximize lifespan, but the design criterion remains that the membrane should possess good chemical stability (e.g. chlorine resistance and pH tolerance) and adequate thermal stability to handle real-world feed streams and cleaning protocols. These properties ensure the membrane can be deployed in diverse applications (from seawater desalination to industrial wastewater treatment) without rapid degradation.

*Mechanical Strength and Module Integration*: While FO membranes do not need to withstand high hydraulic pressures, they must still have sufficient mechanical strength to be manufactured, handled, and operated in large-scale modules. FO membrane sheets are typically assembled into modules (such as spiral-wound, plate-and-frame, or tubular configurations) similar to those used in pressure-driven processes. To facilitate this, the membrane often includes a support fabric or backing layer that imparts tear strength and dimensional stability during module winding and operation. The inclusion of a fabric backing (common in TFC membranes) makes the membrane more robust for module fabrication, but it introduces a trade-off by increasing thickness and potentially impeding mass transfer (as discussed earlier, the backing can reduce overall hydrophilicity and add to the structural parameter S). An important design consideration is thus module compatibility: The membrane must be designed in a format that can be readily integrated into standard module designs without excessive performance loss. For flat-sheet FO membranes, spiral-wound modules are widely used due to their high packing density and proven manufacturing techniques. TFC-FO membranes have been successfully deployed in spiral-wound configurations analogous to RO systems, demonstrating that FO membranes can achieve scalable module integration. However, long membrane envelopes in spiral-wound FO modules can experience a gradual decline in osmotic driving force from inlet to outlet (a phenomenon related to draw dilution and ICP along the flow path), which requires careful design of feed flow and spacer geometry to mitigate. Hollow-fibre FO (HFFO) membranes offer an alternative, inherently self-supported format with very high packing density, but they demand excellent fibre tensile strength to avoid breakage and can be prone to shell-side fouling if densely packed (54). In HF configurations, the pressure drop within the lumen increases as the length of the HF extends. Therefore, a minimum strength is necessary to withstand the pressure drop on the lumen side. Regardless of configuration, the membrane must endure operational stresses such as cross-flow shear, occasional hydraulic pressure during cleaning or testing, and repeated handling for module maintenance. Therefore, an ideal FO membrane achieves a balance between mechanical durability and performance: Its SL is robust enough for large-scale deployment (withstanding handling, cleaning, and long-term operation) yet as thin and porous as possible to preserve mass transfer efficiency. Innovations in support materials—for example, woven or non-woven supports with open structures, or mechanically reinforced ultrathin substrates—are continually being explored to meet this criterion.

In summary, high-performance FO membranes must not only excel in lab-scale metrics (water flux and solute rejection) but also be engineered for scalability, exhibiting the mechanical strength and module compatibility required for real-world FO applications.

The above design criteria are interrelated, and optimizing one aspect often impacts others. For instance, increasing membrane porosity and hydrophilicity can boost water flux but may slightly reduce mechanical strength or require chemical modifications; similarly, enhancing selectivity by tightening the AL might lower permeability. Chapter 2 highlights the trade-offs among permeability, selectivity, and durability using the example of TFC versus CA membranes. TFC membranes achieve higher flux and better selectivity (low B/Lp ratio) than CA, but CA offers greater longevity in challenging chemical environments. Likewise, a very low S (thin support) improves flux but can compromise the membrane's physical integrity unless new support structures are adopted. Engineering decisions in FO membrane design thus require a holistic approach: The optimal membrane finds a sweet spot that maximizes water permeability and minimizes salt passage and ICP, without sacrificing the structural strength and stability needed for operation. The ongoing advancements in FO membrane technology—from novel hydrophilic materials and anti-fouling coatings to nanostructured supports—are driven by these fundamental design principles. By meeting the outlined criteria for high Lp, low B, low S, favourable surface chemistry, and robust stability, next-generation FO membranes aim to significantly enhance performance and reliability, thereby accelerating the adoption of FO in sustainable water treatment and resource recovery.

With these foundational design criteria in mind, the following section on FO membrane fabrication methods will delve into the specific techniques and innovations that enable the practical realization of these performance goals in modern membrane development.

### 2.2.2 Membrane Fabrication Methods

Membranes specifically manufactured for FO use can be generally classified based on materials into cellulose, TFC, and biomimetic membranes, as discussed below.

#### 2.2.2.1 Cellulose Membranes

CTA membranes were the first commercially available membranes. Due to its unique structure and properties, it has been widely used in membrane technology. However, low pH, stability with free chlorine, rejection rate, and permeability were found to be the drawbacks of the CTA membranes.

CTA membranes are mostly used in domestic desalination applications and less concentrated industrial wastewater applications due to their good mechanical strength and lower fouling propensity. In the 2000s, a CTA-FO membrane with an improved asymmetric structure was developed and able to provide almost 13 times higher water flux than a commercial RO membrane (55). Although the CTA membrane presented improved performance due to hydrophilic properties, it was found to be susceptible to hydrolysis. In addition, the magnitude of the water flux achievable by the CTA membrane is relatively lower with respect to biomimetic membranes due to low water permeability (56).

#### 2.2.2.2 Thin-Film Composite Membranes

TFC membranes typically comprise three layers: a thin PA AL formed by IP, a porous polymeric SL that provides mechanical stability, and a non-woven fabric backing layer for additional strength and handling. Compared to TFC-RO, TFC-FO membranes are generally thinner, more porous, and more hydrophilic; however, the thick SL in FO membranes can exacerbate ICP, while the non-woven fabric may limit overall hydrophilicity by hindering water penetration. The best-known commercial vendors for cross-flow TFC-FO membranes are Porifera, Oasys Water, Sterlitech Corporation, Aquaporin, and Osmotic Engineering. The performance of these membranes is significantly higher than that of cellulose membranes; however, these types of membranes have certain drawbacks: The thickness of the SL is relatively high and requires modification by new SL structures. Furthermore, the surfaces of these membranes typically suffer from fouling over long periods. Moreover, the surfaces of these membranes are instantly polluted by bacterial growth and require antibacterial solutions for storage after testing.

The synthesis of the AL in TFC membranes via IP is often accompanied by challenges in completing the polymerization reaction. Incomplete reactions can compromise the integrity of the selective layer, leading to significant reductions in overall membrane performance.

A membrane's efficiency in FO is largely determined by its transport parameters, namely water permeability ($L_p$), salt permeability (B), and the selectivity ratio (B / $L_p$). These parameters provide a quantitative basis for evaluating and comparing membrane types under different operational conditions.

When assessing FO membranes, comparisons between TFC and CA membranes offer valuable insights. TFC membranes are generally characterized by high water permeability, which makes them highly effective for achieving elevated water flux. Conversely, CA membranes, although typically exhibiting lower permeability, demonstrate excellent structural stability and broad chemical compatibility, making them suitable for a wide range of applications.

By analysing these parameters, it is possible to evaluate the relative effectiveness of TFC and CA membranes not only in desalination but also in broader sustainable water treatment contexts. Such comparative assessments help identify trade-offs among permeability, selectivity, and durability, thereby guiding the advancement of FO technology for diverse industrial and environmental applications. Table 2.5 summarizes recent studies comparing the performance of TFC and CA membranes in FO processes.

Table 2.5 presents a distinct pattern where most TFC-FO membranes exhibit higher water permeability ($L_p$) and lower $B / L_p$ ratios than CA-FO membranes. This implies that TFC membranes allow water molecules to pass through them more readily compared to the CA membranes. The improved water permeability and reduced salt permeability in TFC membranes can be attributed to their unique dual-layer structure, which combines a highly porous SL with an ultrathin selective layer, allowing efficient water transport while limiting salt passage. Although recent studies

**TABLE 2.5**
**TFC-FO membrane performance and transport parameters**

| Membrane Type | Water Permeability "$L_p$", L m$^{-2}$ h$^{-1}$ bar$^{-1}$ | Solute Permeability "B", L m$^{-2}$ h$^{-1}$ | Water Flux, $J_w$ L m$^{-2}$ h$^{-1}$ | Provide Ratio | Draw Solution | Ref. |
|---|---|---|---|---|---|---|
| TFC-CNFs | 1.84 | 0.12 | 13.08 | | 1 M NaCl | (57) |
| TFC polyacrylonitrile | 2.12 | 0.027 | 37 | | 2 M NaCl | (58) |
| TFC-polyimide | 5.5 | 5.3 | 11.6 | | 1 M NaCl | (59) |
| TFC-silver-MOF nanoparticles | 2.24 | 0.410 | 42 | | 2 M NaCl | (60) |

that combined CA membranes with covalent organic frameworks (COF) particles have shown promising results (high water permeability and low salt permeability), TFC membranes have been generally highlighted as more promising structures for advancing FO technologies, offering more sustainable solutions for water purification and desalination compared to CA membranes.

The most important part of all the membranes is the chemical structure of the different synthesized membrane layers. Application of the most reactive materials to form a stabilized polymeric layer can significantly improve the performance of membranes. Through IP techniques, almost all TFC-FO membranes have been synthesized using two common monomers, MPD and TMC. Monomers with hydrophilic functional groups, such as carboxylic, hydroxyl groups, and sulphur trioxide, can facilitate polymerization and the change in hydrophilic structure of FO membranes.

#### 2.2.2.3 Biomimetic Membranes

The term "biomimetics" refers to the adaptation of naturally occurring systems and functional principles for the design of advanced engineering solutions (62,63). In membrane science, biomimetic membranes are broadly classified into unimolecular channels and self-assembled channels, both aiming to replicate the selective and efficient transport functions observed in biological molecules (64).

Unimolecular channels are typically based on cyclic macromolecules that provide well-defined pore structures. The most widely studied examples are pillar[n] arenes, composed of hydroquinone units interconnected by methylene bridges (65). These structures offer tunable cavities that can emulate nature-like transport pathways. In contrast, self-assembled channels are formed by monomers linked through non-covalent interactions, creating dynamic transport structures. Representative examples include zwitterionic polymers, dendritic dipeptides, and block co-polymers (BCPs), which mimic the functionality of AQPs. For instance, in zwitterionic polymers, $ZnBr(H_2O)$ and dicarboxylate anions organize into helical macrostructures, while dipeptides assemble into one-dimensional channels. Dipeptide-based channels, in particular, are regarded as promising alternatives for replicating AQP-like selectivity in biomimetic membranes.

BCP (block co-polymers) are a type of polymer consisting of two or more distinct polymer segments (blocks) that are covalently bonded together. In the context of biomimetic membranes, BCPs can self-assemble to form structures resembling biological membranes, such as polymerosomes. These polymerosomes are vesicle-like structures that mimic the properties of natural cell membranes, making them useful for applications in filtration and drug delivery and as models for studying membrane dynamics. BCPs can also mimic the cell membrane by producing polymerosomes.

BCPs reinforce the stability and durability of the membranes, as the properties of the membrane can be changed by increasing the chain length and introducing different functional groups in the polymers. In recent years, the advantages and disadvantages of using BCPs and lipid bilayers have been investigated, where BCPs have exhibited great performance by forming stable membranes with high durability and low synthetic cost (61). Reportedly, since the thickness of both lipid and polymeric membranes is low, the membrane becomes fragile. Thus, an SL is used to improve the stability of biomimetic membranes. Furthermore, based on the substrate structure and characteristic properties, different synthetic methods including vesicle fusion, cross-linking by chemical materials, vesicle adsorption, magnetic adsorption, and polymerization can be applied to synthesize membranes (62–64). Overall, biomimetic FO membranes typically consist of a thin, selective layer that mimics the structure and properties of biological membranes, supported by a porous substrate. The selective layer is engineered to have a highly ordered and uniform pore structure, with precisely tuned pore sizes and surface chemistry, enabling efficient water transport while rejecting other molecules and contaminants. The incorporation of biomimetic FO membranes in advanced commercial RO applications offers the following advantages:

- Improved Water Flux: Biomimetic membranes demonstrate significantly higher water permeability and selectivity compared to conventional membranes, resulting in increased water flux and enhanced system throughput (66).
- Enhanced Fouling Resistance: The biomimetic design enables the incorporation of surface modifications or self-cleaning mechanisms inspired by natural systems, which reduce fouling tendencies and extend membrane lifespan.
- Reduced RSF: A key advantage of biomimetic membranes lies in their ability to deliver high water flux while maintaining low RSF. The reduced RSF per unit volume of water transported minimizes contamination of the FS and shortens the required operating time, thus improving the overall separation efficiency.
- Energy Considerations: Although biomimetic membranes facilitate high water flux, the overall energy consumption in FO processes is not substantially impacted, as the process inherently requires minimal energy input—primarily for FS and DS circulation. Thus, the primary benefit is not in direct energy savings but rather in achieving higher water flux with comparatively lower RSF, enhancing process selectivity and reducing solute leakage.

- Sustainability: The application of biomimetic materials and green chemistry principles in membrane fabrication can reduce the environmental impact associated with membrane production and operation, supporting the broader goals of sustainable engineering.

Commercial AQP-embedded TFC membrane was made of AQP water channels incorporated into the selective layer of HF membranes (67,68). They exhibit excellent chemical resistance to cleaning agent such as sodium hypochlorite (NaOCl), chelating 5 mM disodium ethylenediaminetetraacetic acid (EDTA), HCl, SDS, and anionic detergent (Alconox) with a pH ranging from 1.4 to 11.7. After chemical cleaning, a considerable increase in the water flux and salt rejection was achievable (69). The presence of AQP protein was shown to be more effective for Alconox cleaning processes, increasing water flux by up to 75% after cleaning as compared to unmodified membranes, thereby suggesting that AQPs help the detergent enlarge the membrane pores, leading to higher flux performance by the bio-membrane (70). However, over long-term operation, chemical cleaning may negatively impact the AQP protein channels, causing reduced membrane performance (in terms of flux and solute rejection efficiency) (71,72).

Biomaterials have also been used in HF membranes as a new technique for measuring the water flux of these membranes. Ren and McCutcheon (2018) (67) used AQP inside HF membranes to improve FO membrane performance in terms of water permeability and selectivity. The morphology analysis of the AQP membrane revealed a dense structure of HF with reduced surface roughness. It was observed that the membrane has a uniform porous structure with high aspect ratio and pore density compared to conventional polymer substrates while exhibiting a higher water flux of 27.6 ± 3.6 L $m^{-2}$ $h^{-1}$ and superior membrane selectivity of 0.01 g $L^{-1}$. With respect to the flat sheet AQP membrane, the HF AQP membrane had a comparatively lower structural parameter value (~210 nm). This may be because of the dense middle zone of the flat sheet AQP, which was absent in the case of the HF counterpart (2,67).

A ceramic-based AQP biomimetic membrane utilizing anodic aluminium oxide (AAO) as a substrate was reported (66). In this study, the AAO substrate was functionalized with aminosilane followed by polydopamine (PDA) coating; then, AQP vesicles were immobilized on the functionalized AAO substrate surface using an electrokinetic method.

AQPs have also been used as nanocomposites of FO membranes, where the functional groups of AQP are mixed with other materials and added to the membrane synthesis process to form new bio-membranes. A thin-film nanocomposite (TFN) membrane incorporating GO, aquaporin Z (AQP-z), and a reconstituted nanostructure within the AL was prepared using IP techniques (73). In this study, GO was first added to the MPD solution and sonicated for 30 min, after which AQN was slowly added to the solution. Reportedly, the presence of GO flakes and AQP significantly reduced surface roughness as the membrane roughness decreased from 51.2 nm to 32.9 nm. The contact angle also decreases from 83° to 55°. Owing to the new surface structure, the synthesized TFC-AQP/GO membrane exhibited about three times higher water flux (24.1 L $m^{-2}$ $h^{-1}$) and approximately 27% lower RSF than the TFC membrane.

Biomimetic membranes can be used in wastewater treatment for rejection of pollutants and trace organic contaminants (66,74,75). A recent study investigated the impact of various AQP-based biomimetic membranes on greywater treatment efficiency (76). The study revealed that the rough membrane surface and low zeta potential facilitate the adsorption of linear alkylbenzene sulfonates (LAS), thus improving the water flux and rejection of calcium/magnesium ions. In recent studies, biomimetic membranes were also utilized in liquid food processing industries for the concentration of sugarcane juice and plant extracts (such as tea) (2,77,78).

Overall, biomimetic membranes push the boundaries in FO because of their capability of effectively marrying biological ingenuity with synthetic innovation. Though many of these challenges remain, their prospect of delivering high-performance and sustainable solutions to various global issues with water and energy places biomimetic membranes on the frontier of membrane science to well illustrate the transformative power of designs from nature.

#### 2.2.2.4 Zwitterionic-Based FO Membranes

Zwitterions are known to be valuable materials for membrane modification due to their strong hydrophilicity, bipolar properties, and non-toxicity, which enhance water flux, contaminant removal and anti-fouling properties (79). Zwitterions are generally considered a specific material with a similar number of anions and cations in their polymer chains. Zwitterionic materials represent a unique structure in membranes, adding hydrophilic groups to the surface of a membrane and improving membrane stability with cross-linked structures.

Recent efforts in the surface modification of FO membranes with novel zwitterion nanocomposites to improve anti-fouling and anti-bacterial properties have shown good results. A recent study demonstrated the application of two novel nanocomposites—3-bromopropionic acid and 1,3-propane sulfone—chemically linked to silver-based metal–organic frameworks (MOFs) and grafted onto the top layer of FO membranes using 2 wt.% *N,N*-diethylethylenediamine (DEDA), 0.03 wt.% potassium persulfate, and 0.02 wt.% sodium metabisulphite (80). Subsequently, an aqueous 1,3-propane sulfone solution (5 wt.%, pH 5.0) or a BPA solution (5 wt.%, pH 5.0) was mixed for 20 h at 40 °C to graft sulphobetaine zwitterions on the membrane. Reportedly, the synthesized FO membrane showed that the dynamic behaviour of water molecules passing through the membrane was considerably changed by introducing novel modifiers on the membrane surface. It was assumed that the presence of zwitterions in the PS solution significantly increased water permeability because of their stronger interaction with water molecules. The RSF also improved with the introduction of new functionalized materials. This was attributed to the lower negative charge on the surface of the membrane after functionalization. This, in turn, decreased the electrostatic repulsion between the FS and the AL, resulting in a higher solute flux. The transport performance of FO membranes was enhanced by the newly introduced materials. Regarding antibacterial properties, FO membranes containing zwitterionic composites showed a high removal rate of *E. coli* as a surrogate bacterium for biofouling organisms. Over 99% of bacterial elimination resulted from the use of silver-based MOFs. Table 2.6 shows zwitterion derivatives applied to FO membranes based on previous studies.

**TABLE 2.6**
**Literature overview of the use of zwitterionic materials by different types of forward osmosis membranes**

| Zwitterion Composite | Method of Membrane Modification | Membrane Performance | Ref. |
|---|---|---|---|
| 4-Bis (3-aminopropyl)-piperazine propane carboxylate with silver nanoparticles | TFC active layer was synthesized using MPD and TMC monomers. Post-fabrication, the membrane was immersed in a zwitterionic solution for surface modification | - Achieved long-term anti-fouling performance with only 8% flux decline (vs. 50% in the unmodified membrane).<br>- Demonstrated high regeneration potential.<br>- Salt rejection up to 97%.<br>- Water flux increased to >30 $L{\cdot}m^{-2}{\cdot}h^{-1}$ (vs. 22 $L{\cdot}m^{-2}{\cdot}h^{-1}$ for the unmodified FO membrane). | (81) |
| Z-CNT | Z-CNT solution was uniformly coated onto the TFC FO membrane surface. Functional groups facilitated attachment via the van der Waals forces | - Significantly reduced RSF: NaCl (55.5%), $NH_4Cl$ (70.8%), and $NH_4HCO_3$ (61.9%).<br>- Low flux decline over 12-day operation: 14.9% (vs. 54.5% in control TFC FO membranes).<br>- Enhanced bidirectional solute flux control. | (82) |
| $P{-}SO_3{-}NH_2$ and $P{-}2SO_3{-}2NH_2$ | Zwitterions were grafted onto the active layer surface of the nascent membrane post-polymerization. | - Increased water flux by 47%.<br>- Flux recovery of 94% (vs. 82% for unmodified membrane).<br>- Exhibited excellent fouling resistance. | (83) |

Z-CNT: zwitterion-functionalized carbon nanotubes; $P{-}SO_3{-}NH_2$: n-amino-ethyl piperazine propanesulfonate; $P{-}2SO_3{-}2NH_2$: 1,4-bis (3-aminopropyl) piperazine propanesulfonate; RSF: reverse solute flux

Overall, based on the literature, using zwitterions in FO membranes has proven to be promising. However, further studies are needed to fully understand the performance of zwitterionic membranes in different conditions. FO zwitterionic-based membranes can be effective with hydrophilic and amine-shaped structure zwitterions. The presence of sulphur in zwitterionic sulphobetaine can also facilitate the polymerization process and the formation of a well-connected membrane with high stability. However, the use of zwitterion derivatives in FO membranes is limited, and this gap needs to be addressed in future studies. This promising material is widely available for synthetic processes. Zwitterions can also be used with other materials via coating or grafting and then attached to the AL or dispersed through polymerization processes in the aqueous phase. Since zwitterions are highly hydrophilic, they enhance the water permeability of FO membranes. The charges applied by zwitterions can increase the anti-fouling propensity of FO membranes, as foulants settle less on the surface of the membrane because of the Donnan effect in zwitterion-modified membranes, thereby suggesting that zwitterions are highly effective materials for developing high-efficiency FO membranes.

#### 2.2.2.5 Metal–Organic Frameworks FO Membranes

MOFs are an emerging type of structurally flexible porous material known as porous co-ordination polymers, consisting of inorganic metal ions co-ordinated to organic units by strong bonds. The channel-like structure of MOFs has been successfully effective in the enhancement of different membranes.

Zeolitic imidazolate framework-8 (ZIF-8) was synthesized and modified with PDA to enhance the dispersion efficiency of nanoparticles, yielding a composite designated as PDA@ZIF-8 (84). The modified FO membrane was prepared by coating a gel solution containing PDA@ZIF-8 and PVA onto a PES substrate. The performance and characteristics of the modified membrane were compared with those of a conventional PA TFC prepared via IP. The presence of ZIF-8 on the membrane surface substantially enhanced the flux of the novel prepared membrane (from 8.7 to 14.2 L $m^{-2}$ $h^{-1}$) while maintaining its selectivity (SRSF = 0.44 to 0.4 g $L^{-1}$). Moreover, the prepared membrane displayed stronger anti-fouling properties to organic foulants than the PA TFC membrane because of its improved hydrophilicity and smoother surface.

UiO-66-$NH_2$ MOF was synthesized and subsequently modified with graphene quantum dots (GQDs) to enhance water affinity on the MOF surface and to improve both the hydrophilicity and porosity of FO membranes integrated with these modified MOFs (85). The membrane modified with MOFs (0.25 wt.%) exhibited higher water permeability of 4.88 L $m^{-2}$ $h^{-1}$ $bar^{-1}$ compared to the control PA TFC, with water permeability of 2.23 L $m^{-2}$ $h^{-1}$ $bar^{-1}$ while maintaining selectivity. The modified membrane exhibited a lower fouling propensity against bovine serum albumin (BSA), which was attributed to the more hydrophilic and smoother surface of the modified membrane.

A PA TFC membrane was modified with PDA and MOF-801 to enhance salt rejection and improve heavy metal ion removal (86). The modified membrane was prepared via IP of an MPD/PDA aqueous solution and TMC/MOF-801 organic solution on a polyacrylonitrile (PAN) UF membrane. The modified membrane exhibited a higher pore tortuosity of the PA layer, which increased the salt and heavy metal rejection of the modified membrane compared to the pristine PA TFC. In addition, the membrane modified with MOFs (0.01 wt.%) had higher water permeability (2.12 L $m^{-2}$ $h^{-1}$ $bar^{-1}$) and lower reverse salt permeability (0.88 L $m^{-2}$ $h^{-1}$ $bar^{-1}$) compared to the pristine PA TFC, with water permeability of 1.81 L $m^{-2}$ $h^{-1}$ $bar^{-1}$ and reverse salt permeability of 1.49 L $m^{-2}$ $h^{-1}$ $bar^{-1}$.

A facile deposition method was developed for covalently bonding silver-based MOFs (Ag-MOFs) onto a PA TFC membrane, aiming to enhance its anti-fouling and antimicrobial properties (87). The modified membrane had higher hydrophilicity, roughness, and negative surface charges owing to the anchoring of the Ag-MOFs on the membrane surface. Water permeability of the modified membrane (0.94 L $m^{-2}$ $h^{-1}$ $bar^{-1}$) was slightly lower than that of the pristine PA-TFC (1.1 L $m^{-2}$ $h^{-1}$ $bar^{-1}$), as the anchored Ag-MOFs on the membrane surface reduced the accessible membrane surface for water permeation. Contrastingly, the reverse salt flux of the modified membrane decreased from 0.33 L $m^{-2}$ $h^{-1}$ to 0.27 L $m^{-2}$ $h^{-1}$ compared to the pristine PA TFC due to the higher negative charge of the modified membrane rejecting ions. More importantly, the modified membrane caused a 100% reduction in live bacteria on the membrane surface, indicating the excellent antibacterial property of the modified membrane.

A novel method involving interlayer decoration with MOFs during the IP process was developed for fabricating TFC membranes (88). Their study compared the performance of control TFC and TFC modified by MOFs through the conventional method of incorporating MOFs in TMC/hexane solution (noted as the conventional method) and TFC modified by introducing an interlayer of MOFs before adding TMC/hexane solution (denoted as the decorating method). The results showed that the decorating methods caused a greater improvement in water permeability, a higher enhancement of salt rejection, and a lower structural parameter. Therefore, decorating an interlayer inside the PA layer can be an effective strategy for developing high-performance TFC membranes for FO processes. A PA TFC membrane was modified by incorporating MOF nanocrystals composed of silver (I) and 1,3,5-benzene tricarboxylic acid (3HBTC) into the PA layer (89). The prepared membrane showed improved water permeability from 2 to 7.1 L $m^{-2}$ $h^{-1}$ $bar^{-1}$ compared to the control PA TFC membrane, without detrimental effects on membrane selectivity. In addition, testing real seawater caused flux declines of 7% and 18% for the modified and unmodified membranes, respectively, which confirmed the improved performance stability of the modified membrane.

Table 2.7 presents a summary of the studies that have modified the PA layer of TFC FO membranes. Most studies have reported higher water permeability and anti-fouling properties of membranes modified with MOFs than control PA TFC membranes. UiO-66 MOFs have been used more than other types of MOFs because of their excellent water stability, which improves the performance of membranes without the potentially undesired leaching of components (90,91). Further studies are suggested to investigate the stability of MOFs in the PA layer during long-term operation and their impact on pollutant removal. A few studies have investigated the use of MOFs for improving mass transfer in FO substrates to control the ICP effect in FO processes. The incorporation of UiO-66 nanoparticles into the polymer dope of a polysulphone (PSf) membrane during the phase inversion process was investigated (91). The presence of MOFs in the FO substrates reduced the calculated structural parameter (S) of the membrane compared with that of the TFC control, thereby mitigating the ICP effect and improving the water flux. Similar results were reported using TFC substrates modified with MOF-5 (Zn) during the phase inversion process (92).

#### 2.2.2.6 Other New Additives

In recent years, considerable efforts have been made to modify the surfaces of novel materials, especially those with unique structures, to significantly improve membrane permeability and fouling. Novel additives with specific functional groups can potentially strengthen membrane structure and introduce hydrophilic cross-linked

**TABLE 2.7**
**Polyamide thin-film composite forward osmosis membranes modified via metal–organic frameworks**

| MOF Composite | Method of Modification | Membrane Performance | Ref. |
|---|---|---|---|
| Aspartic acid-modified MIL-53-$NH_2$ | MP nanoparticles were added to the MPD and 2,6-diaminopyridine solution of the IP process | • Improved selectivity<br>• Higher water flux<br>• Good resistance to fouling | (93) |
| MIL-53 | MOF nanocrystals added to an organic solution (consisting of TMC in n-hexane solution) of the IP process | • Improved tortuosity<br>• Higher water permeability<br>• High rejection rate (99.73%) for doxycycline (an antibiotic) | (94) |
| CuBTC MOF | Blended with PES support layer; CS/TMC IP | • Water flux improved by 2.8 times<br>• Improved anti-fouling properties | (95) |
| MIL-101 (Cr) | Incorporated into PA layer via IP | • Enhanced water flux and selectivity<br>• 81% of heavy metal (As and Cr) rejection | (96) |
| ZIF-8 | In situ hydrophilization using tannic acid | • Increased water flux from 12.2 L $m^{-2}$ $h^{-1}$<br>• Improved anti-fouling properties<br>• High lithium recovery | (97) |

groups on the membrane surface. For the introduction of new additives into the membrane structure, different methods were explored. The two main methods followed are as follows: firstly, the incorporation of porous additives into the membrane SL through the phase inversion method and secondly, the incorporation of additives in the AL through mixing with the organic/aqueous phase or grafting/coating to change the surface roughness. Using additives that can create a selective surface and fully interact with the polymer-based structure of the molecule through these two methods reinforces membrane effectiveness and therefore is highly recommended for the modification of the FO membrane AL. For example, COFs are perfect nanofillers that increase membrane selectivity and hydrophilicity because of their tunable structures, porosities, and hydrophilic groups. They can also be applied to the SL and prevent RSF because they have aromatic rings that are highly selective to water molecules while efficiently rejecting salts from permeating from DS to FS. For example, an FO membrane with a CA-based structure was synthesized using sulfonated COFs. The presence of $SO_3H$ group increases hydrophilicity, resulting in improved permeation rate. The mixed CA and sulfonated COF were cast onto a nylon support using a casting knife (thickness, 200 μm). The sulfonated COFs created a dense surface, which prevents the RSF. In addition, the COFs created a negatively charged surface, which increased the probability of salt rejection by ionic repulsion.

Surface modifications of macromolecules have emerged as a promising strategy to enhance the anti-fouling properties of membranes. In FO membranes, the application of surface-modified macromolecules as polymer additives offers a unique opportunity to tackle the fouling issues effectively. For instance, PEG is known for its high hydrophilicity. Hence, when PEG-based additives are incorporated in the membrane, they inhibit the adhesion of the majority of organic and inorganic foulants by creating a hydration layer on the membrane surface (98). Overall, the mechanisms underlying the anti-fouling properties of surface-modified macromolecules involve steric hindrance, electrostatic repulsion, and hydration effects. Steric hindrance occurs when the bulky macromolecules physically block the foulant, thereby preventing it from adhering to the membrane surface (99). Electrostatic repulsion occurs due to the charged nature of macromolecules; the presence of like charges on the membrane surface and foulants generates repulsive forces, thereby preventing adhesion of foulants (100). Additionally, the hydration effects of hydrophilic additives create a water-rich environment that further deters the interaction of the membrane surface with hydrophobic foulants (101).

Carbon quantum dots (CQDs), also known as carbon dots, are zero-dimensional, super-hydrophilic nanomaterials that enhance the surface hydrophilicity of FO membranes due to their abundance of epoxy, hydroxyl, and carboxyl functional groups. Their nanoscale size allows for uniform dispersion and integration into the membrane matrix, facilitating the formation of a defect-free and highly hydrophilic AL. Unlike tubular and sheet-like nanomaterials, CQDs can be easily incorporated onto the membrane surface without compromising its integrity. TFC membranes fabricated using carbon polyetherketone (PEK-C), polyethyleneimine, and CQDs have demonstrated high water flux, and improved RSF performance (102) synthesized PEK-C-based FO membranes by applying polyethyleneimine and CQDs. The resulting membrane exhibited excellent water flux and RSF. CQDs derived from palm oil have also been successfully integrated into the PSf substrate via phase inversion techniques, resulting in FO membranes that not only exhibit enhanced separation efficiency but also possess antimicrobial properties, contributing to extended membrane lifespan in food processing applications (103).

Overall, the incorporation of novel modifiers such as MOFs and zwitterionic materials has shown significant promise in enhancing the efficiency of FO membranes. Nevertheless, further efforts are needed to improve their performance and anti-fouling properties against both conventional and emerging contaminants while maintaining cost-effectiveness. The following section discusses the respective roles of the AL and SL in determining FO membrane performance.

Role of Active and Support Layers on FO Performance:

The TFC membrane is composed of a resiliently bound layer with exceptional mechanical strength and a compact AL with remarkable salt rejection capabilities (104). The unrivalled permeability, chemical stability, selectivity, and design versatility have made the TFC membrane the preferred choice over other FO membranes. The performance of TFC-FO membranes is critically governed by the structure and properties of both the AL and the SL. While the AL primarily dictates water permeability, solute rejection, and RSF, the SL plays a pivotal role in ICP, provides mechanical stability, and facilitates flow pathways for mass transfer.

The SL of TFC-FO membranes is predominantly produced via phase inversion, in which a polymer solution (dope) is applied onto either woven or non-woven SLs and then immersed in a coagulation bath. The choice of support fabric plays a critical role in defining both mechanical performance and transport behaviour. Woven substrates, owing to their tight, interlaced structure, deliver superior mechanical strength, whereas non-woven substrates, being thinner and more porous, minimize ICP, leading to enhanced water flux.

The porous substrate structure is further tailored by manipulating several fabrication parameters, including:

- Polymer Concentration: Higher polymer content increases solution viscosity, slows phase separation, and leads to thicker top layers with lower porosity and fewer micro-voids. Conversely, low polymer concentrations (e.g. 10 wt.%) result in thinner skin layers, wider channels with thinner walls, lower resistance, and higher porosity—ultimately enhancing water permeability.
- Solvent Choice: Solvents such as dimethylformamide (DMF) and *N*-methyl-2-pyrrolidone (NMP) significantly influence the substrate's morphology. Membranes cast with DMF generally exhibit greater surface roughness, larger pore sizes, and higher porosity compared to those cast with NMP. While DMF-based supports promote greater water and salt permeability, NMP-based membranes demonstrate improved resistance to organic fouling. For example, under fouling conditions, NMP-based TFC membranes showed a flux decline of 7.47 ± 0.15%, compared to 12.7 ± 2.62% for their DMF-based counterparts.
- Additives: The incorporation of additives like PEG-400 or lithium chloride (LiCl) significantly alters the final membrane morphology. For instance, 6 wt.% PEG-400 improves both water flux and salt rejection, while 3 wt.% LiCl not only enhances water flux but also increases RSF under AL-FS orientation.
- Environmental Conditions: High air humidity during fabrication (30–50%) leads to slower solvent evaporation, resulting in a denser, less porous support structure with larger micro-voids in the upper layer.

Optimizing the SL is inherently complex due to the interplay of multiple parameters such as pore size, surface hydrophilicity, roughness, and even the morphology of the bottom surface. The chemical structure of the polymer also influences support properties; for example, non-sulfonated supports often result in dense bottom layers, while sulfonated substrates exhibit higher porosity. Incorporating sulfonated polymers like sulfonated polyetherketone (SPEK) into a polysulphide (PSU) matrix has been shown to form sponge-like porous structures, enhance hydrophilicity, and reduce the structural parameter (S) (104).

The AL of membranes is typically formed via IP, most commonly using MPD as the aqueous-phase monomer and TMC as the organic-phase monomer. This reaction produces a cross-linked PA selective layer over the porous support. Although alternative monomer combinations have been investigated, the MPD-TMC system remains the industry standard due to its favourable separation performance and robustness.

The structure and properties of the PA layer can be tuned by adjusting:

- MPD/TMC Ratio: A high MPD/TMC ratio promotes dense cross-linking, leading to improved salt rejection but reduced water permeability. Conversely, a lower MPD/TMC ratio favours higher permeability due to a looser network but may compromise solute rejection.
- Monomer Concentration: Studies have shown that a 2 wt.% MPD solution can result in a denser, less permeable rejection layer with enhanced solute retention, helping to mitigate ICP. Meanwhile, reducing TMC concentration to 0.1 wt.% has been associated with improved water flux and salt rejection, likely due to reduced membrane resistance and the predominance of ICP-reducing effects.
- IP Reaction Time: The duration of the IP step also significantly affects the membrane's AL thickness. A short IP time (e.g. 15 seconds) may result in incomplete layer formation and poor rejection, while excessively long IP times (e.g. 45 seconds) produce a thicker PA layer, increasing transport resistance and decreasing water flux.

The performance of TFC-FO membranes is a function of not only the individual layers but also the interfacial compatibility and synergy between the SL and the AL. A well-matched interface enhances layer adhesion, prevents delamination, and promotes uniform PA formation. Furthermore, a hydrophilic and smooth support facilitates better monomer diffusion during IP, leading to a more uniform and defect-free AL.

The performance of TFC-FO membranes is intricately linked to the structural and physicochemical properties of both the SL and the AL. Through careful control of fabrication parameters—such as polymer concentration, solvent selection, additive inclusion, environmental conditions, and IP reaction settings—membrane developers can finely tune the balance among water flux, salt rejection, and resistance to fouling or ICP. While considerable progress has been made, continued innovation in material science and membrane engineering remains essential for advancing the efficiency and scalability of TFC-FO membranes for real-world water treatment and resource recovery applications.

In recent years, numerous research studies on membrane fabrication and modification have been conducted using different polymers and additives to modify the SL structure and to alter the dense-layer characteristics. The membrane development generally goes through three processes: CTA-flat-sheet membrane, TFC-flat-sheet membrane, and TFC-HF membrane. Phase inversion, IP, electrospinning, and layer-by-layer (LbL) deposition are the typical fabrication methods of double-skin membranes (105).

### *2.2.2.6.1 Interfacial Polymerization*

IP is the chemical reaction between two liquids or liquid air, which leads to polymer synthesis. IP is an important method for TFC membrane fabrication for FO, permitting fine adjustments of crucial properties of the AL, including thickness, permeability, and selectivity (106–111). This method involves the interaction of two monomers at the interface of two immiscible phases (usually an aqueous solution containing amine monomers such as MPD and an organic solvent (e.g. hexane) containing acyl chloride monomers such as TMC). IP has emerged as the TFC membrane industry standard due to its ease of process implementation and commercial scale, especially in challenging separation applications such as desalination and wastewater treatment, where selectivity and stability are essential.

Recently, IP-based innovations have sought to mitigate inherent challenges, like fouling due to surface roughness and the trade-off between permeability and selectivity. In addition, traditional PA layers with rough nodular structures favour the accumulation of foulants, contributing to accelerated performance decline. To overcome this problem, researchers report about the addition of hydrophilic additives (such as zwitterionic polymers and silica nanoparticles) in the aqueous phase during IP. These modifications yield smoother, more fouling-resistant surfaces (biofilm formation reduction by up to 60% at no flux cost). Additionally, incorporation of nanomaterials such as GO and MOFs into the PA matrix has resulted in more competent TFN membranes. For instance, GO-doped PA layers exhibit 30–50% higher water flux, as the incorporation of GO leads to the creation of nanochannels conducive to fast transport, while MOFs like ZIF-8 add molecular sieving functionality that improves selectivity towards monovalent species (112–114).

Reportedly, the fatty acid concentration and other IP parameters are important in tailoring membrane performance. Higher MPD gives thicker and more selectivity layers with declining permeability. Reaction durations and solvent extraction methods can also modulate pore structure and eliminate defects in a reversal manner upon fine-tuning (115). Strategies for polymerization of interfacial polymer that led to improved PA morphology have enabled the development of ultrathin and defect-free ALs with high permeability (up to 40 L $m^{-2}$ $h^{-1}$) (116).

Recent research validated the significance of controlling viscosity in IP and showcased how the precise addition of additives can enhance membrane performance by regulating permeability, improving selectivity, and controlling structural stability. The research illustrates how viscosity modulation caused by additives can regulate PA layer growth to yield membranes with improved permeability, selectivity, and structural integrity. It was revealed that PVP increases solution viscosity, thereby slowing the diffusion rate of amine monomers. Due to slow diffusion, a thinner (~30 nm), smoother, and compactly cross-linked PA layer was created. The reduced PA layer enhances water transport, which is accountable for the higher water flux in the PVA-modified membranes by 25% compared to the original membranes. Additionally, the denser and smoother cross-linked network suppressed pore flaws and free volume, resulting in reduced specific salt flux by 43%.

Due to its versatility, IP can be integrated with other fabrication techniques. For instance, incorporating LbL assembly preceding IP allows for precise tuning of surface charge and hydrophilicity, resulting in compatibility with a wide array of FSs.

Despite these advantages, some critical challenges need to be addressed for the efficient scaling up of the IP process for industrial applications of the FO process. The employment of harmful solvents such as hexane also brings ecological concerns, and there has been extensive investigation on more environmentally benign alternatives (such as ILs) and solvent-free portrayals of IP (117).

#### 2.2.2.6.2 *Layer-by-Layer Assembly*

LbL assembly is a technique used to synthesize multiple layers on the surface of a membrane. These layers are usually assembled by depositing different layers with different surface charges onto each other to form a multilayer structure for the purification process. LbL assembly, a more complex manufacturing technique, builds an alternating nanoscale layer of positively and negatively charged material on a substrate, facilitating fine, ordered control of the membrane structure (109). Reportedly, by utilizing diverse molecular interactions, this membrane fabrication technique can be utilized to construct ultrathin ALs with tailored properties to improve FO performance.

This is one of the most effective methods for synthesizing membranes with thin layers that exhibit unique properties of water selectivity in separation systems. The LbL approach was initially developed by Decher in 1997 to prepare thin-film structures with different layers using polyelectrolyte spacer layers to create PE multilayers. In these techniques, polycations are absorbed on the surface of the polyanions and consequently form a multilayer membrane. After each LbL assembly, the membrane is rinsed with DI water to eliminate weak polychain structures on the surface layers. One of the main benefits of the LbL techniques is the synthesis of membranes with nanosized layers. Accordingly, the thickness of each layer can be adjusted during synthesis. In membrane technology, the LbL technique can form defect-free layers with nanosized thicknesses to improve membrane performance, especially fouling. The synthesized multilayer membranes with tunable structures obtained by the LbL method can significantly improve membrane selectivity and flux. Therefore, it can be concluded that using the LbL technique to fabricate different types of membranes can significantly change the structure of the membranes and create a unique structure to transfer water molecules without fouling.

#### 2.2.2.6.3 *Phase Inversion and Nanomaterial Incorporation*

Phase inversion plays a pivotal role in developing asymmetric polymeric membranes used for applications of FO. This allows them to create a porous SL containing a dense selective skin. The membranes are created by dissolution of polymers (such as PSf, polyethersulphone (PES), or CA) in a solvent, casting it into a film, and then immersing the film in a non-solvent bath. The resulting rapid solvent/non-solvent exchange initiates phase separation, resulting in a microstructure with interconnected pores in the SL and a thin, dense top layer required for solute rejection. Due to their established common usage, conventional phase inversion membranes have shortcomings including moderate hydrophilicity, mechanical fragility, and fouling propensity, and much research has focused on integrating nanomaterials to augment performance.

The integration of nanomaterials into phase inversion matrices represents a major step forward, making use of the properties of such nanoparticle materials to address FO-specific challenges such as ICP and poor water flux. Phase inversion can be combined with $TiO_2$, GO, CNTs, COFs, and MOFs mixed into a polymer solution. These additives change the thermodynamics and kinetics of phase separation to improve pore structure and membrane performance. For example, in the FO desalination process, hydrophilic $TiO_2$ nanoparticles improved surface wettability and decreased contact angles by up to 40% and increased water flux by 20–30% (111). The oxygen-functional-rich GO nanosheets can generate nanochannels in the polymer matrix, increasing the rate of water transport with > 95% salt rejection (118). They function as molecular sieves, allowing ions and organic contaminants to be excluded at the molecular level due to their unique porosity and tunable pore sizes. For example, ZIF-8-incorporated membranes were shown to achieve 99% rejection of divalent ions in hypersaline feeds.

The mechanical strength and anti-fouling ability of the composite are further enhanced by the introduction of nanomaterials. CNTs with excellent tensile strength can provide reinforcement for polymer matrices and significantly increase membrane stability against hydraulic pressure by 50%. However, more work is needed towards achieving an even distribution of nanomaterials. However, clumping during phase inversion can form defects that undermine selectivity. In this regard, sonication, surface modification, and covalent attachment are used to enhance the dispersion of nanoparticles. $TiO_2$ modified with silane is a good example where compatibility with PSf is enhanced, thus limiting $TiO_2$ aggregation. Scaling up production and cost management remain challenges, as GO or MOFs are expensive, high-grade nanomaterials to manufacture.

These hybrid membranes have applications in numerous FO systems. The $TiO_2$-embedded PES membranes may provide high flux with fouling resistance, thus being beneficial for treating oily or organic-rich industrial effluents. In nutrient reclamation, MOF-embedded membranes extract phosphate and ammonium selectively from agricultural drainage for the purpose of nutrient reuse, aligning with the principles of the circular economy (52).

Future research activities may focus on green synthesis approaches such as water-based phase inversion systems and multifunctional nanocomposites responsive to stimuli including pH and/or temperature. Advances in machine learning (ML) for optimizing interactions between nanoparticles and polymers combined with scalable fabrication methods such as roll-to-roll casting will bridge the gap between lab-scale discoveries and fungible, industrial-scale realization. This strategy merges the adaptability of phase inversion with the game-changing revelations of nanomaterials, seeking to drive the advancement of FO membranes as practicable remediation tools for global water and energy challenges. Table 2.8 provides a comparative overview of various FO membrane fabrication techniques, detailing their advantages, limitations, and key recommendations for enhancing scalability, performance, and industrial feasibility.

In all FO membrane fabrication methods discussed so far, TFC-FO membranes remain the most promising membranes for potential FO applications, owing to their high rejection capabilities.

Hydrophilic synthetic polymers and monomers such as CA, CTA, polyvinylidene fluoride (PVDF), PES, PSf, and PAN, PES, PSf, and PAN were used for the fabrication of layers (Table 2.9). Additives such as silica, graphene, carbon nanotubes, and zeolites are employed to enhance the structural parameter and membrane performance.

**TABLE 2.8**
**Overview of different fabrication techniques for forward osmosis membranes (110)**

| Fabrication Method | Advantages | Limitations | Recommendations |
|---|---|---|---|
| Interfacial polymerization for active layer | • Industry standard method for TFC membranes<br>• High selectivity and stability<br>• Compatible with nanomaterial modifications | • Requires a strict process to control to avoid defects<br>• Uses hazardous organic solvents (e.g. hexane)<br>• Surface roughness leads to fouling | • Optimize monomer diffusion for uniform layer formation<br>• Uses green solvents or solvent-free approaches<br>• Integrates nanomaterials to improve selectivity and permeability |
| Electrospinning for support layer | • Produces highly porous, nanofibre-based support layers<br>• Reduces ICP and enhances water flux.<br>• Allows for tunable membrane properties | • High production cost and scalability challenges<br>• Requires additional processing for active layer deposition<br>• Weak mechanical strength | • Develop scalable electrospinning techniques (such as roll-to-roll electrospinning)<br>• Enhance mechanical strength via cross-linking or hybrid structures<br>• Improve cost-effectiveness through material selection |
| Layer-by-layer (LbL) assembly for the support layer | • Precise control over membrane thickness and surface properties<br>• Enables multifunctional coatings (e.g. anti-fouling responsive behaviour) | • Labour-intensive and time-consuming process<br>• Scaling up for industrial applications is difficult | • Implement automated LbL deposition methods (such as spray LbL and roll-to-roll coating)<br>• Explore hybrid approaches integrating LbL with other fabrication techniques |
| Phase inversion for support layer | • Simple and cost-effective fabrication<br>• Produces robust support layers<br>• Compatible with nanomaterial incorporation for performance enhancement | • Moderate hydrophilicity and fouling tendency<br>• Less tunability compared to electrospinning and LbL<br>• Uneven nanomaterial distribution can cause defects | • Optimize polymer concentration and solvent exchange rates to improve porosity.<br>• Use surface modifications (such as plasma treatment) for better wettability<br>• Improve nanomaterial dispersion techniques |

Apart from SL, the selective layer also plays a crucial role in rejecting different contaminants, which determines the water permeability and selectivity. Polymers such as PA and piperazine are commonly used for the synthesis of the FO membrane selective layer. Modification of the selective layer by grafting or coating various nanomaterial additives can improve the selectivity, hydrophilicity, and anti-fouling properties of the membrane. Incorporation of CNTs in the PSf double-skinned membrane by IP exhibited enhanced water flux (about 54%) compared to the TFC membrane (52). Reportedly, the amount of CNTs incorporated in the double-skinned membrane can improve the porosity and densify the selective layer, resulting in higher anti-fouling capacity and osmotic water flux.

### 2.2.2.7 Membrane Surface Modifications

To date, most FO applications utilize RO membrane, and on a commercial scale, only a handful of companies manufacture specific membranes for FO applications. Generally, two types of membranes are applied for the FO process, that is asymmetric CA/CTA and TFC PA membranes. Asymmetric CA membranes are typically produced with the phase inversion method. The CA membrane has a smoother surface than the TFC membrane, making it less susceptible to fouling.

Hydration Technology Innovations (HTI) manufactured the first commercial CA/CTA thin-film and HF membranes for the FO processes. The HTI-CTA membrane has a thickness of less than 50 µm embedded in a polyester mesh to provide mechanical support.

#### *2.2.2.7.1 Chemical Cross-Linking*

Chemical modification can significantly alter membrane properties, enhancing hydrophilicity, providing anti-fouling characteristics, and reducing ICP. One example of chemical modification is the incorporation of chitosan (CS) via IP, using TMC as the cross-linking agent (125). The SL in this case is a sulfonated PES/PES (SPES/PES) composite. A CS solution ranging from 0.5% to 2.0% by weight was applied to the SL, both with and without NaOH treatment.

The highest water flux, 19.7 L m$^{-2}$ h$^{-1}$, was achieved with a 1% CS concentration without NaOH treatment when using 1 M NaCl as the DS. This flux was 8.2% higher than the unmodified membrane and 71.3% higher than the 0.5% CS-modified TFC

**TABLE 2.9**
**Summary of commonly used materials for FO substrate**

| Materials | Advantages | Disadvantages |
|---|---|---|
| CA/CTA | Hydrophilic (resistant to fouling) | Narrow pH tolerance (pH 3–8) |
| PES/PS | Good chemical resistance and mechanical stability | Relatively hydrophobic (prone to fouling) |
| PEI | Good durability; excellent solvent resistance | Reactive with amine-containing chemicals at elevated temperatures |
| PAN | Hydrophilic | Nitrile group is sensitive to alkaline |
| Nylon | Good mechanical strength; hydrophilic nature | Poor resistance to acids and bases; limited chemical stability |
| Inorganic materials | Outstanding thermal and chemical stability | High cost; complex preparation process |

CA/CTA: cellulose acetate/cellulose triacetate; PES/PS: polyethersulphone/polysulphone; PEI: polyetherimide; PAN: polyacrylonitrile

membrane. However, a 2.0% CS solution resulted in a lower flux of 8.8 L $m^{-2}$ $h^{-1}$. The increased CS concentration led to a thicker AL due to a higher amine-to-acyl chloride ratio, which enhanced the amine group's accessibility and increased the rate of internal polarization reactions. As a result, the thicker AL led to higher cross-link density on the membrane surface, creating a barrier that hindered water transport.

Interestingly, NaOH treatment further improved reverse salt rejection by forming a denser AL, although it resulted in a lower overall water flux. In basic conditions, the deprotonation of cationic moieties ($NH_3^+$) to $NH_2$ facilitated a stronger reaction with TMC's acyl chloride groups, thus increasing the degree of cross-linking and the AL thickness. The final reverse salt flux was lower when CS was treated with NaOH, as the flux dropped from 71.8 to 6.1 g $m^{-2}$ $h^{-1}$, compared to a decrease from 96.3 to 13 g $m^{-2}$ $h^{-1}$ for the CS treated without NaOH. This phenomenon can be attributed to the presence of $NH_3^+$, which influences the structure and properties of the CS-TFC membrane's AL.

(ii) Polymer and/or Mineral Coating:

Polymer coatings are widely used to manage the surface characteristics of membranes, particularly to create a dense layer on the surface of CA membranes. One approach involves incorporating PVA along with a cross-linking agent and PDA, followed by an accelerated polymerization process. This combination resulted in a TFC membrane with a water flux of 16.72 L $m^{-2}$ $h^{-1}$ and an RSF of 0.14 mol $m^{-2}$ $h^{-1}$, showing a three-times lower specific reverse salt flux compared to the pristine CA membrane (119). The application of PDA coating to TFC membranes increases hydrophilicity and reduces surface roughness from 48.9 nm to 36.1 nm. Interestingly, the water permeability remained similar to that of uncoated TFC membranes ($2.2 \times 10^{-5}$ L $m^{-2}$ $h^{-1}$ $Pa^{-1}$), despite a longer coating duration (4 hours), which resulted in a reduction of water flux to approximately $1.9 \times 10^{-5}$ L $m^{-2}$ $h^{-1}$ $Pa^{-1}$. It was suggested that the extended coating time increased hydraulic resistance. Nevertheless, the selectivity towards NaCl remained relatively low (0.3–0.7 L $m^{-2}$ $h^{-1}$), compared to the control TFC membrane. PDA coatings are regarded as promising anti-fouling agents due to their ability to weaken the adhesion between fouling gels and the membrane surface, with added stability during cleaning cycles (120). While longer coating times improve selectivity, they tend to reduce water flux, highlighting the need for further exploration of alternative coating materials, such as zwitterionic, carbon-based, or other functional coatings (121).

Microporous 2D materials, such as GO, are emerging as promising candidates for enhancing the permeability and selectivity of desalination membranes. GO's ultrathin thickness, specific pore size, and superior water permeability make it an attractive material for FO applications. A PDA-coated rGO membrane was prepared, exhibiting a high-water flux of 36.6 L $m^{-2}$ $h^{-1}$, a relatively low reverse solute diffusion of 0.042 mol $L^{-1}$ $m^{-2}$, and a salt rejection rate of 92.0% (122). Additionally, a recent study explored the performance of a geopolymeric membrane coated with GO for orange juice concentration. Using a 10 M NaCl solution, the membrane achieved a water flux of 1.053 L $m^{-2}$ $h^{-1}$, significantly outperforming the pristine membrane, and concentrated the orange juice to up to 26 Brix (123).

A PA membrane was modified by in situ surface grafting of amine-terminated sulfonated poly(arylene ether sulfone) ($NH_2$-BPSH100), resulting in enhanced anti-fouling performance during the filtration of a soybean oil/water emulsion (40 $g{\cdot}L^{-1}$) (125). The modified membrane allowed for easy flux recovery through simple hydraulic flushing and demonstrated a significantly higher flux recovery factor (69.8%) compared to the unmodified membrane (11%) (124).

A dip-coating process using a PVA solution was introduced to enhance the hydrophilicity of PVDF nanofibre supports (126). The modified TFC membrane exhibited excellent water flux of 34.2 L $m^{-2}$ $h^{-1}$ when using 1 M NaCl as the DS and DI water as the FS (126). The hydrophilicity of a PES-based support was enhanced by coating the pore surfaces with calcium carbonate ($CaCO_3$). This increased hydrophilicity reduced ICP, enabling the resulting TFC membrane to achieve a water flux of 52 L $m^{-2}$ $h^{-1}$ when operated in the AL facing the FS (AL-FS) mode, using DI water as the FS and 2 M NaCl as the DS (127). While the addition of nanoparticles has improved membrane performance, challenges remain, particularly with nanoparticle aggregation at high concentrations due to reduced miscibility, which can lead to inconsistencies in membrane properties (128).

The modification of TFC membranes was explored using a zwitterionic monomer, *N*-aminoethyl piperazine propane sulfonate (AEPPS), either by incorporating it into the aqueous phase before IP or by grafting it onto the membrane surface after the polymerization process (129). The modified membranes demonstrated superior fouling resistance in long-term FO tests using grey wastewater as the FS. In contrast, the membranes fabricated using the first method (before polymerization) showed no significant reduction in fouling, indicating the potential advantage of post-polymerization grafting in enhancing membrane stability and fouling resistance.

(iii) Grafting Modifications:

A photo-bacterial TFC FO membrane was synthesized by grafting porphyrin onto a TFC membrane. The grafting process involved a coupling reaction between the amine groups of porphyrin and the acyl chloride groups on the TFC membrane (130). With regard to sunlight radiation, grafted porphyrin TFC can generate reactive oxygen species, which are bactericidal and thus diminish biofouling propensities in FO. Lower concentration of porphyrin was apparently insufficient to tackle microbial growth, whereas porphyrin aggregation at higher concentrations can significantly reduce its effectiveness. The porphyrin concentration was varied from 0 to $10^{-4}$ M, and after grafting, the ethanol and water were used to clean up the remaining porphyrin attached to the TFC surface. The highest water flux (36 L $m^{-2}$ $h^{-1}$) was reported for $10^{-5}$ M porphyrin concentration against 2 M NaCl as DS in FO mode. This experiment revealed a 50% water flux improvement compared to the non-ethanol-treated pristine membrane, but the SRSF was also increased. Ethanol treatment exhibited better water flux for the extraction of non-reacted monomers and short chain from the membrane. Conversely, the physical swelling and wetting emerged due to the hydrogen bonding and non-polar interaction between ethanol and PA, leading to higher water and salt permeability as the chain–chain interaction was weakened, chain–chain distance was enlarged, and chain flexibility in the PA layer was increased. Such modification may be useful for fruit-juice concentration for preventing biofouling on the FO membrane due to the presence of pathogenic bacteria in the FS.

(iv) Organic Acid Insertion:

The surface of a TFC membrane was modified through a three-step process involving polyethyleneimine (PEI) grafting—via a reaction between the acyl chloride groups of the TFC and the amine groups of PEI—followed by organic acid grafting and subsequent mineralization. The modified membrane incorporating diethylenetriaminepentakis (methylphosphonic acid) ("DTPMP", 50 wt.% in water) demonstrated superior anti-fouling properties against BSA and sodium alginate due to increased hydrophilicity. Additionally, it exhibited enhanced water flux along with reduced RSF.

#### 2.2.2.8 Membrane Bulk Modifications

In the given section, modifications related to sulfonation, additive blending, intercalation of hydrophilic polymer, and intercalation of nanomaterials/nanoparticles (zeolite, GO, zinc oxide, carbon nanotubes, silica, electro-spun membrane, and titanium oxide) are discussed in detail.

##### *2.2.2.8.1 Sulfonation*

A chemical modification was carried out to initiate the covalent bonding of aniline sulfonate/bisulfonate through IP, involving reactions between amine and acyl chloride groups, on a PES layers (131). This modification induced a smoother surface with more hydrophilic features. Compared to the pristine membrane, membranes modified with 2,5-disulfonanilin disodium salt exhibited enhanced anti-fouling capabilities and a 52% increase in water flux. Additionally, when seawater was used as the FS and 2 M $MgCl_2$ as the DS, the modified membrane achieved superior water flux (16.7 L $m^{-2}$ $h^{-1}$) compared to the pristine membrane (12.3 L $m^{-2}$ $h^{-1}$).

##### *2.2.2.8.2 Polymer Blending*

Polymer blending is a straightforward and effective approach for enhancing membrane properties. In one method, FO membrane supports were fabricated by blending PSf with SPEK (132,133). The incorporation of SPEK improved membrane hydrophilicity and porosity, resulting in a water flux of 28.3 L $m^{-2}$ $h^{-1}$. Additionally, a disulfonated poly(arylene ether sulfone) hydrophilic–hydrophobic multiblock co-polymer was integrated into the PSf matrix to function as an SL for TFC membranes, further enhancing performance (134).

The synthesis of a porous aromatic framework within a TFN membrane was investigated by carbonizing and pre-impregnating fullerene onto the PA layer in the presence of hydrophobic, porous membrane additives, resulting in the formation of C60@PAF900 (135). The modification led to a less dense, cross-linked network, facilitated by the IP process and the repulsive interaction between C60@PAF900 and PA. When compared to the pristine membrane, the incorporation of 0.01 w/v% C60@PAF900 resulted in an increase in water flux from 7.4 to 12.4 L $m^{-2}$ $h^{-1}$, as demonstrated in simulations using 2 M NaCl as the DS and 10 mM NaCl as the FS.

#### 2.2.2.8.3 *Hydrophilic Material Intercalations*

The slow progress in utilizing GO laminates in FO membrane is mainly due to their unstable properties and hydrophobic nature in water, which hinder their potential as additives in membrane applications (136). The addition of hydrophilic PDA adhesive offers a solution by chemically reducing GO laminates. This modification improves the selectivity for hydrated ions through compact 3.45 Å nanochannels and accelerates water adsorption. As a result, the modified membrane achieved a superior FO performance (water flux: 36.6 L $m^{-2}$ $h^{-1}$; RSF: 0.042 g $m^{-2}$ $h^{-1}$; salt rejection rate: 92%).

To address issues such as ICP and weak mechanical strength, a self-standing FO membrane was developed that does not require an SL (137). The membrane was synthesized from -COOH-derived polyoxadiazole co-polymer with 5–15 μm thickness. The membrane exhibited high $Na_2SO_4$ rejection and acceptable water permeability (8.37 × 10–12 L $m^{-2}$ $h^{-1}$ $Pa^{-1}$). The given design highlights the advantage of a support-free membrane in mitigating ICP.

In another study, a membrane matrix that was incorporated by CS nanobiopolymer and prepared using a phase inversion method (138). This modification was aimed at enhancing the surface hydrophilicity and porosity, leading to a reduction in the structural parameter (S-value) between 0.30 to 1.30 mm. This resulted in a fourfold increase in water flux (31.2 L $m^{-2}$ $h^{-1}$), along with a significantly lower RSF (0.09 g $m^{-2}$ $h^{-1}$) compared to the pristine membrane. Furthermore, a novel hydrophilic CS-based thin AL membrane was synthesized. In this study, the CS was interracially polymerized with TMC monomer on the surface of SPES-PES SL (139). Reportedly, compared to the commercial TFC membrane, the modified membrane with optimal CS concentration demonstrated remarkably higher hydrophilicity and water permeation.

A novel FO membrane was fabricated using an LbL assembly of CS and GO nanosheets, deposited via electrostatic interactions onto a porous SL (140). This SL was prepared by blending hydrophilic SPES into the PES matrix using a wet phase inversion process. Reportedly, the abundant functional groups on CS and GO structures improved final membrane surface hydrophilicity, being a vital factor to improve membrane performance and anti-fouling properties.

#### 2.2.2.8.4 *Nanomaterial/Nanoparticle Intercalations*

The incorporation of nanoparticles into the membrane matrix has emerged as an effective strategy to enhance membrane properties by combining the advantages of both inorganic and organic materials. Integrating inorganic nanoparticles into the membrane support has proven effective in reducing ICP in the FO process. For instance, silica nanoparticles have been added to a polyetherimide (PEI) nanofibrous support to improve the performance of PA TFC membranes by mitigating ICP. Incorporation of silica was found to reduce porosity during heat-press treatments due to its higher heat resistance. The highest concentration of silica (1.6 wt.%) resulted in the highest substrate porosity (83%), the smallest S-value (174 μm), and a water flux of 42 L $m^{-2}$ $h^{-1}$ in AL-FS mode of operation with DI water as FS against 1.0 M NaCl as DS.

Porous zeolite nanoparticles are embedded into a porous PSf SL via IP to create a PSf nanocomposite with reduced ICP. For instance, with 0.5% zeolite, the TFC membrane achieved a significant reduction in S-value (from 0.96 mm to 0.34 mm) and a doubled water flux (~ 80 L $m^{-2}$ $h^{-1}$ in AL-DS mode and ~40 L $m^{-2}$ $h^{-1}$ in AL-FS mode) against 2 M NaCl as DS and (0.0–0.01) M NaCl as FS (141).

Titanium oxide ($TiO_2$) nanoparticles were incorporated onto CTA and AQP membranes through grafting methods using a monomer chain precursor, 3-(trimethoxysilyl) propyl methacrylate–polymethyl methacrylate–bromide (MEMO–PMMA–Br) (142). The addition of $TiO_2$ nanoparticles reduced the water contact angle on the modified CTA membrane surface, improving both water flux and fouling resistance. The water flux increases by 73.4% for the modified CTA membrane and 13.6% for the AQP membrane. Furthermore, $TiO_2$ nanoparticles enhanced anti-fouling properties by catalysing foulant photodegradation at the membrane interface while also preventing functional water channel blockage. However, the increased hydrophilicity of the modified AQP membrane led to water transport obstruction due to the coverage of MEMO–PMMA–Br monomers and $TiO_2$ nanoparticles.

Nanoparticles, such as GO nanosheets incorporated into the PA layer of TFC membranes, exhibited higher water permeability ($2.20\times10^{-5}$ L $m^{-2}$ $h^{-1}$ $Pa^{-1}$) and water flux (35.4 L $m^{-2}$ $h^{-1}$) (143). This improvement was attributed to the increased hydrophilicity and decreased thickness of the AL. The presence of GO accelerated the reaction rate between MPD and TMC, leading to lower water resistance, smoother membrane surface, and enhanced salt rejection. However, when GO loading exceeded 50–800 mg $L^{-1}$, membrane agglomeration occurred, increasing salt permeability due to defect formation in the membrane.

Amine-functionalized multi-walled carbon nanotubes (F-MWCNTs) were also used as additives in the aqueous phase of MPD to enhance TFN membranes. These nanotubes acted as nanochannels in the top surface of the membrane, facilitating water transport. The addition of F-MWCNT increases surface hydrophilicity, and the membrane with 0.1 % F-MWCNTs exhibited 160% higher water flux than the TFC membrane and an acceptable salt rejection of 73% (144).

Incorporation of $MoS_2$-Ag nanofillers into the TFC membrane resulted in a smoother membrane surface, which can be represented by a decrease in roughness from 40.09 ± 1.39 nm to 15.53 ± 0.17 nm. Furthermore, the contact angle of the membrane also decreases from 79.0 ± 2.0° to 52.2 ± 0.7°, along with improved FO performance in terms of water flux by 35.2% (145).

While nanoparticle incorporation offers significant improvements, challenges such as nanoparticle aggregation due to incompatibility with the polymer matrix can lead to reduced mechanical stability and selectivity. Thus, enhancing compatibility between nanoparticles and polymers, through chemical modifications or improved dispersion techniques, is essential for optimizing membrane performance.

Thereby concluding, although nanoparticles have been successfully integrated into FO membranes to enhance properties such as water flux, hydrophilicity, and anti-fouling performance, further optimization is needed. Membranes with improved flux, low fouling tendencies, and reduced ICP are particularly promising for applications such as desalination and food concentration. The development of such membranes with high hydrophilicity, low S values, and enhanced selectivity is expected to offer higher productivity and better resistance to fouling, making them ideal for producing fruit-juice concentrates with minimal salt contamination.

### 2.2.3 Forward Osmosis Membrane Characterization

FO membranes are usually characterized by their water flux, RSF, and structural parameter. The fundamental transport properties, water permeability and solute permeability of a membrane material, are critical to fully understand the transport properties and mechanisms. The three parameters that fully describe the FO performances are water permeability ($L_p$), salt permeability ($B$), and the structural parameter ($S$). The following subsections summarize the experimental techniques reported in the literature for evaluating these parameters. While several of these methods are based purely on transport phenomena, some analytical approaches are also included for discussion.

#### 2.2.3.1 Indirect Techniques

##### *2.2.3.1.1 RO–FO-Based Method*

Pressurized RO–FO-based methods are the most common techniques to evaluate transport parameters of FO membranes. In this method, water permeability and solute (or salt) permeability are measured using an RO system, whereas the S-value is calculated using an FO mode. A standardized methodology was introduced to evaluate osmotic-driven membrane processes with high permeability, utilizing a combined RO and FO-based approach for calculating transport parameters. These standard conditions were used to create a "*round-robin*" experiment with seven independent participating laboratories, and the results were compared. The following relations were used to evaluate the water and salt permeability coefficients based on the solution–diffusion (SD) model for describing the mass transport across the membrane; the following relation was used to estimate pure water permeability and solute permeability (146):

$$\text{Pure water permeability coefficient, } L_p = \frac{J_w}{\Delta P} \tag{2.1}$$

where the term $J_w$ is water flux and $P$ refers to the applied hydraulic pressure.

$$\text{Solute permeability coefficient based on SD model, } B = J_w\left(\frac{1-R}{R}\right)\exp\left(-\frac{J_w}{k}\right) \tag{2.2}$$

where membrane rejection "$R$" is expressed as: $R = \left(1 - \frac{c_p}{c_b}\right)$, where the terms $c_p$ and $c_b$ denote permeate and bulk concentration, respectively.

The structural parameter (S) is a quantitative measure used to describe the internal structure of FO membrane SLs, particularly their thickness, porosity, and tortuosity.

It is determined by integrating the SD model with film theory to account for ICP, which can significantly impact membrane performance. The S parameter essentially reflects how easily water and solutes can move through the membrane's porous SL—lower S values indicate a support structure that minimizes ICP and thus enhances water flux, while higher S values are associated with greater resistance to mass transfer within the membrane. The structural parameter S is determined as given below:

$$S = \frac{D}{J_w} \ln \left[ \frac{B + L_p \pi_{D,b}}{B + J_w + L_p \pi_{F,m}} \right] \tag{2.3}$$

where the term D refers to the diffusivity of the draw solute and the terms $\pi_{D,b}$ and $\pi_{F,m}$ refer to the osmotic pressure of FS and DS, respectively.

"*Round-robin*" experiments by seven different laboratories were conducted using two commercial membranes—a TFC PA membrane and an asymmetric cellulose-based membrane. The $L_p$ value for both these membranes was consistent across these laboratories (i.e. ~9–10 $Lm^{-2}h^{-1}bar^{-1}$ for TFC and ~8–1.2 $Lm^{-2}h^{-1}bar^{-1}$ for cellulose-based membrane). However, the B-value of the TFC membrane showed more variability (~5×10$^{-7}$ to ~3×10$^{-6}$ m s$^{-1}$) versus the cellulose-based membrane (~1.0×10$^{-7}$ to ~3.0×10$^{-7}$ m s$^{-1}$). The S-value for both TFC and cellulose-based membranes remained consistent (for TFC, 350–400 μm, and for cellulose-based, 500–550 μm) under FO mode. This study demonstrated that the "pressurized RO-FO method", when performed under carefully controlled experimental conditions, produces very reproducible results.

*2.2.3.1.1.1 Ultra-Low-Pressure RO/FO-Based Method* In FO membrane, SLs are thin and highly porous; therefore, a high-pressure RO method to obtain $L_p$ and $B$ values may not be ideal for such membranes. Parameter evaluation was proposed using an ultra-low-pressure (25 psi) RO method based on Equation (2.1), while the parameters B and S were estimated using the standard FO process (132). The following equations were solved simultaneously to evaluate using $B$ and $S$:

$$J_w = L_p \left[ \frac{\pi_{D,b} \exp\left(-\frac{J_w S}{D}\right) - \pi_{F,b} \left(\frac{J_w}{k}\right)}{1 + \frac{B}{J_w} \left\{ \exp\left(\frac{J_w}{k}\right) - \exp\left(-\frac{J_w S}{D}\right) \right\}} \right] \tag{2.4}$$

$$J_s = B \left[ \frac{C_{D,b} \exp\left(-\frac{J_w S}{D}\right) - C_{F,b} \left(\frac{J_w}{k}\right)}{1 + \frac{B}{J_w} \left\{ \exp\left(\frac{J_w}{k}\right) - \exp\left(-\frac{J_w S}{D}\right) \right\}} \right] \tag{2.5}$$

where $C_{D,b}$ and $C_{F,b}$ are the concentrations of DS and FS, respectively.

FO membranes are operated at low osmotic pressures; thus, a high hydraulic pressure for determining $L_p$ and B would not be ideal and, in some cases, could lead to erroneous results. This ultra-low-pressure RO/FO method is preferred in that case. However, prior to widespread adoption of this method, it is recommended to compare and analyse the values obtained from this method versus the pressurized RO–FO method.

#### 2.2.3.1.2 FO-Based Methods for Determining Transport Properties of FO Membranes

A comprehensive approach for determining the key transport parameters of FO membranes—namely water permeability ($L_p$), solute permeability (B), and structural parameter (S)—can be achieved using FO mode alone, eliminating the need for RO testing. This methodology involves a sequential experimental design in which DI water is employed as the FS, and an NaCl-based DS is progressively concentrated over four stages. At each stage, the system is allowed to reach a steady state, and the corresponding water and RSF are measured. These values are then used to compute transport parameters using analytical expressions derived from equations governing steady-state mass transfer, specifically those equivalent to Equations (2.4) and (2.5) (133).

For each of the four stages, two sets of values for $J_w$ and $J_s$ were obtained, having a set of eight equations with three unknown parameters ($L_p$, $B$, and $S$). A rigorous least squares non-linear regression analysis was then performed to determine the parameters. The results obtained from this technique were directly compared with the more prevalent "combined RO-FO" method, and significant discrepancies were found—for example, a 46% average deviation was found in the membrane $B$-values. The cause of this deviation was primarily attributed to the fact that, in the case of the combined RO–FO method, hydraulic pressure was applied for the determination of $L_p$ and $B$, which possibly altered the properties of the membranes as well as introduced errors in the comparison to the pure-FO method described earlier. While this method can yield highly accurate estimates, it is labour-intensive, and its prerequisites are often too challenging to fulfil under standard laboratory conditions (132).

#### 2.2.3.1.3 Analytical Approach-Based "Direct" Techniques

*2.2.3.1.3.1 Using Tortuosity Calculation Models* An analytical method has been developed to estimate the structural parameter (S) of FO membranes directly, utilizing values of hydraulic permeability (Lp) and solute permeability (B) obtained from RO experiments. Unlike indirect approaches, where S is back-calculated from water and solute flux data, this method determines tortuosity, porosity, and thickness independently to compute the S-value. Thickness and porosity are measured using conventional techniques such as scanning electron microscopy (SEM) and gravimetric analysis. Tortuosity is estimated through a fractal-theory-based model that correlates it with the measured porosity of the SL. The structural parameter is then calculated from these measured values. However, significant deviations have been observed when comparing this direct method with flux-based indirect estimation techniques. These discrepancies may arise from inherent assumptions in the empirical models, as well as experimental uncertainties in porosity and thickness measurements. Without further validation, the reliability of this approach as a substitute for established estimation methods remains uncertain (134).

*2.2.3.1.3.2 Using Mercury Intrusion Porosity and X-Ray Microscopy* Two direct analytical techniques have been employed to estimate the structural parameter (S) of osmotic membranes: mercury intrusion porosimetry (MIP) combined with micrometric measurements, and micro-X-ray microscopy (micro-XRM). The MIP-based approach enables the calculation of average porosity and membrane thickness, while micro-XRM provides simultaneous measurements of thickness, porosity, and tortuosity. However, S-values derived from both methods were found to be one to two orders of magnitude lower than those obtained using conventional indirect techniques. These discrepancies were attributed to several factors, including mass transfer limitations, inadequate wetting of hydrophobic materials, and variations in hydrodynamic conditions. Overall, these findings underscore the significant uncertainty associated with direct analytical methods for determining the S-value, suggesting that such techniques may be limited in reliability without further refinement and validation (135).

### 2.2.4 Forward Osmosis Membrane Module

FO membrane modules serve as the operational backbone of FO systems, combining membrane sheets with supportive hardware to facilitate osmotic transport under real-world conditions. In contrast to pressure-driven RO systems, FO modules operate based on osmotic gradients, requiring specialized design strategies to enhance performance, reduce hydraulic complexity, minimize RSF, and optimize energy efficiency.

From an industrial standpoint, FO modules—especially those based on flat-sheet configurations—have regained interest due to their ease of assembly, lower replacement costs, and simplified operator training. Similar to plate-and-frame heat exchangers, such as those used in Alfa Laval systems, FO modules can employ a sandwich-like flat-sheet structure that enhances scalability, user-friendliness, and ease of manufacturing.

FO membrane modules are generally available in four main configurations: flat-sheet (plate-and-frame), tubular, spiral-wound, and HF configurations (Figure 2.6). Each design offers distinct advantages and limitations in terms of packing density, operational control, fouling resistance, and cleanability. Selecting the appropriate configuration requires careful consideration of system goals, feedwater characteristics, energy efficiency requirements, and maintenance practicality (*Table 2.11*).

- Flat-Sheet (Plate-and-Frame) Membrane Module: These modules consist of flat membrane sheets alternately stacked with spacers and housed within a rigid frame. While traditionally characterized by relatively low packing densities (~300 $m^2$ $m^{-3}$), advancements in manufacturing have enabled densities that rival those of spiral-wound modules. Their key industrial advantages include quick membrane replacement via simple disassembly

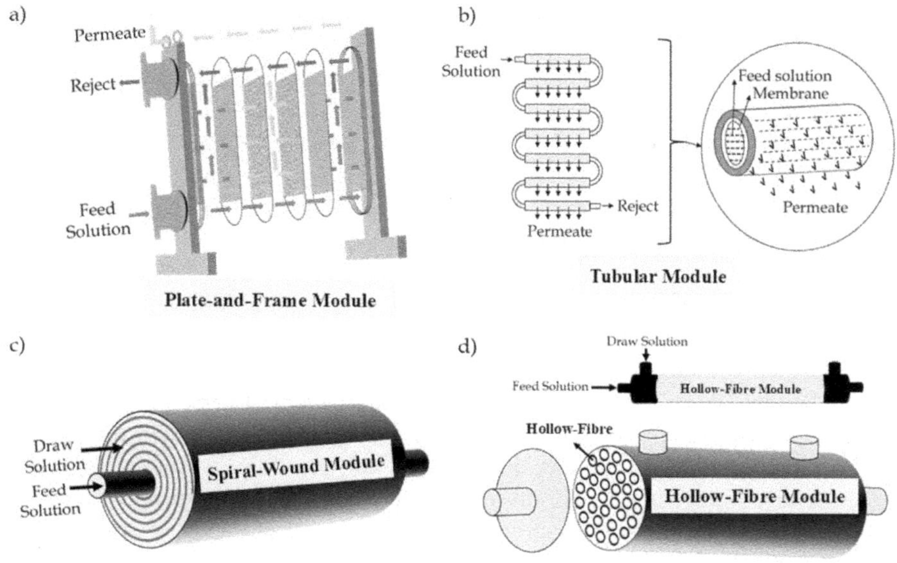


**FIGURE 2.6** Schematic representation of FO membrane modules

and reassembly, ease of operator training, minimal internal dead zones, and accessible flow channels that support effective cleaning. As a result, flat-sheet modules are particularly well-suited for small- to medium-scale applications in food and water treatment, where system uptime and operational simplicity are paramount.

- Tubular Membrane Modules: Tubular membranes are commonly operated in high-velocity cross-flow mode, making them ideal for treating viscous or particulate-laden feed streams. Their wide flow channels allow for effective mechanical and chemical cleaning, contributing to strong fouling resistance and ease of maintenance. However, their relatively low packing density (~60 $m^2$ $m^{-3}$) limits their suitability for space-constrained applications. Although they offer straightforward reassembly and durability, their larger footprint and higher energy consumption have curtailed broader adoption. As such, tubular modules are best suited for processes involving high solids or viscous feeds, where operational robustness is prioritized over compactness.
- Spiral-Wound Membrane Modules: Spiral-wound FO modules are widely used due to their moderate-to-high packing density (~600 $m^2$ $m^{-3}$) and compatibility with established manufacturing techniques. However, performance limitations arise with increasing membrane leaf length. The U-shaped internal flow path effectively doubles the flow distance within each leaf, resulting in significant pressure drops. This mismatch between internal and external flow resistance can lead to uneven flow distribution, localized CP, and elevated energy consumption—challenges that complicate efficient operation, particularly under low-pressure and high-recovery conditions. Recent research highlights these limitations, revealing performance

degradation in long-leaf modules, particularly when using sugarcane juice as the FS. The study proposes mitigation strategies such as optimizing membrane leaf length, tailoring channel height design, and ensuring uniform inlet flow distribution (147).

- HF Membrane Modules: HF membranes provide the highest packing densities—reaching up to 1200 $m^2 m^{-3}$—and are widely employed in large-scale water treatment and food concentration processes. They are typically available in two main configurations, namely axial and radial flow configurations.

  In axial configurations, the feed flows longitudinally through the fibre lumens, allowing for easier cleaning due to more accessible flow paths and reduced risk of irreversible fouling. However, this design typically results in slightly lower packing densities compared to radial flow configurations.

  In radial configurations, the feed flows perpendicularly across the HFs, enabling maximum packing densities and uniform flow distribution across the membrane surface. However, the tightly packed nature of radial modules increases the risk of fouling and complicates cleaning due to restricted channel access. These trade-offs are evident in the challenges associated with cleaning efficiency and maintaining stable flux during intermittent operation in radial-flow modules (147).

  As such, the choice between axial and radial configurations should be guided by factors such as ease of cleaning, membrane durability during backflushing, and spatial constraints.

Achieving optimal performance in FO modules requires careful attention to flow channel design, support materials, and operational parameters. The following strategies are essential:

1. Channel Geometry Optimization: Spacer configurations and channel heights play a crucial role in influencing shear rates and mitigating CP. Computational fluid dynamics (CFD) simulations are valuable tools for modelling internal flow behaviour, enabling the identification of geometries that enhance mass transfer and minimize dead zones.
2. Ease of Membrane Replacement: Especially in flat-sheet configurations, modules should be designed to allow for straightforward and rapid membrane sheet replacement, facilitating maintenance and reducing downtime.
3. Leaf Length Optimization: In spiral-wound modules, balanced flow distribution is critical to minimizing internal pressure drops and osmotic imbalances that can compromise performance.
4. Material Compatibility: Housing materials and seals must be chemically and thermally compatible with both FS and DS, particularly in hybrid systems incorporating RO or MD for DS regeneration.
5. Accessibility for Cleaning and Maintenance: For feed streams prone to fouling or rich in organics, module designs such as tubular or axial-flow HF configurations are advantageous, as they support effective mechanical and chemical cleaning protocols.

The performance of FO systems is deeply influenced by module design (Table 2.10). An optimal module strikes a balance among packing density, energy efficiency, cleanability, and operational resilience. Design choices should be driven by the specific application, feed water characteristics, and long-term cost considerations.

#### 2.2.4.1 System Design

(i) Flow Configurations: In the FO process, the selection of flow configuration (co-current and countercurrent) is a fundamental aspect of FO system design. It significantly impacts the mass transfer efficiency, overall water recovery, RSF, and energy consumption.

In co-current flow systems, the FS and DS travel in the same direction along the membrane. While this configuration simplifies design and reduces potential mechanical stress on the membrane module, it suffers from a gradual reduction in the osmotic pressure difference along the membrane length. This decline limits water flux and overall efficiency, particularly for high-recovery applications where maintaining a consistent driving force is critical. This configuration is used in low-recovery systems, where the operational simplicity is more critical than recovery. Conversely, it is used in countercurrent flow systems, where FS and DS flow in opposite directions.

**TABLE 2.10**
**Comparison of the forward osmosis membrane module**

| Membrane Module | Packing Density, $m^2\ m^{-3}$ | Energy Cost (Pumping) | Particulate Plugging | Ease of Cleaning | Hold Up Volume |
|---|---|---|---|---|---|
| Flat sheet | 300 | Moderate | Moderate | Moderate | Low |
| Tubular | 60 | High | Low | Excellent | High |
| Hollow-fibre Axial flow | 1200 | Low | High | Fair | Low |
| Hollow-fibre radial flow | ~1000 | Low | Moderate to high | Fair | Low |
| Spiral wound | 600 | Low | Very high | Moderate | Low |

**TABLE 2.11**
**Comparative analysis of co-current and countercurrent flows**

| | Co-current Flow | Countercurrent Flow |
|---|---|---|
| Osmotic pressure gradient | Declines along the membrane | Relatively constant throughout |
| Water recovery | Lower | Higher |
| Reverse solute flux | Higher | Lower |
| System complexity | Simpler | More complex |
| Shear stress | Lower | Higher |
| Application suitability | Low-efficiency or simple systems | High-efficiency and high-recovery systems |

This configuration leads to higher water flux and better recovery rates, making it ideal for applications such as wastewater treatment or desalination where performance optimization is critical. Despite these advantages, countercurrent configurations involve more complex designs and require precise flow control to prevent cross-contamination.

Overall, in an FO system, the flow configuration is integral to its success, impacting every major performance metric. Countercurrent flow provides superior efficiency and recovery, making it the preferred choice for most high-demand applications. However, the co-current flow's simplicity can be advantageous in specific scenarios (Table 2.11). A thorough understanding of system requirements and operational goals is essential when selecting the appropriate flow configuration for an FO process.

## 2.3 OPERATIONAL PARAMETERS: FACTORS INFLUENCING PERFORMANCE AND OPTIMIZATION

The performance of FO significantly depends on various operating parameters, which include operating temperature, flow rate, and concentration (148). The optimization of operating conditions such as flow velocity, concentration, pH, temperature, and membrane orientation is essential to make the FO process more efficient.

### 2.3.1 Feed and Draw Solution Velocity

Flow velocity significantly influences various operation parameters such as CP, water flux, and overall system efficiency (149). Higher flow velocity reduces the CP, which in turn impacts the water flux and recovery of the FO system. This variation in flow velocity also depends on membrane orientation (i.e. AL-FS or AL-DS). For instance, in AL-FS orientation, increasing the DS flow rate can reduce water flux. This effect is generally attributed to increased turbulence in the membrane's SL and enhanced back diffusion of salt. In contrast, AL-DS orientation typically shows improved water flux at higher DS flow rates, due to reduced ECP at the membrane surface. Similarly, increasing the FS flow rate enhances water flux by mitigating ICP and reducing fouling in the SL (150). When both DS and FS flow rates are increased simultaneously in AL-DS orientation, a synergistic improvement in water flux has been reported—experimental data suggest flux increases on the order of ~76% under these conditions (151).

Typically, an increase in flow velocity enhances the mass transfer coefficient, which enhances the FO performance by reducing ECP. Conventionally, an FO system is considered efficient when the final DS concentration is low and the water recovery is high. A software tool was developed to analyse and optimize the performance of an FO system (152). The detailed parametric study was performed to determine the optimum operating condition. The study revealed that a higher recovery rate can be achieved by maximizing the DS flow rate.

A one-dimensional mathematical model was developed to investigate the influence of operating flow rate on overall FO performance, aiming to maximize water flux while minimizing specific RSF (SRSF) (78). The study revealed that reducing the DS flow rate while increasing the FS flow rate enhanced water flux, achieving optimal dilution of the final DS. Although the slower DS velocity is favourable, it can cause membrane fouling. As a result, a higher FS flow rate is required to reduce the effect of CP and to mitigate membrane fouling (153). However, the change in mass transfer and ECP caused by a change in flow rate was not observed to be significant when the flow rate was varied by a little (154).

Even though increasing the flow rate may mitigate CP and enhance water flux, this may augment the energy consumption for the pumping process. For example, reducing the DS flow rate from 100 ml $min^{-1}$ to 10 ml $min^{-1}$ decreases consumption from 1.86 kWh $m^{-3}$ to 0.02 kWh $m^{-3}$. In summary, increasing both DS and FS flow rates simultaneously was 40% more efficient than increasing either of the flow rates individually. This may be attributed to the simultaneous reduction of ICP and ECP.

### 2.3.2 Concentration

A difference must be maintained between the concentration of FS and DS to maintain the osmotic pressure gradient, which drives the high water flux. The type of DS and its concentration are important factors to get higher water flux. With an increase in DS concentration, the osmotic pressure difference ($\pi_{DS} - \pi_{FS}$) is between the two sides of the membrane (155). Water flux has both linear and non-linear relations with the DS concentration. The type of draw solute used and its concentration influence water flux. For example, inorganic salts such as NaCl and $MgCl_2$ showed a linear relation to 13 L $m^{-2}$ $h^{-1}$, whereas draw solutes such as $ZnSO_4$ and EDTA exhibited a non-linear relation with water flux (156,157). For example, with EDTA as DS, the water flux increases from 4 L $m^{-2}$ $h^{-1}$ to 13 L $m^{-2}$ $h^{-1}$ when DS concentration increases from 0.1 mol $L^{-1}$ to 1 mol $L^{-1}$. The water flux varied linearly up to 0.7 mol $L^{-1}$ EDTA, beyond 0.7 mol $L^{-1}$, about a 3% increase in water flux was observed when EDTA concentration was increased from 0.7 mol $L^{-1}$ to 1.0 mol $L^{-1}$. Furthermore, when EDTA sodium salt concentration exceeded 1.0 mol $L^{-1}$, the change in water flux remained unchanged because the increased water flux aggravated the dilute ICP in SL (168). In the case of NaCl and $MgCl_{2,}$ with an increase in DS concentration, the water flux increases, but RSF also increases. However, RSF increases when EDTA sodium salt concentration increases from 0.1 mol $L^{-1}$ to 0.5 mol $L^{-1}$. However, RSF became stable when the concentration exceeded 0.5 mol $L^{-1}$ (157). Furthermore, increasing DS concentration also increases the viscosity, which increases the pumping energy (169). In the DS, which has a higher diffusion coefficient, higher RSF was one of the major limitations while increasing DS concentration. Thus, the selection of DS concentration should be done based on a high diffusion coefficient and a low RSF for achieving a higher water flux (158). However, the increase in FS concentration decreases the osmotic pressure difference across the membrane on both sides of the membrane, which eventually reduces the water flux (159).

In summary, FO performance is typically optimized by increasing DS concentration rather than FS concentration. This is because increasing DS enhances the osmotic driving force for water flux, while FS concentration is process-dependent and tends to reduce the driving force if increased. The optimal DS concentration depends on the type of draw solute and its associated transport properties, like RSF and diffusion coefficient.

### 2.3.3 Operating Temperature and pH

Temperature significantly affects water flux, RSF, and membrane fouling (160–162). The increasing temperature improves the water flux by reducing CP (163). As the temperature increases, the viscosity decreases, which in turn increases the water permeability across the membrane. Thus, higher temperature may provide higher initial flux; however, after a few hours, membrane fouling and scaling may significantly influence the efficiency of the system. The increase in DS temperature could significantly increase the osmotic pressure, thereby facilitating the FO performance (164).

As the pH of the solution increases, the cross-linked polymer network of the FO membrane becomes more hydrophilic and acquires a greater negative surface charge due to deprotonation of functional groups. This pH elevation also influences the speciation of pollutants in the FS; for example, sulfamethoxazole transitions to its anionic form above pH 5.8. Concurrently, the increasing negative charge on the membrane's AL enhances electrostatic repulsion between the membrane and anionic species like sulfamethoxazole, thereby improving their rejection (165).

pH also influences the hydrated radius and charge of pollutants in FS to affect retention. Boric acid can produce a larger hydrated radius of borate anion ($B(OH)_4^-$) when the pH is above the pKa of 9.24. The electrostatic repulsion between the negatively charged membrane and the borate anion increased boric acid retention (154). Moreover, pH has an influence on RSF by changing the chemical forms of the draw solute. When EDTA sodium was used as DS, $EDTA_4^-$ combined with $Na^+$ to form the complex formation ($Na\,(EDTA)_3^-$) at higher pH, which reduced the RSF (157).

To summarize, water flux can be improved by maintaining a constant temperature in both DS and FS due to enhanced mass transfer across the FO membrane. Increasing flow rates of DS and FS can simultaneously reduce the CP across the FO membrane and improve the water flux. Modifying the flow rates of DS and FS reduces the ECP, whereas modifying the SL structure and properties can reduce the ICP in FO membranes. Maintaining a constant temperature of FS and DS is one of the main challenges for applications in temperature climate zones, where the ambient temperature will be less than 15 °C. Heating may be needed to maintain the temperature of FS and DS, leading to additional energy consumption. Hence, further research must focus on overall process optimization with respect to effectiveness and energy consumption.

## 2.4 CHALLENGES AND LIMITATIONS

Although the FO process has attracted attention, it has not been widely accepted. Membrane fouling, RSF, and CP are the major limitations affecting the overall membrane performance and water flux.

### 2.4.1 Concentration Polarization

In both osmotically driven and pressure-driven membrane processes, CP is one of the major challenges limiting large-scale application. In FO, CP significantly reduces water flux by lowering the effective osmotic driving force across the membrane. During operation, solutes accumulate on the membrane surface or within its porous structure, forming a mass transfer boundary layer that restricts solvent transport. Unlike pressure-driven processes, where CP mainly occurs externally, FO suffers from both ECP and ICP. ECP develops on the AL surface, while ICP occurs within the porous SL of the membrane, making it especially severe in FO due to the membrane's asymmetric structure. Depending on orientation, CP can manifest as dilutive polarization on the feed side or concentrative polarization on the draw side, both of which diminish osmotic efficiency.

Among these, ICP is considered the most detrimental because it occurs inside the SL, where solute accumulation cannot be easily mitigated by hydrodynamic conditions. This leads to reduced water flux and hinders the practical scalability of FO. In contrast, ECP can be partially alleviated through flow design, turbulence promoters, or spacers.

The severity of CP is strongly influenced by the structural properties of the SL, including thickness, porosity, and tortuosity. A thick or poorly designed SL exacerbates ICP, whereas highly porous and thin supports reduce its impact. CP not only lowers efficiency but also increases the energy demand for draw solute regeneration in hybrid FO systems.

In summary, CP remains a key bottleneck for FO commercialization. While advances in TFC membrane design, novel support structures, and hydrodynamic optimization have mitigated some effects, overcoming ICP remains essential to unlock the full potential of FO for industrial-scale water and food processing applications.

### 2.4.2 Membrane Fouling

Membrane fouling is inevitable for most membrane technologies. The deposition of contaminants on the surface of the membrane may result in fouling and loss of performance (166). This occurs from the interplay between foulants and the membrane surface, driven by the convective forces that propel foulant deposition onto the membrane surface. Over time, fouling reduced membrane permeability, resulting in increased operational costs, higher demand for chemical cleaning, and shorter membrane life (167,168). Similar to the RO process, fouling in the FO process is influenced by several factors including FS properties, membrane characteristics, and operating conditions. Membrane fouling can be reversible or irreversible in nature, and to mitigate it, these strategies such as optimizing membrane structure, feed water pretreatment, and effective cleaning protocols are essential.

Based on different types of foulants in the FS and fouling mechanisms, membrane fouling can be classified into four major categories, namely organic fouling, inorganic fouling, biofouling, and colloidal fouling (169).

Organic fouling—Typical organic fouling is attributed to the presence of organic macromolecules, including SA, BSA, and humic acid. They represent the polysaccharides, protein, and natural organic compounds, respectively. The adhesion of organic foulant to the FO membrane can significantly impact the FO performance (in terms of water flux). Organic pollutants are the most commonly selected model foulants. Organic foulants contribute to pore blocking, cake layer formation, or gel layer adsorption, leading to rapid flux decline. Reportedly, the most common method of mitigating organic fouling on FO membrane is to increase the hydrophilicity of the membrane surface, thereby preventing the adhesion of organic macromolecules on the membrane. For example, the modification of an FO membrane with GO enhanced its hydrophilicity, leading to reduced adhesion SA (170). Furthermore, the contamination of the aquatic environment by plastic waste (particularly microplastics "MPs", particle size smaller than 5 mm) has emerged as a significant global issue. The presence of MPs has posed a significant challenge to the long-term stability of the FO membrane. The development of FO membrane represents a crucial strategy for mitigating the accumulation of MPs. Two primary modification strategies emerged to address this issue: by increasing the electronegativity of the FO membrane to repel negatively charged MPs and by enhancing the hydrophilicity of FO membranes to resist the hydrophobicity of the MPs.

Inorganic Fouling—Typical inorganic fouling occurs when a supersaturated salt solution has been observed to precipitate readily on the membrane surface, resulting in membrane scaling. In comparison to CTA membranes, the surface of the conventional TFC-FO membrane is more susceptible to scaling. This is due to the fact that carboxylic groups on the surface of the TFC membranes can act as binding sites for calcium ions and silica gel. Also, the ridge-and-valley morphology of the PA AL also facilitates the accumulation and formation of scale by salts. Thus, to improve the anti-scaling properties of TFC-FO membranes, two key aspects must be considered: (i) incorporation of scale inhibitors into PA AL to salt-ion deposition and (ii) loading of material with hydrophilic functional groups to enhance the hydrophilicity of the TFC membrane and to prevent accumulation of scale ions.

Biological fouling—Biological fouling or biofouling refers to the undesirable accumulation of microorganisms such as bacteria, algae, fungi, and protozoa on the membrane surface. While other forms of fouling can be controlled with a variety of pretreatments, biofouling is the most ubiquitous type and is difficult to control. Understanding the mechanisms of biofilm formation is vital for tackling biological fouling. As the biofilm develops, it obstructs the membrane pores and increases hydraulic resistance, reducing permeate flux and impairing separation efficiency. Biofouling can also lead to pore-clogging and assist with other types of fouling such as inorganic fouling, and these channelling matters can lead to precipitation of soluble salts and eventually scaling. Biofouling mitigation strategies include pretreatment of feed water to reduce microbial load, the use of biocides, and the application of anti-fouling membrane coatings. Membrane cleaning methods such as backwashing and chemical cleaning with disinfectants or enzymes are essential to remove accumulated biofilm and restore membrane performance.

Colloidal fouling—Colloids are generally defined as nanoscale particles ranging in size from 1 to 1000 nm, such as colloidal silica particles. Their suspension exhibits reversible transitions between liquid-like flow and solid-like behaviours under specific conditions. As the colloidal concentration increases, this dispersed phase may undergo a phase transition from a disordered state to an ordered glassy state. This phase transition promotes colloid accumulation on the membrane surface, forming a dense layer. It would elevate the hydraulic resistance and compromise FO performance. Two strategies have been explored to alleviate colloidal fouling on FO membranes: (i) increasing the roughness of the membrane surface and (ii) enhancing the surface potential of the FO membranes. The influence of FO membrane roughness on colloidal fouling depends on surface hydrophilicity. Enhanced roughness generally suppresses colloidal deposition on hydrophilic membranes, whereas hydrophobic membranes exhibit an opposing trend with increased roughness exacerbating fouling.

Understanding the critical factors influencing membrane fouling in the FO process is crucial for optimizing performance and extending membrane lifespan. Key factors include membrane material properties, membrane orientation, operational parameters, and FS/DS characteristics.

Membrane material plays a vital role in FO fouling behaviour. TFC membranes suffer severe fouling and lower reversibility due to high surface roughness, hydrophobicity, and abundant carboxyl groups. For instance, the organic fouling and scaling are more pronounced on TFC membranes due to interactions between foulants and negatively charged carboxyl groups. Membrane properties like hydrophilicity, charge, and morphology are critical in fouling behaviour.

Higher cross-flow velocity reduces external fouling by enhancing shear forces that help prevent foulant deposition on the membrane surface but is ineffective against internal/irreversible fouling. To reduce or prevent fouling, operating parameters such as hydrodynamic conditions, temperature, and membrane orientation may be controlled (171,172). For example, regarding hydrodynamic conditions, increasing the DS flow rate from 5 L $h^{-1}$ to 10 L $h^{-1}$ reduces the water flux by 2.0%. Similarly, at elevated temperatures (above 40°C), due to higher water permeability, fouling is higher on the DS side than the FS side and there is more rapid diffusion on the FS side. Temperature also influences fouling; though higher temperature may provide higher initial flux, after a few hours, membrane fouling reduces the efficiency of the system. Temperature further influences fouling by affecting solute solubility and foulant–membrane interactions. For example, higher FS temperature reduces fouling by improving back diffusion and solubility, whereas higher DS temperature exacerbates fouling due to increases in permeation drag.

In terms of membrane orientation, since the hydraulic pressure was more on AL than on SL. Thus, in AL-FS mode, the deposition of foulant on the membrane surface is less (173,174). Membrane orientation significantly impacts fouling in FO. In AL-FS mode, fouling can occur on the AL surface (Figure 2.7 (a)). This external fouling can be mitigated by optimizing the hydrodynamic conditions of FS (such as by enhancing FS flow velocity or by using pulsed flow). AL-FS typically experiences lower fouling than AL-DS because of its smoother surface and the efficient shear force that helps remove foulants. Contrastingly, AL-DS is more prone to fouling as internal pore clogging occurs within the SL, which is challenging to remove easily by shear force alone (Figure 2.7 (b)). For instance, pollutants with a small size can enter

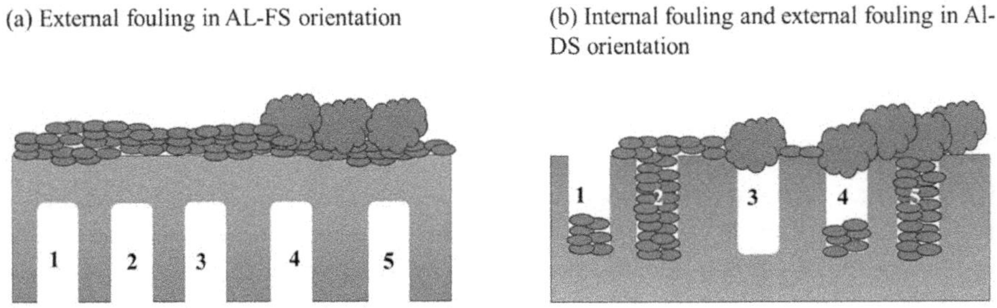


**FIGURE 2.7** External and internal fouling in forward osmosis system at (a) active layer facing feed solution (AL-FS) and (b) active layer facing draw solution (AL-DS) mode

membrane pores to form the internal fouling (refer to scenario 1 in Figure 2.7 (b)). Under more severe fouling conditions, the internal fouling gradually aggravates and forms external fouling on the outer surface of the SL (refer to scenario 2 in Figure 2.7 (b)). Pore blockage or internal fouling caused by a smaller foulant is serious and is extremely difficult to remove (175). Furthermore, if the size of pollutants is large, only the external fouling can occur (refer to scenario 3 in Figure 2.7 (b)), whereas, if the FS consists of different sizes of pollutants, both internal and external fouling can occur simultaneously (refer to scenarios 4–5 in Figure 2.7 (b)).

The membrane fouling behaviour of the FO process during orange juice concentration was systematically investigated (176). The study revealed that the AL-FS mode was suitable for the concentration of orange juice. The increase in the cross-flow rate made it difficult for pollutant to adhere to the membrane surface, thereby reducing the fouling layer so as to improve water flux and concentration efficiency.

Modification of the membrane surface is another approach to mitigate fouling, which includes changing the roughness, hydrophilicity, and characteristic group of the FO membrane. Reducing the roughness of the membrane and increasing the hydrophilicity of the membrane reduces the contact angle and increases the anti-fouling capacity of the membrane (177). To overcome this, several experimental works were focused on developing anti-fouling FO membranes. A dopamine-incorporated TFC-FO membrane was fabricated, exhibiting enhanced surface hydrophilicity and smoothness, which contributed to improved anti-fouling performance (178). Functionalization of CNTs with PDA exhibited good anti-fouling performance in the FO process by achieving 81.4% water flux. Table 2.12 summarizes the literature available for FO membranes and their anti-fouling performance. In recent years, a significant effort has been put into developing novel and innovative membrane materials to mitigate fouling in the FO membrane process. A TFN FO membrane was developed by incorporating carbon nanoparticles to mitigate fouling through enhanced electrostatic repulsion between foulants and the membrane surface (179). Additionally, an electrically assisted FO process was introduced using carbon nanotube HFs to enhance fouling resistance in a TFC-FO membrane. Application of a 2.0 V potential was found to be effective in significantly reducing organic fouling during operation (180).

**TABLE 2.12**
**Forward osmosis membranes and their anti-fouling performance**

| Membrane Type | Model Fouling Tests: FS Foulant | Model Fouling Tests: DS | Membrane Orientation | Cleaning Strategy | Anti-fouling Performance | Ref. |
|---|---|---|---|---|---|---|
| (CS)-AlFu MoF-CA membrane | Synthetic wastewater<br>Real wastewater | 1 M $MgCl_2$ | AL-FS | Backwashing and backflushing | Over 80% water recovery | (181) |
| Mxene-CA membranes | Treated sewage effluent | Seawater | AL-FS | - | Water flux decline: 10.7% | (182) |
| TFC (dopamine)-PSf support layer | Seawater | 0.5–4.0 M NaCl | AL-FS | Physical cleaning | Water flux decline: 14.4% | (178) |
| PDA-SWCNTS-PES MF membrane | BSA | 1 M NaCl | AL-DS | - | Water flux decline: 19.0% | (183) |
| TFC FO membrane (from Porifera) | Wastewater | Synthetic seawater (35g/L NaCl) | AL-FS | Physical and chemical cleaning | Water flux recovery: 110% | (184) |
| CTA membrane (from Sterlitech) | Landfill leachate wastewater | 1 M NaCl | AL-FS and AL-DS | Chemical cleaning (sodium docusate) | Water flux recovery: 99% | (185) |
| CTA membrane (FTS) | Landfill leachate wastewater | 0.6 M NaCl | AL-FS and AL-DS | Chemical cleaning ($H_2O_2$) | Water flux recovery: 100% | (185) |

The feed properties, including pH, ionic strength, and divalent cations, strongly influence fouling. Higher pH reduces fouling by increasing repulsion, while lower pH promotes hydrophobic interactions. However, at acidic pH, foulants may form a loose foulant layer, whereas at basic pH, severe fouling can result. Ionic strength enhances fouling by weakening electrostatic repulsion, promoting foulant aggregation. Higher ionic strength can lower the electrostatic repulsion between foulants, promoting foulant accumulation on the membrane surface. RSF, an inherent challenge in the FO process, occurs when solutes from DS diffuse to FS. This phenomenon significantly impacts fouling, causing solute accumulation in the FS, which alters its physico-chemical property, ultimately influencing the fouling behaviour. RSF is affected by certain factors such as DS concentration, membrane properties, and operating conditions with RSF rates often exacerbating fouling severity. The DS concentration also affects fouling by providing the osmotic driving force. High DS concentration increases permeate flux by accelerating foulant transport to the membrane surface. Studies revealed that even at low concentrations, NaCl induces more severe irreversible fouling than dextrose due to higher ionic strength, pH changes, and interactions with foulants. Additionally, divalent ions in NaCl can strengthen electrostatic interactions with negatively charged foulants and the membrane, promoting persistent fouling deposits.

Other than employing low-fouling membrane materials, researchers suggested various strategies in the FO process to mitigate FO membrane fouling such as the implementation of different cleaning techniques and optimization of systems' operational parameters.

Similar to any membrane process, fouling is unavoidable in the FO process as well, thereby suggesting that immediate fouling detection ensures and restores membrane performance. Thus, periodic rinsing and cleaning may be used to reverse the fouling (187). Membrane cleaning involves removing deposits/particulates (also known as "foulants") attached to the membrane surface and within pores. Ideally, the cleaning method should (i) completely remove foulants and restore full membrane performance, (ii) avoid damaging the membrane or altering its surface properties, and (iii) be cost-effective. Membrane cleaning can be classified into physical, chemical, physico-chemical, and biological cleaning methods. The membrane cleaning efficiencies are mainly indicated by flux recovery.

As the fouling in FO is less dense compared to the RO process, around 80–95% of water flux can be recovered through periodic rinsing and cleaning (188). In the FO process, cleaning strategies include chemical treatment, biological cleaning, physical cleaning (such as surface flushing, ultrasonication, and osmotic backwash), increased cross-flow velocity, and alteration of membrane orientation.

(i) Physical cleaning method involves applying mechanical forces to loosen and remove fouling layers from the membrane surface. They are primarily effective against foulants loosely attached to the membrane (i.e. reversible fouling). Surface flushing, osmotic backwashing, and air scouring are the most commonly used physical cleaning methods.

Surface flushing involves flushing the membrane with DI water at cross-flow velocity. Due to the high shear rate generated by increased flow turbulence, foulants on the surface are easily dislodged and washed away. Many studies revealed surface flushing is highly effective in restoring 95–100% water flux of membranes fouled by microplastic, alginate, silica, and gypsum. However, surface flushing is insufficient to remove stubborn foulants such as biofilms that strongly adhere to the membrane surface or within the membrane pores. Surface flushing increases the shear force and cleans the foulant on the membrane surface (i.e. AL). Moreover, the surface flushing method is ineffective for cleaning internal fouling that occurs in AL-DS mode.

Ultrasonication cleaning utilizes sound waves to generate tiny bubbles through a process known as cavitation. The collapse of these bubbles generates heat and pressure, thereby dislodging foulants from the membrane surface and within the membrane pores. The combined application of ultrasonication and cross-flow velocity (CFV) was investigated as an effective method for cleaning fouled membranes (189). It was revealed that employing low-frequency ultrasonication (30 kHz) combined with a high cross-flow velocity (42 cm $s^{-1}$) effectively restored water flux by efficiently removing mineral crystals and particulate fouling layers. A recent study revealed that the inclusion of 33 kHz ultrasonication in the FO process can partially mitigate the ICP effect, enhancing transmembrane flux and reducing concentration time (by 2 h) (190). Although transmembrane flux was improved, ultrasonication led to an increased RSF across the membrane into the FS and alteration of nutritional quality parameters of concentrated beetroot juice. Additionally, extended exposure to ultrasonication can damage membrane pores, leading to expansion or rupture (191). Further research is needed to optimize ultrasonic frequency and intensity while minimizing potential long-term damage to membrane durability during repeated cleaning cycles.

Osmotic backwashing is another physical cleaning method used for FO membranes. This involves reversing the flow from the DS to the FS, thereby forcing foulants embedded within the membrane pores and membrane surface (Table 2.13). Studies revealed that osmotic backwash can achieve moderate to high efficiency

**TABLE 2.13**
**Effectiveness of osmotic backwashing in cleaning FO membranes**

| Feed Solution | Composition of Osmotic Backwash | Cleaning Condition | Duration | Membrane Type | Water Flux Recovery, % | Ref. |
|---|---|---|---|---|---|---|
| Skimmed goat milk | 2 M NaCl | 30 min | 5 cycles × 480 min | TFC | 100% | Cao et al. (2024) (192) |
| Skim milk | 0.5% NaCl | 30 min; CFV: 0.6 m $s^{-1}$ | | CTA and TFC | 77–95% | Zhou et al. (2024) (193) |
| | 2 M NaCl | 30 min; CFV: 600 ml $min^{-1}$ | | | >90% | |
| Green fluorescent protein | 0.1–3 M NaCl | 1 min | No data available | Aquaporin-based TFC-FO | 93% | Daly et al. (2021) (194) |
| | 0.5–3 M $CaCl_2$ | | | | 39% | |

(40–100%) in restoring water flux across various FO membrane foulants. Overall, improved cleaning efficiency was achieved with optimized osmotic backwash conditions, which included higher NaCl concentration, increased CFV, and longer cleaning durations.

(ii) Although physical cleaning can efficiently reduce FO membrane fouling, it is unable to efficiently recover water flux if the foulant has strong adhesion with FO membranes. In such a case, chemical agents such as $NaCl/HNO_3$ can be used to treat the membrane. Chemical cleaning weakens the adhesive bond between foulant and membrane, thereby mitigating fouling (195). Chemical cleaning reagents work by breaking the strong adhesion forces between foulants and membrane surface using various mechanisms such as hydrolysis, saponification, solubilization, chelation, oxidation, disinfection, and dispersion.

Membrane chemical technique involves five stages: (i) The cleaning agent is applied to the fouled membrane surface, (ii) it then penetrates the fouling layer to reach the membrane, (iii) chemical reaction between the cleaning agent and foulants on the membrane facilitates the detachment of foulants, (iv) the waste cleaning agent is subsequently moved to the interface, and (v) finally, the resulting by-products are transported into the bulk solution.

The selection of chemicals and cleaning cycle frequency is based on the type and severity of fouling. It is crucial to select a cleaning agent that can efficiently remove complex foulants, prevent secondary fouling, and protect the membrane from further damage. Common cleaning agents used in FO membrane cleaning include alkalis, chelating agents, oxidants, acids, and surfactants.

Acidic reagents, including mineral and organic acids, are commonly used to remove inorganic fouling. In a recent study, HCl cleaning outperformed both SDS and EDTA, achieving a flux recovery rate of 89.91% in an FO process used for concentrating skimmed goat milk (SGM) (192). However, its performance is not consistent across all applications, and reportedly, the severe fouling developed over time during a long-term operation can make HCl cleaning ineffective, as it struggles to remove the accumulated foulants from the membrane surface. Cleaning with organic acids (such as citric acid) is ineffective in removing foulants, particularly when the fouling is more complex. However, citric acid cleaning was ineffective for restoring fouled membranes in polymer-flooding water treatment, achieving only 56% flux recovery. Overall, acid cleaning alone is typically ineffective, especially when there is complex fouling, as it primarily targets inorganic fouling. It is generally more efficient when combined with other cleaning agents.

Alkaline cleaning agents are specifically used to remove organic fouling (including biofouling) from membranes. Sodium hydroxide (NaOH) is a commonly used cleaning agent in the FO process. Cleaning with NaOH is more effective due to the excellent efficiency of the alkaline solution in the cleaning and removal of organic materials. It was demonstrated that sodium hydroxide (NaOH) cleaning could fully restore flux in TFC membranes following the processing of SGM (192). However, in certain cases, NaOH can fail to effectively remove organic foulants, as the swelling of membrane pores in alkaline conditions may enable smaller foulant fragments to enter the pores, intensifying fouling.

Oxidizing agents such as sodium hypochlorite and hydrogen peroxide are primarily used to remove organic and biological foulants through processes of oxidation and disinfection.

Although chemical cleaning is efficient in mitigating fouling, it may affect the membrane integrity and membrane lifetime. Moreover, the chemical cleaning increases chemical load and causes secondary pollution to the environment.

In addition to the cleaning methods, studies suggested that membrane fouling can be controlled by optimizing operating conditions. For example, compared to AL-DS, AL-FS is generally more prone to fouling. The increased fouling propensity of AL-FS (FO mode) is attributed to the accumulation of contaminants within the porous SL, leading to severe internal fouling. For example, osmotic backwashing of the FO membrane after wastewater treatment restored 95% to 91% of the declined water flux, compared to 90% to 68% in the AL-DS configuration. Pretreatment of FS can also impact the chances of membrane fouling. For instance, prefiltration of FS with a sand filter resulted in a 35% increase in water flux compared to the FO experiment without prefiltration.

In summary, membrane fouling is still a significant operational issue, hurdling the commercial/large-scale application of the FO process. Though anti-fouling membrane exhibited good resistance to organic and inorganic fouling, high fabrication cost and lack of long-term and pilot-plant performance data make them unreliable for commercial application. Chemical cleaning not only generates secondary pollution but also compromises the integrity of membranes AL. For example, most FO membranes are sensitive to oxidants (such as chlorine) that could damage the AL of the membrane. When compared to chemical cleaning, physical cleaning methods have more potential for fouling mitigation in the FO process. Physical cleaning methods (such as hot water and osmotic backwash) are cost-effective methods, with no impact on the AL, thereby eliminating chemical wastewater generation as in the chemical cleaning methods. Alteration of membrane orientation and flow rate can reduce the FO membrane fouling but cannot eliminated it. Pretreatment methods can reduce membrane fouling; however, they increase the overall treatment cost.

### 2.4.3 Reverse Solute Flux

The diffusion of draw solute through the membrane into the FS is defined as RSF. This results in a gradual loss of draw solute, reduces the concentration gradient, and increases the fouling propensity. While concentrating on liquid food, RSF is an undesirable process. As the diffusion of draw solute into the FS can alter the chemistry and composition of FS, similar to membrane fouling, RSF is inevitable in the FO process, which is affected by DS physiochemical properties (viscosity, ion/molecule size, ion charge, and diffusivity), flow velocity, membrane structural parameters (thickness, porosity, and tortuosity), and CP. RSF can aggravate the biofouling and scaling, whereas the cake layer deposited on the surface of the membrane in turn promotes the solute leakage. Effective control and mitigation of RSF are pivotal for maintaining the efficiency and sustainability of FO operations. Despite its significance, RSD has received comparatively less research attention than membrane fouling. Complete elimination of RSF remains impractical; however, its reduction and mitigation can be achieved by utilizing less permeable draw solutes, optimizing operational strategies, or developing advanced membranes with enhanced selectivity and structural properties; studies on FO systems lack detailed RSF data, hindering a comprehensive understanding of strategies to reduce RSF. Addressing this research gap requires a stronger focus on the design and synthesis of large-sized solutes, such as stimuli-responsive polymers, which can potentially lower diffusion rates. Additionally, developing robust mathematical models to predict solute diffusion and investigating the interplay between RSD and fouling will provide valuable insights. Future research should also prioritize the development of cost-effective methods for mitigating RSD, ensuring scalable and economically viable FO applications.

According to the above discussion, membrane fouling, CP, and RSF are closely interrelated in the FO process. Severe CP not only decreases the osmotic pressure but also increases the foulant deposition on the membrane surface, resulting in a decrease in water flux. Membrane fouling has an influence on the water flux by changing the membrane property, introducing additional hydraulic resistance, and introducing a fouling-enhanced CP or ICP self-compensation effect. RSD could increase the CP and change the chemistry and composition of FS, further impacting the fouling behaviour.

In summary, severe CP can decrease reverse solute diffusion (RSD) while exacerbating membrane fouling. Conversely, membrane fouling may intensify CP and further suppress RSD. At the same time, RSD can contribute to both enhanced membrane fouling and CP, creating a complex interplay among these factors. Future research should prioritize the development of innovative membranes and advanced DS to effectively manage RSD, CP, and membrane fouling. Additionally, robust mathematical models and comprehensive experimental investigations are essential to uncover the underlying mechanisms governing these interactions.

### 2.4.4 Strategies to Mitigate Concentration Polarization and Membrane Fouling

#### 2.4.4.1 Cleaning Protocols

Various cleaning strategies have been developed and applied in FO processes to mitigate membrane fouling, scaling, and CP, remove foulants, and recover water flux (196). Membrane fouling not only reduces water permeability but also compromises selectivity. Effective cleaning can therefore lower transport resistance and control the severity of CP (197). Commonly used cleaning methods in FO include physical, chemical, and biological approaches (198,199).

Physical cleaning methods, such as circulating clean water at high flow rates or applying backwashing, have been shown to effectively remove surface-deposited foulants (210). Additionally, reversing the water flow across the membrane has been demonstrated to fully restore membrane permeability (200).

Chemical cleaning is another effective method for fouling reversal and involves flushing membranes with chemical agents during or after FO operation (198,199).

These agents react with foulants to weaken their adhesion and increase their solubility (201). For instance, alkaline agents, surfactants, and oxidants are used to remove organic foulants, while acids and chelating agents can effectively dissolve inorganic scale. Chemical cleaning protocols were investigated for TFC FO membranes employed in wastewater treatment (202). Their study showed that a combination of 0.8% EDTA and 1% Alconox improved both water and solute fluxes. While alkaline treatment led to modest improvements, acid cleaning alone reduced performance. The most effective sequence was a two-step hybrid treatment involving 0.1% NaOH and 0.1% SDS, followed by an acid rinse using either 2% citric acid or 0.5% HCl. This approach achieved significant flux recovery and restored membrane selectivity. However, there remains a concern that chemical penetration into the sublayer may alter membrane properties and shorten its operational lifespan (201).

Biological cleaning has gained attention, particularly in membrane bioreactor (MBR) applications, and is categorized into enzymatic cleaning, energy uncoupling, and quorum quenching (197,199). Enzymatic methods use specific enzymes to degrade protein- or lipid-based biofilms, while energy uncoupling reduces microbial ATP production, thereby inhibiting biofilm formation. Quorum quenching disrupts microbial communication pathways (quorum sensing), preventing biofilm maturation on membrane surfaces. To preserve membrane integrity, it is critical to select biological agents that are compatible with the membrane material. For example, proteases have been used to remove proteinaceous foulants, and alginate lyase has been applied to target polysaccharide fouling via the β-elimination mechanism.

Amylase has been utilized to break down humic substances on membrane surfaces, and biofilm control has been demonstrated using energy uncouplers such as 2,4-dinitrophenol (DNP). Xu and Liu (2011) (203) found that DNP effectively inhibited microbial activity and suppressed the production of autoinducer-2 (AI-2), a key molecule in quorum sensing. Consequently, membranes cleaned with DNP showed enhanced water flux due to reduced biofilm accumulation after wastewater treatment.

While numerous cleaning protocols have been developed, they must be non-damaging, operationally feasible, and time-efficient to be viable for long-term FO operation (201). Each cleaning strategy—physical, chemical, or biological—has specific advantages and limitations, and the optimal approach should be tailored to the fouling characteristics and operating conditions of the FO system.

#### 2.4.4.2 Using Additives and Anti-Scaling Agents in the Draw Solution

Among the various strategies developed to mitigate fouling, scaling, and CP (both internal and external) in FO processes, the incorporation of anti-scaling agents into the DS has emerged as a promising approach (204). This method reduces inorganic scaling on the membrane surface by allowing the reverse diffusion of anti-scalants from the DS to the FS, where they inhibit scale formation.

Poly(aspartic acid sodium salt) (PaspNa) was proposed as an environmentally friendly and highly water-soluble anti-scalant for use in FO systems (205). Owing to its high molecular weight and large hydrated radius, PaspNa generates high osmotic pressure while exhibiting low RSF. Scaling tests using real wastewater as the FS demonstrated that PaspNa resulted in lower reverse salt flux compared to conventional inorganic DS solutes. At an optimal concentration of 0.20 g $L^{-1}$, PaspNa achieved a water flux of approximately 10 L $m^{-2}$ $h^{-1}$ with an RSF of just 10 g $m^{-2}$ $h^{-1}$. Moreover, the membrane experienced reduced ICP due to lower salt intrusion into the porous SL.

However, increasing the concentration of PaspNa beyond optimal levels resulted in higher water flux accompanied by increased solute diffusion, a common trade-off in membrane processes. At excessively high concentrations, severe ICP effects were observed, offsetting the benefits of reduced ECP. The thick, porous SL facilitated salt accumulation within the substructure, compromising hydrodynamic flow and reducing membrane performance. These findings highlight the importance of optimizing DS concentration to balance water permeability and anti-fouling performance.

A blended DS for seawater desalination was developed by combining TERGITOL NP7 and NP9—non-ionic surfactants with long hydrophobic chains and low critical micelle concentrations—with highly charged ethylenediaminetetraacetic acid disodium salt (EDTA-2Na) (206). This formulation aimed to reduce both RSF and membrane fouling. The presence of EDTA-2Na hindered NP7 adsorption onto the membrane surface due to hydrophobic interactions between the surfactant tails and the hydrophobic membrane matrix. As a result, the average membrane pore size decreased, thereby limiting the diffusion of $Na^+$, $H2[EDTA]^{2-}$, and $H[EDTA]^{3-}$ ions.

Using synthetic brackish water containing 5, 10, and 20 g $L^{-1}$ NaCl as FS, the FO system achieved water fluxes of 7.68 L $m^{-2}$ $h^{-1}$, 6.78 L $m^{-2}$ $h^{-1}$, and 5.95 L $m^{-2}$ $h^{-1}$, respectively, under 1 mol $L^{-1}$ EDTA-2Na and 15 mmol $L^{-1}$ NP7 as DS. For seawater (35 g $L^{-1}$ NaCl), the water flux declined to 3.81 L $m^{-2}$ $h^{-1}$. While increasing the EDTA-2Na concentration improved osmotic driving force, it also elevated RSF flux, exacerbating ICP and reducing overall water permeability. An optimum condition was identified using 0.1 mol/L EDTA-2Na and 15 mmol $L^{-1}$ NP7 as DS with DI water as FS, yielding a minimal RSF of 0.067 g $m^{-2}$ $h^{-1}$. This performance enhancement was attributed to the formation of a hydrophobic layer on the membrane surface, which contracted pores and limited foulant intrusion.

A novel anti-scaling strategy was recently introduced by incorporating ethylenediamine tetra(methylene phosphonic acid) sodium salt (EDTMPA-Na) into DI water to formulate an organophosphate-based DS (207). Compared to 1 mol $L^{-1}$ NaCl, a 0.69 mol $L^{-1}$ EDTMPA-Na DS provided comparable water flux while significantly reducing reverse solute diffusion and gypsum scaling. After 48 hours of FO operation, a water recovery rate of 60.6% was achieved, attributed to the reverse diffusion of EDTMPA-Na into the FS, where it inhibited calcium sulphate (gypsum) crystal formation. The favourable performance of EDTMPA-Na is associated with its high molecular weight, large hydrated radius, and strong negative charge, which induces electrostatic repulsion with the negatively charged TFC membrane, further lowering reverse solute transport. Nonetheless, the complex synthesis and high cost of EDTMPA-Na present challenges for large-scale application. Furthermore, introducing such additives may alter DS properties and membrane compatibility.

#### 2.4.4.3 Membrane Modification

Extensive research has focused on modifying the selective layer's properties and architecture to enhance membrane performance and reduce fouling. A key strategy involves surface modification of both the selective layer and SL through chemical treatments and the integration of additives or nanomaterials, which collectively improve the membrane's physicochemical characteristics and structural integrity (see Section 2.1.3, FO membrane surface modification).

One effective approach for mitigating fouling and CP is the chemical functionalization of the selective layer, often through nanomaterial incorporation. For example, Jeffamine—a PEG-based amine—was chemically grafted onto the selective layer by reacting it with acyl chloride groups (208). This modification significantly improved anti-fouling performance against synthetic wastewater rich in alginate (250 mg $L^{-1}$). Under fouling conditions, the water flux declined by only 2.8% in the modified membrane, compared to a 14.3% reduction in the pristine counterpart. The enhanced anti-fouling behaviour was attributed to reduced adhesion forces between carboxyl-modified latex particles and the membrane surface. While the modification did not affect salt permeability, it resulted in a decreased water permeability coefficient from 1.71 to 0.89 L $m^{-2}$ $h^{-1}$.

Similarly, a novel grafting approach was employed to introduce polyzwitterions (PZs) onto PA chains through SN2 nucleophilic substitution (209). This approach leverages the abundance of amide groups in PA to improve hydrophilicity and impart anti-fouling resistance. The resulting PSf-based FO membrane demonstrated superior wettability and fouling resistance, achieving a slightly higher water flux than the unmodified membrane when using 4 mol $L^{-1}$ NaCl as a DS and 20 mmol $L^{-1}$ NaCl as an FS, without significantly compromising salt rejection.

Conventional IP using MPD and TMC tends to yield rough, hydrophobic surfaces that are prone to fouling. Reportedly, the given issue was resolved by incorporating tris(2-aminoethyl) amine (TAEA) into the MPD phase to accelerate the IP reaction, forming a highly cross-linked, semi-aromatic PA structure (210). This was addressed by incorporating TAEA into the MPD phase to accelerate the IP reaction, forming a highly cross-linked, semi-aromatic PA structure. This led to a denser, smoother membrane surface with improved anti-fouling properties. However, variations in pH during synthesis influenced membrane morphology: Lowering the pH to 8–9 resulted in looser, thicker, and more hydrophilic layers, enhancing water flux but reducing salt rejection. Conversely, increasing TAEA concentration or pH improved both water flux and salt retention while significantly enhancing anti-fouling performance against sodium alginate.

A recent study emphasized the importance of selecting SLs with high hydrophilicity and low swelling potential. For example, nylon 6,6 exhibited lower swelling than CA and served as a robust SL for FO membranes (211). PVDF, a commonly used support material, was modified using 1,6-hexanediamine and dipoyl chloride through IP to improve chemical resistance and minimize swelling. This treatment produced nylon 6,6 nanofibres with enhanced strength and wettability. The resulting membranes achieved water fluxes of 22 L $m^{-2}$ $h^{-1}$ and 31 L $m^{-2}$ $h^{-1}$ in FO and PRO modes, respectively, using DI water as FS and 1 mol $L^{-1}$ NaCl as DS. The salt permeability was limited to 0.25 L $m^{-2}$ $h^{-1}$, while the structural parameter dropped to 193 μm, among the lowest recorded for FO membranes.

This work was further advanced by incorporating perfluorosulfonic acid (PFSA) into PVDF substrates. The sulfonic acid groups in PFSA enhanced hydrophilicity and facilitated phase inversion during membrane fabrication, resulting in porous structures characterized by straight, narrow finger-like pores and decreased tortuosity (212). The SL's structural parameter decreased to 334 μm, mitigating ICP effects. A PA layer synthesized via IP further contributed to a well-defined, uniform selective surface. The membrane achieved a water flux of ~27.0 L $m^{-2}$ $h^{-1}$ with DI water as FS and 1 mol $L^{-1}$ NaCl as DS, with salt leakage of 8.4 g $m^{-2}$ $h^{-1}$.

Incorporating nanofibre additives is another promising avenue. Carboxylated carbon nanofibres (CNFs), at various concentrations ranging from 0 to 1 wt.%, were deposited onto CTA-based FO membranes (213). The modified membranes exhibited a nanoporous sponge-like morphology, enhanced pore density, reduced roughness, and a lower structural parameter (0.6 mm), indicative of improved mass transfer and reduced ICP. Hydrophilicity increased, evidenced by a lower contact angle (57° vs. 70° for the control), and water flux rose to 15.6 L $m^{-2}$ $h^{-1}$ (from 8.7 L $m^{-2}$ $h^{-1}$ for the control) using 10 mmol $L^{-1}$ NaCl as FS and 1 mol $L^{-1}$ NaCl as DS. Salt leakage was minimized (~1 g $m^{-2}$ $h^{-1}$), and mechanical strength improved compared to the pristine CTA membranes. An emerging enhancement strategy involves coupling FO membranes with ultrasonic devices to further mitigate fouling and CP.

#### 2.4.4.4 Ultrasound Applications

ICP presents a persistent challenge in the FO system and cannot be effectively mitigated through turbulent flow enhancement or by adjusting hydrodynamic conditions alone. However, ultrasound treatment has emerged as a promising method to alleviate ICP without the need for toxic cleaning agents. Ultrasound devices induce acoustic streaming and generate cavitation bubbles in the liquid medium, which increases turbulence and enhances mass transfer across the membrane (214).

By applying ultrasonic vibrations, solute accumulation near the membrane can be minimized, thus reducing both ICP and ECP. For example, the performance of FO membranes was evaluated under 22 kHz ultrasonic irradiation (215). The application of ultrasound to the SL improved water flux from 30 L $m^{-2}$ $h^{-1}$ to 35 L $m^{-2}$ $h^{-1}$ when using pure water as the FS and sodium sulphate as DS. When using polyphenolic tannins FS, the water flux nearly doubled from 14 L $m^{-2}$ $h^{-1}$ to 25 L $m^{-2}$ $h^{-1}$. However, ultrasound exposure to the selective layer resulted in elevated RSF in both scenarios, indicating the importance of targeted application. Additionally, higher frequencies (573 kHz and 1136 kHz) adversely impacted membrane performance, underscoring the need for frequency optimization.

In another study, the effects of solute molecular weight, membrane orientation, and ultrasound treatment were explored in the concentration of fruit juices (215). Ultrasound facilitated intense mixing and stabilized solute concentration, effectively mitigating dilutive ICP. Substantial water flux enhancement was observed with sweet lime juice and anthocyanin-rich rose extract. While anthocyanin loss was minimal (~1.82%) at 30 kHz, ultrasound was ineffective against gel layer formation when 0.5% pectin was present in FS, leading to severe ECP and increased RSF.

Further investigation was conducted on the application of 40 kHz ultrasonication in FO systems treating synthetic brackish water and seawater, using magnesium sulphate and copper sulphate as DS (216). Water flux improvements of 43.9%, 40.5%, and 12.5% were recorded for DS concentrations of 0.5, 0.75, and 1.0 mol

$L^{-1}$, respectively. Overall, ultrasound-enhanced water fluxes of 34.6% (magnesium sulphate) and 43.9% (copper sulphate) were achieved. Notably, the enhancement was more pronounced at lower DS concentrations. Despite the benefit, RSF increases by 5% and 11.7% for magnesium and copper sulphate DSs, respectively.

In summary, ultrasound-assisted FO can significantly reduce CP and fouling, thereby improving membrane efficiency. Nevertheless, limitations related to scalability, equipment cost, energy demand, and the availability of suitable transducers hinder its application in large-scale operations.

### 2.4.5 Advanced Techniques for Detecting Concentration Polarization

Primarily, CP is influenced by the concentration difference between the membrane/solution interface. To characterize CP and fouling issues, most studies utilized experimental measurements of macroscopic and indirect operating conditions (217). However, accurate quantification of fouling behaviour and cake layer structure formed during the filtration process is of considerable interest in verifying current theories and modelling approaches. Macroscopic boundary layer measurements at different regions can help determine the water mass transfer rate, the concentration of permeate, and relevant properties of the membrane material. Generally, the CP and fouling mechanisms in the FO membrane depend on the solution properties, membrane materials, and operating conditions (218).

So far, fouling and CP have not been widely studied using real-time monitoring of long-term filtration experiments. Most studies focused on characterization of foulants forming on membrane surfaces and the structure of fouled membranes. One non-invasive characterization method is to monitor the mass transport impedance at the membrane interface by electrochemical impedance spectroscopy (EIS).

EIS was employed to reliably assess the effects of fouling and CP on ion mass transport at the anion exchange membrane interface, both before and after cleaning, assuming the EIS method could determine the different contributions of fouling and CP to ion mass transport in an electrodialysis system. In the given study, it was observed that organic matter acted as an anionic foulant and caused severe fouling on an anion exchange membrane. Due to fouling, when mass solute transfer was low, the solute concentration close to the membrane was higher than that of the bulk solution. This induced the growth of the cake layer on the membrane surface, leading to a gradual increase in CP over the anion exchange membrane. Online EIS analysis indicated that the chemical cleaning of an inverted electrode was effective and almost completely restored the ion flux (> 99%). This method appears to be effective in analysing the ion mass transfer during the desalination process and in monitoring fouling and CP.

Raman microspectroscopy (RM) is a well-established measurement technique to perform in situ measurements of the concentration distribution within the feed channels of a membrane separation lab set-up. A proof-of-concept laboratory study combined with numerical modelling was designed to explore the potential of RM for measuring CP in the presence of biofouling (219). Additionally, RM was employed to measure sulphate concentration boundary layers in an NF process (220). The concentration boundary layers were visualized at cross-flow velocities of 0.04 m $s^{-1}$ and

0.2 m $s^{-1}$. At cross-flow velocities of 0.04 and 0.2 m $s^{-1}$, RM visualized the sulphate concentration boundary layer. The results indicated that higher cross-flow velocity leads to a thinner CP layer and lower membrane wall concentration. Besides, when decreasing the cross-flow velocity, the CP effects are double those at the bulk concentration. However, the signal intensity and depth resolution are affected by refraction, thereby influencing quantitative analysis.

Time-resolved in situ small-angle X-ray scattering and time-resolved in situ velocity microparticle image velocimetry (PIV) techniques were employed to characterize membrane surfaces and foulants, respectively, during the filtration process (221). Both strategies were employed to obtain the velocity field's concentration profiles related to the distance from the membrane surface in the fouling layer. The fouling mechanism was reported to depend on the type of electrostatic interactions occurring among the various foulants in the fouling layer. For instance, when a Laponite solution was used in the experiment, the fouling layer formed over the membrane surface was thicker and more concentrated than when using a Peptizer solution. Furthermore, the stress field of the concentrated fouling layers could be calculated by integrating in situ shear rate fields (measured by micro-PIV) into rheological measurements. These data indicate that the FS flow mechanism is correlated to the solution's rheological behaviour. The flow accelerates due to the impacts of operating conditions, such as velocity and pressure, during the filtration process. Also, an increase in the volume fraction in the FS was correlated to a drop in the velocity in the fouling layer during the filtration test. Researchers identified three areas close to the membrane surface with different shear stress/shear rate behaviours. In the area where the shear rate and shear stress peaked, a fouling layer was formed with a thickness of a few hundred micrometres.

To understand the developmental stages and formation of fouling layers over a membrane, real-time measurement tools such as laser-induced fluorescence are powerful (222). LIF is an advanced non-destructive technique that can image a polarized layer and quantify a CP concentration profile by tracking the fouling process, from initial CP to the formation of a compacted fouling layer. The impact of operating conditions on membrane performance was investigated using advanced LIF to characterize CP and quantify the solute concentration field over a membrane during an NF cross-flow experiment. The fouling layer undergoes two phases, namely rapid formation and consolidation. The final fouling was dense and strongly adhered to the membrane surface, with a thickness of around 41 μm. The concentration of solute in the fouling layer was 1.5-fold higher than that of the FS. These numerical data confirm that the flow rate significantly affects the development of the fouling layer.

Due to the scarcity of non-invasive characterization instruments, few studies have assessed the microscopic process of the fouling layer and the growth mechanism of the polarized layer during FO filtration experiments. Advanced optical coherence tomography (OCT) was used to observe the fouling layer formed during FO using clay suspensions as the FS and various divalent salts, including sodium sulphate and magnesium chloride ($MgCl_2$), as DS. They also used numerical analysis to study the complicated interaction between the suspension and the membrane surface during a desalination test. Visual observations together with numerical analysis revealed that

RSF adversely contributed to the accumulation of clay components on the membrane surface. Cake layer formation was slow when $MaCl_2$ was used as DS, causing a low critical point in the initial flux. The divalent salts induced the formation of strong metal–ligand complexation when clay components interacted with the membrane surface. In this case, the OCT method is determining fouling development in the boundary layer, making it easy to optimize the operating parameters during a filtration experiment.

The structural and transport properties of FO membranes were investigated using magnetic resonance imaging (MRI) (223). This technique demonstrated a non-invasive technique for both ECP and ICP detection, with results that were consistent with theoretical data, thereby confirming that $^{23}Na$ MRI is a promising tool for determining the CP mechanism in FO and RO membranes during desalination processes.

Overall, determining the characteristics of a complex CP mechanism using quantitative analysis is necessary to validate the feasibility of using advanced optical instruments. However, there is a lack of useful reports on using advanced techniques to measure ICP in FO processes, which has hampered the development of effective fouling mitigation methods.

## REFERENCES

1. Zhang L, Sun X, Wu S. A comprehensive review of responsive draw solutes in forward osmosis: Categories, characteristics, mechanisms and modifications. Desalination. 2024 Aug;583:117676.
2. Bardhan A, Senthilmurugan S, Mohanty K. Optimisation of multi-component inorganic salt composition as draw solute for preparation of concentrated tea extract using forward osmosis process. Food and Bioproducts Processing [Internet]. 2023;138:126–38. Available from: https://doi.org/10.1016/j.fbp.2023.01.007
3. Ge Q, Su J, Amy GL, Chung TS. Exploration of polyelectrolytes as draw solutes in forward osmosis processes. Water Res [Internet]. 2012;46(4):1318–26. Available from: http://dx.doi.org/10.1016/j.watres.2011.12.043
4. Gadelha G, Nawaz MS, Hankins NP, Khan SJ, Wang R, Tang CY. Assessment of micellar solutions as draw solutions for forward osmosis. Desalination. 2014 Dec;354:97–106.
5. Nawaz MS, Parveen F, Gadelha G, Khan SJ, Wang R, Hankins NP. Reverse solute transport, microbial toxicity, membrane cleaning and flux of regenerated draw in the FO-MBR using a micellar draw solution. Desalination. 2016 Aug;391:105–11.
6. Nguyen HT, Nguyen NC, Chen SS, Li CW, Hsu HT, Wu SY. Innovation in draw solute for practical zero salt reverse in forward osmosis desalination. Ind Eng Chem Res. 2015 Jun 17;54(23):6067–74.
7. Nguyen HT, Nguyen NC, Chen SS, Duong HC, Nguyen ML, Tran CS, et al. Exploration of a cost-effective draw solution based on mixing surfactant and sodium chloride for forward osmosis desalination process. Environ Technol Innov. 2023 May;30:103088.
8. Dutta S, Nath K. Prospect of ionic liquids and deep eutectic solvents as new generation draw solution in forward osmosis process. J Water Process Eng. 2018 Feb;21:163–76.
9. Chen Q, Ge Q, Xu W, Pan W. Functionalized imidazolium ionic liquids promote seawater desalination through forward osmosis. J Memb Sci. 2019 Mar;574:10–6.
10. Haddad AZ, Menon AK, Kang H, Urban JJ, Prasher RS, Kostecki R. Solar desalination using thermally responsive ionic liquids regenerated with a photonic heater. Environ Sci Technol. 2021 Mar 2;55(5):3260–9.

11. Chen Q, Ge Q, Xu W, Pan W. Functionalized imidazolium ionic liquids promote seawater desalination through forward osmosis. J Memb Sci. 2019 Mar;574:10–6.
12. Chen R, Liao X, Ge Q. A novel multinuclear zinc complex Zn-Bet-Tf2N for electroplating wastewater treatment using forward osmosis technique. Chem Eng J. 2021 Jan;404:126569.
13. Pan Z, Guo H, Huang T, Wen G, Qu F, He J, et al. Effect of sewage sludge ash contents on the performance of thermo-sensitive hydrogel as draw agent for forward osmosis application. J Clean Prod. 2021 Sep;313:127941.
14. Zhang K, Li F, Wu Y, Feng L, Zhang L. Construction of ionic thermo-responsive PNIPAM/γ-PGA/PEG hydrogel as a draw agent for enhanced forward-osmosis desalination. Desalination. 2020 Dec;495:114667.
15. Vafaei MA, Shakeri A, Salehi H, Razavi SR, Salari N. The effect of nanosheets on polymer hydrogels performance in Rhodamine B dye removal by forward osmosis process. J Water Process Eng. 2021 Dec;44:102351.
16. Hartanto Y, Yun S, Jin B, Dai S. Functionalized thermo-responsive microgels for high performance forward osmosis desalination. Water Res. 2015 Mar;70:385–93
17. Zufía-Rivas J, Morales P, Veintemillas-Verdaguer S. Effect of the sodium polyacrylate on the magnetite nanoparticles produced by green chemistry routes: Applicability in forward osmosis. Nanomaterials. 2018 Jun 27;8(7):470.
18. Kim J, Kang H, Choi YS, Yu YAH, Lee JC. Thermo-responsive oligomeric poly(tetrabutylphosphonium styrenesulfonate)s as draw solutes for forward osmosis (FO) applications. Desalination. 2016 Mar;381:84–94.
19. Inada A, Yumiya K, Kumagai K, Matsuyama H. Effect of branch structure of thermoresponsive oligomers on draw solution performance in forward osmosis process. Colloids Surf A Physicochem Eng Asp. 2021 Jan;609:125659.
20. Kim YC, Han S, Hong SA. Feasibility study of magnetic separation of magnetic nanoparticle for forward osmosis. Water Sci Technol. 2011;64(2):469–76. https://doi.org/10.2166/wst.2011.566
21. Khazaie F, Shokrollahzadeh S, Bide Y, Sheshmani S, Shahvelayati AS. Forward osmosis using highly water dispersible sodium alginate sulfate coated-Fe3O4 nanoparticles as innovative draw solution for water desalination. Process Saf Environ Prot. 2021 Feb;146:789–99.
22. Bide Y, Shokrollahzadeh S. Toward tailoring of a new draw solute for forward osmosis process: Branched poly (deep eutectic solvent)-decorated magnetic nanoparticles. J Mol Liq. 2020 Dec;320:114409.
23. Park SY, Ahn HW, Chung JW, Kwak SY. Magnetic core-hydrophilic shell nanosphere as stability-enhanced draw solute for forward osmosis (FO) application. Desalination. 2016 Nov;397:22–9.
24. Na Y, Yang S, Lee S. Evaluation of citrate-coated magnetic nanoparticles as draw solute for forward osmosis. Desalination. 2014 Aug;347:34–42.
25. Shakeri A, Salehi H, Khankeshipour N, Nakhjiri MT, Ghorbani F. Magnetic nanoparticle-crosslinked ferrohydrogel as a novel class of forward osmosis draw agent. J Nanopart Res. 2018 Dec 7;20(12):325.
26. Nazari R, Aghababaie M, Razmjou A, Landarani-Isfahani A, Amini M, Hajjari M, et al. Multifunctional hyperbranched polyglycerol-grafted silica-encapsulated superparamagnetic iron oxide nanoparticles as novel and reusable draw agents in forward osmosis process. Desalination Water Treat. 2017 Feb;64:81–9.
27. Yang HM, Park CW, Han MJ, Seo BK, Moon JK, Lee KW. Hyperbranched polyglycerol carboxylate-coated magnetic nanoparticles as a draw solute in a combined forward osmosis and ultrafiltration process. J Nanosci Nanotechnol. 2016 Oct 1;16(10):10858–63.

28. Yang HM, Choi HM, Jang SC, Han MJ, Seo BK, Moon JK, et al. Succinate functionalization of hyperbranched polyglycerol-coated magnetic nanoparticles as a draw solute during forward osmosis. J Nanosci Nanotechnol. 2015 Oct 1;15(10):8279–84.
29. Ge Q, Yang L, Cai J, Xu W, Chen Q, Liu M. Hydroacid magnetic nanoparticles in forward osmosis for seawater desalination and efficient regeneration via integrated magnetic and membrane separations. J Memb Sci. 2016 Dec;520:550–9.
30. Tayel A, Nasr P, Sewilam H. Forward osmosis desalination using pectin-coated magnetic nanoparticles as a draw solution. Clean Technol Environ Policy. 2019 Oct 2;21(8): 1617–28.
31. Ma D, Tian Y, He T, Zhu X. Preparation of novel magnetic nanoparticles as draw solutes in forward osmosis desalination. Chin J Chem Eng. 2022 Jun;46:223–30.
32. Zhao Q, Chen N, Zhao D, Lu X. Thermoresponsive magnetic nanoparticles for seawater desalination. ACS Appl Mater Interfaces. 2013 Nov 13;5(21):11453–61.
33. Ban I, Markuš S, Gyergyek S, Drofenik M, Korenak J, Helix-Nielsen C, et al. Synthesis of Poly-Sodium-Acrylate (PSA)-coated magnetic nanoparticles for use in forward osmosis draw solutions. Nanomaterials. 2019 Aug 31;9(9):1238.
34. Hartanto Y, Yun S, Jin B, Dai S. Functionalized thermo-responsive microgels for high performance forward osmosis desalination. Water Res. 2015 Mar;70:385–93.
35. Cai Y, Shen W, Wang R, Krantz WB, Fane AG, Hu X. CO2 switchable dual responsive polymers as draw solutes for forward osmosis desalination. Chem Comm. 2013; 49(75):8377.
36. Joafshan M, Shakeri A, Razavi SR, Salehi H. Gas responsive magnetic nanoparticle as novel draw agent for removal of Rhodamine B via forward osmosis: High water flux and easy regeneration. Sep Purif Technol. 2022 Feb;282:119998.
37. Kadhim RM, Al-Abodi EE, Al-Alawy AF. Citrate-coated magnetite nanoparticles as osmotic agent in a forward osmosis process. Desalination Water Treat. 2018 May;115:45–52.
38. Shabani Z, Rahimpour A. Chitosan- and dehydroascorbic acid-coated Fe3O4 nanoparticles: preparation, characterization and their potential as draw solute in forward osmosis process. Iran Polym J. 2016 Oct 7;25(10):887–95.
39. Razmjou A, Barati MR, Simon GP, Suzuki K, Wang H. Fast deswelling of nanocomposite polymer hydrogels via magnetic field-induced heating for emerging FO desalination. Environ Sci Technol. 2013;47(12):6297–305
40. Cai Y, Shen W, Wang R, Krantz WB, Fane AG, Hu X. CO2 switchable dual responsive polymers as draw solutes for forward osmosis desalination. Chem Comm. 2013; 49(75):8377
41. Ellis SN, Cunningham MF, Jessop PG. A forward osmosis hydrogel draw agent that responds to both heat and CO2. Desalination. 2021 Aug;510:115074.
42. Cruz-Tato P, Richardson TMJ, Romero-Mangado J, Flynn M, Nicolau E. Performance evaluation of 1-cyclohexylpiperidine as a draw solute for forward osmosis water separation and CO2 recovery. ACS Omega. 2020 Oct 13;5(40):25919–26.
43. Amjad M, Gardy J, Hassanpour A, Wen D. Novel draw solution for forward osmosis based solar desalination. Appl Energy. 2018 Nov;230:220–31.
44. Zeng Y, Qiu L, Wang K, Yao J, Li D, Simon GP, et al. Significantly enhanced water flux in forward osmosis desalination with polymer-graphene composite hydrogels as a draw agent. RSC Adv. 2013;3(3):887–94.
45. Wu S, An Y, Lu J, Yu Q, He Z. EDTA-Na2 as a recoverable draw solute for water extraction in forward osmosis. Environ Res. 2022 Apr;205:112521.
46. Chen Y, Ge Q. A bifunctional zwitterion that serves as both a membrane modifier and a draw solute for forward osmosis wastewater treatment. ACS Appl Mater Interfaces. 2019 Oct 2;11(39):36118–29.

47. Wu Y, Liu Y, Chen R, Zhang WH, Ge Q. A pH-responsive supramolecular draw solute that achieves high-performance in arsenic removal via forward osmosis. Water Res. 2019 Nov;165:114993.
48. Shi Y, Liao X, Chen R, Ge Q. pH-responsive polyoxometalates that achieve efficient wastewater reclamation and source recovery via forward osmosis. Environ Sci Technol. 2021 Sep 21;55(18):12664–71.
49. Lee S, Boo C, Elimelech M, Hong S. Comparison of fouling behavior in forward osmosis (FO) and reverse osmosis (RO). J Memb Sci. 2010 Dec;365(1–2):34–9.
50. Piash KS, Sanyal O. design strategies for forward osmosis membrane substrates with low structural parameters—a review. Membranes (Basel). 2023 Jan 7;13(1):73.
51. Zhang Q, Zhou R, Peng X, Li N, Dai Z. development of support layers and their impact on the performance of thin film composite membranes (TFC) for water treatment. Polymers (Basel). 2023 Aug 3;15(15):3290.
52. Jamadi-Torab J, Malekzadeh A, Rahbari-Sisakht M, Ismail AF. State-of-the-Art in forward osmosis membrane process: materials, fabrication, and applications. Chemical Engineering and Processing—Process Intensification. 2025 May;110385.
53. Eyvaz M, Arslan S, İmer D, Yüksel E, Koyuncu İ. Forward osmosis membranes—a review: Part I. In: Du H, Thompson A, Wang X, editors. Osmotically driven membrane processes—approach, development and current status [Internet]. INTECH. p. 11–40. Available from: https://www.intechopen.com/chapters/59495
54. Wan CF, Yang T, Lipscomb GG, Stookey DJ, Chung TS. design and fabrication of hollow fiber membrane modules. J Memb Sci. 2017 Sep;538:96–107.
55. Garcia-Castello EM, McCutcheon JR, Elimelech M. Performance evaluation of sucrose concentration using forward osmosis. J Memb Sci. 2009 Aug;338(1–2):61–6.
56. Altaee A, Sharif A. Pressure retarded osmosis: Advancement in the process applications for power generation and desalination. Desalination. 2015;356:31–46. https://doi.org/10.1016/j.desal.2014.09.028
57. Kim WJ, Heldman DR. A mathematical estimation of the structural parameter for prediction of Forward Osmosis (FO) performance. J Water Process Eng. 2021 Feb;39:101719.
58. Tavakol I, Hadadpour S, Shabani Z, Ahmadzadeh Tofighy M, Mohammadi T, Sahebi S. Synthesis of novel thin film composite (TFC) forward osmosis (FO) membranes incorporated with carboxylated carbon nanofibers (CNFs). J Environ Chem Eng. 2020 Dec;8(6):104614.
59. Shen L, Zhang X, Tian L, Li Z, Ding C, Yi M, et al. Constructing substrate of low structural parameter by salt induction for high-performance TFC-FO membranes. J Memb Sci. 2020 Apr;600:117866.
60. Chi XY, Zhang PY, Guo XJ, Xu ZL. A novel TFC forward osmosis (FO) membrane supported by polyimide (PI) microporous nanofiber membrane. Appl Surf Sci. 2018 Jan;427:1–9.
61. Kriegesmann J. Processing of silicon carbide-based ceramics. In: Comprehensive hard materials. Elsevier; 2014. p. 89–175.
62. Kim YR, Jung S, Ryu H, Yoo YE, Kim SM, Jeon TJ. Synthetic biomimetic membranes and their sensor applications. Sensors. 2012 Jul 11;12(7):9530–50.
63. Shen Y, Saboe PO, Sines IT, Erbakan M, Kumar M. Biomimetic membranes: A review. J Memb Sci. 2014 Mar;454:359–81
64. Fuwad A, Ryu H, Malmstadt N, Kim SM, Jeon TJ. Biomimetic membranes as potential tools for water purification: Preceding and future avenues. Desalination. 2019 May;458:97–115.
65. Sathiyajith C, Shaikh RR, Han Q, Zhang Y, Meguellati K, Yang YW. Biological and related applications of pillar[n]arenes. Chem Comm. 2017;53(4):677–96.

66. Fuwad A, Ryu H, Han ED, Lee JH, Malmstadt N, Kim YR, et al. Highly permeable and shelf-stable aquaporin biomimetic membrane based on an anodic aluminum oxide substrate. NPJ Clean Water. 2024 Feb 15;7(1):11.
67. Ren J, McCutcheon JR. A new commercial biomimetic hollow fiber membrane for forward osmosis. Desalination. 2018;442(October 2017):44–50.
68. Abaie E, Xu L, Shen Y. Bioinspired and biomimetic membranes for water purification and chemical separation: A review. Front Environ Sci Eng. 2021 Dec 10;15(6):124.
69. Revanur R, Roh I, Klare JE, Noy A, Bakajin O. Thin film composite membranes for forward osmosis and their preparation methods. United States Patent; USOO89 20654B2, 2012
70. Jeong G, Jang D, Jeong S, Jang A. Enhancing modification of forward osmosisrich wastewater. Chem Eng J. 2025 Jan;504:158936 membrane for zwitterion-based surface using chemical structure of arginine to treat ammonium-
71. Li Z, Valladares Linares R, Bucs S, Fortunato L, Hélix-Nielsen C, Vrouwenvelder JS, et al. Aquaporin based biomimetic membrane in forward osmosis: Chemical cleaning resistance and practical operation. Desalination. 2017 Oct;420:208–15.
72. Xia L, Andersen MF, Hélix-Nielsen C, McCutcheon JR. Novel commercial aquaporin flat-sheet membrane for forward osmosis. Ind Eng Chem Res. 2017 Oct 18;56(41):11919–25
73. Akther N, Sanahuja-Embuena V, Górecki R, Phuntsho S, Helix-Nielsen C, Shon HK. Employing the synergistic effect between aquaporin nanostructures and graphene oxide for enhanced separation performance of thin-film nanocomposite forward osmosis membranes. Desalination. 2021 Jan;498:114795.
74. Madsen HT, Bajraktari N, Hélix-Nielsen C, Van der Bruggen B, Søgaard EG. Use of biomimetic forward osmosis membrane for trace organics removal. J Memb Sci. 2015 Feb;476:469–74.
75. Zhang Y, Wang KT, Jiang WL, He JY, Wang H, Li B, et al. Black odorous water concentrating by forward osmosis (FO) with aquaporin biomimetic membranes: Pollutants concentrating and membrane fouling characteristics. Chem Eng J. 2020 Nov;400:125938.
76. Chen Y, Ren X, Huang M, Li Y. Evaluation of aquaporin based biomimetic forward osmosis membrane in terms of rejection performance for contaminants in greywater and its membrane fouling properties. Chemosphere. 2023 Aug;333:138983.
77. Akhtar A, Singh M, Subbiah S, Mohanty K. Sugarcane juice concentration using a novel aquaporin hollow fiber forward osmosis membrane. Food and Bioproducts Processing. 2021 Mar;126:195–206.
78. Bardhan A, Subbiah S, Mohanty K. Modeling and experimental validation for the preparation of concentrated tea extract using a forward osmosis process using a food-grade inorganic draw solute. Ind Eng Chem Res. 2022;62(7):3315–26.
79. Jeong G, Jang D, Jeong S, Jang A. Enhancing modification of forward osmosis membrane for zwitterion-based surface using chemical structure of arginine to treat ammonium-rich wastewater. Chem Eng J. 2025 Jan;504:158936
80. Pejman M, Firouzjaei MD, Aktij SA, Das P, Zolghadr E, Jafarian H, et al. Improved antifouling and antibacterial properties of forward osmosis membranes through surface modification with zwitterions and silver-based metal organic frameworks. J Memb Sci. 2020 Oct;611:118352.
81. Qiu M, He C. Novel zwitterion-silver nanocomposite modified thin-film composite forward osmosis membrane with simultaneous improved water flux and biofouling resistance property. Appl Surf Sci. 2018 Oct;455:492–501
82. Zou S, Smith ED, Lin S, Martin SM, He Z. Mitigation of bidirectional solute flux in forward osmosis via membrane surface coating of zwitterion functionalized carbon nanotubes. Environ Int. 2019 Oct;131:104970.
83. Li P, Ge Q. Membrane surface engineering with bifunctional zwitterions for efficient oil–water separation. ACS Appl Mater Interfaces. 2019 Aug 28;11(34):31328–37.

84. Fu W, Chen J, Li C, Jiang L, Qiu M, Li X, et al. Enhanced flux and fouling resistance forward osmosis membrane based on a hydrogel/MOF hybrid selective layer. J Colloid Interface Sci. 2021 Mar;585:158–66.
85. Bagherzadeh M, Bayrami A, Amini M. Enhancing forward osmosis (FO) performance of polyethersulfone/polyamide (PES/PA) thin-film composite membrane via the incorporation of GQDs@UiO-66-NH particles. J Water Process Eng. 2020 Feb;33:101107.
86. He M, Wang L, Lv Y, Wang X, Zhu J, Zhang Y, et al. Novel polydopamine/metal organic framework thin film nanocomposite forward osmosis membrane for salt rejection and heavy metal removal. Chem Eng J. 2020 Jun;389:124452.
87. Seyedpour SF, Rahimpour A, Najafpour G. Facile in-situ assembly of silver-based MOFs to surface functionalization of TFC membrane: A novel approach toward long-lasting biofouling mitigation. J Memb Sci [Internet]. 2019;573:257–69. Available from: https://www.sciencedirect.com/science/article/pii/S0376738818330199
88. Wang Y, Li X, Zhao S, Fang Z, Ng D, Xie C, et al. Thin-film composite membrane with interlayer decorated metal–organic framework UiO-66 toward enhanced forward osmosis performance. Ind Eng Chem Res. 2019 Jan 9;58(1):195–206.
89. Zirehpour A, Rahimpour A, Ulbricht M. Nano-sized metal organic framework to improve the structural properties and desalination performance of thin film composite forward osmosis membrane. J Memb Sci. 2017 Jun;531:59–67.
90. Hu Z, Peng Y, Kang Z, Qian Y, Zhao D. A Modulated Hydrothermal (MHT) approach for the facile synthesis of UiO-66-type MOFs. Inorg Chem. 2015 May 18;54(10):4862–8.
91. Ma D, Han G, Peh SB, Chen SB. Water-stable metal–organic framework UiO-66 for performance enhancement of forward osmosis membranes. Ind Eng Chem Res. 2017 Nov 8;56(44):12773–82.
92. Arjmandi M, Peyravi M, Altaee A, Arjmandi A, Pourafshari Chenar M, Jahanshahi M, et al. A state-of-the-art protocol to minimize the internal concentration polarization in forward osmosis membranes. Desalination. 2020 Apr;480:114355
93. Bayrami A, Bagherzadeh M, Navi H, Chegeni M, Hosseinifard M, Amini M. Thin-film nanocomposite membranes containing aspartic acid-modified MIL-53-NH2 (Al) for boosting desalination and anti-fouling performance. Desalination. 2022 Jan;521:115386.
94. Samsami S, Sarrafzadeh MH, Ahmadi A. Surface modification of thin-film nanocomposite forward osmosis membrane with super-hydrophilic MIL-53 (Al) for doxycycline removal as an emerging contaminant and membrane antifouling property enhancement. Chem Eng J. 2022 Mar;431:133469.
95. Lakra R, Bilad MR, Balakrishnan M, Basu S. Development of high-performance CuBTC MOF-based forward osmosis (FO) membranes and their cleaning strategies. Chem Eng Res Des. 2023 Feb;190:566–79.
96. Ali Z, Kiran S, Haq NU, Dahshan A, Iqbal SZ, Nazir A, et al. Fabrication of novel (Cr) based (MIL-101) incorporated mixed matrix forward osmosis membranes for environmental remediation. Mater Chem Phys. 2024 Mar;315:128951.
97. Ahmadalipour S, Shakeri A. Hydrophilization of ZIF-8 with tannic acid to enhance forward osmosis membranes performance in lithium concentration. J Mol Liq. 2025 Apr;127666.
98. Romero-Vargas Castrillón S, Lu X, Shaffer DL, Elimelech M. Amine enrichment and poly(ethylene glycol) (PEG) surface modification of thin-film composite forward osmosis membranes for organic fouling control. J Memb Sci. 2014 Jan;450:331–9.
99. Hakimi G, Shakeri A, Yassari M, Salehi H. Enhancing the permeability and fouling resistance of thin-film composite forward osmosis membranes via poly(ethylene glycol) diamine as a comonomer. ACS Appl Polym Mater. 2023 Jan 13;5(1):567–75.

100. Rana D, Narbaitz RM, Garand-Sheridan AM, Westgate A, Matsuura T, Tabe S, et al. Development of novel charged surface modifying macromolecule blended PES membranes to remove EDCs and PPCPs from drinking water sources. J Mater Chem A. 2014;2(26):10059–72.
101. Liu Y, Mi B. Effects of organic macromolecular conditioning on gypsum scaling of forward osmosis membranes. J Memb Sci. 2014 Jan;450:153–61.
102. Deng YH, Chen JH, Yang Q, Zhuo YZ. Carbon quantum dots (CQDs) and polyethyleneimine (PEI) layer-by-layer (LBL) self-assembly PEK-C-based membranes with high forward osmosis performance. Chem Eng Res Des. 2021 Jun;170:423–33.
103. Mahat NA, Shamsudin SA, Jullok N, Ma'Radzi AH. Carbon quantum dots embedded polysulfone membranes for antibacterial performance in the process of forward osmosis. Desalination. 2020 Nov;493:114618.
104. Wang X, Zhang X, Zhang H, Xiao Y, Wen X, Zhao Y. High performance Fabric-Reinforced sulfonated polysulfone supported thin film composite membrane. Sep Purif Technol. 2023 Nov;325:124695.
105. Rastogi NK. Opportunities and challenges in application of forward osmosis in food processing. Crit Rev Food Sci Nutr. 2016;56(2):266–91.
106. Yadav S, Saleem H, Ibrar I, Naji O, Hawari AA, Alanezi AA, et al. Recent developments in forward osmosis membranes using carbon-based nanomaterials. Desalination. 2020 May;482:114375
107. Darabi RR, Hosseini SP, Peyravi M, Jahanshahi M. Thin-Film nanocomposite forward osmosis membranes incorporated with hydrophilic TiO2/Fe3O4 nanoparticles: Toward alleviated ICP. Arab J Sci Eng. 2025 Mar 5;50(6):3971–86.
108. Wei S, Guo H, Cui X, Ding X, Xia X, de Waele V. Fabrication of ultra-permeable membranes with antibiofouling capability by incorporating bifunctionalized zeolite. Sep Purif Technol. 2025 Feb;354:129258.
109. Baig MI, Hardian R, Alharthi FA, Fellows CM, Szekely G. Brine separation with polyamide and polyimine thin film composite nanofiltration membranes obtained from biobased monomers. J Memb Sci. 2025 Jan;713:123324.
110. Kallem P, Pandey RP, Hegab HM, Patel M, Hasan SW, Banat F. Forward osmosis membranes for desalination and wastewater treatment: Review of recent advances in electrospun nanofiber-based substrates, challenges, and future research recommendations. J Environ Chem Eng. 2023 Jun;11(3):110231.
111. Emadzadeh D, Lau WJ, Matsuura T, Rahbari-Sisakht M, Ismail AF. A novel thin film composite forward osmosis membrane prepared from PSf–TiO2 nanocomposite substrate for water desalination. Chem Eng J. 2014 Feb;237:70–80.
112. Xu GR, An ZH, Min-Wang, Ke-Xu, Zhao HL, Liu Q. Polyamide layer modulation for PA-TFC membranes optimization: Developments, mechanisms, and implications. Sep Purif Technol. 2023 Apr;311:123200
113. Akther N, Yuan Z, Chen Y, Lim S, Phuntsho S, Ghaffour N, et al. Influence of graphene oxide lateral size on the properties and performances of forward osmosis membrane. Desalination. 2020 Jun;484:114421.
114. Lin YL, Zheng NY, Chen YJ, Chang CC. Incorporation of zeolitic imidazolate framework-8 (ZIF-8) in the polyamide and polysulfone layers of forward osmosis membranes for enhancing aquaculture wastewater recovery and PPCPs removal. J Water Process Eng. 2024 Sep;66:105980.
115. Bagherzadeh M, Bayrami A, Amini M. Thin-film nanocomposite forward osmosis membranes modified with Zr-based metal–organic framework to improve desalination performance. Appl Organomet Chem. 2020 Feb 16;34(2).

116. Zhu J, Hou J, Zhang R, Yuan S, Li J, Tian M, et al. Rapid water transport through controllable, ultrathin polyamide nanofilms for high-performance nanofiltration. J Mater Chem A Mater. 2018;6(32):15701–9.
117. Rahimi M, Hassanzadeh Tabrizi SA, Aminsharei F. Fabrication and antibacterial properties of TFC membrane modified with cellulose/copper oxide nanoparticles for removal of cadmium from water. Sep Sci Technol. 2022 Jul 24;57(11):1762–74.
118. Karkooti A, Yazdi AZ, Chen P, McGregor M, Nazemifard N, Sadrzadeh M. development of advanced nanocomposite membranes using graphene nanoribbons and nanosheets for water treatment. J Memb Sci. 2018 Aug;560:97–107.
119. Shakeri A, Salehi H, Rastgar M. Chitosan-based thin active layer membrane for forward osmosis desalination. Carbohydr Polym. 2017 Oct;174:658–68.
120. Song H, Zhu L, Zeng Z, Xue Q. High performance forward osmosis cellulose acetate (CA) membrane modified by polyvinyl alcohol and polydopamine. J Polym Res. 2018 Jul 29;25(7):159.
121. Guo H, Yao Z, Wang J, Yang Z, Ma X, Tang CY. Polydopamine coating on a thin film composite forward osmosis membrane for enhanced mass transport and antifouling performance. J Memb Sci. 2018 Apr;551:234–42.
122. Wang J, Xiao T, Bao R, Li T, Wang Y, Li D, et al. Zwitterionic surface modification of forward osmosis membranes using N-aminoethyl piperazine propane sulfonate for grey water treatment. Process Saf Environ Prot. 2018 May;116:632–9.
123. Yang E, Kim CM, Song J, Ki H, Ham MH, Kim IS. Enhanced desalination performance of forward osmosis membranes based on reduced graphene oxide laminates coated with hydrophilic polydopamine. Carbon N Y. 2017 Jun;117:293–300.
124. Naveed A, Gul S, Amin NU, Khraisheh M, Gul R. Graphene Oxide (GO) based coated geopolymeric membrane for concentrating orange juice through forward osmosis. Int J Fruit Sci. 2020 Sep 14;20(sup2):S636–49.
125. Zhang X, Tian J, Gao S, Zhang Z, Cui F, Tang CY. In situ surface modification of thin film composite forward osmosis membranes with sulfonated poly(arylene ether sulfone) for anti-fouling in emulsified oil/water separation. J Memb Sci. 2017 Apr;527:26–34.
126. Park MJ, Gonzales RR, Abdel-Wahab A, Phuntsho S, Shon HK. Hydrophilic polyvinyl alcohol coating on hydrophobic electrospun nanofiber membrane for high performance thin film composite forward osmosis membrane. Desalination. 2018 Jan;426:50–9.
127. Liu Q, Li J, Zhou Z, Xie J, Lee JY. Hydrophilic mineral coating of membrane substrate for reducing Internal Concentration Polarization (ICP) in forward osmosis. Sci Rep. 2016 Jan 22;6(1):19593.
128. Suwaileh WA, Johnson DJ, Sarp S, Hilal N. Advances in forward osmosis membranes: Altering the sub-layer structure via recent fabrication and chemical modification approaches. Desalination. 2018 Jun;436:176–201
129. Wang J, Xiao T, Bao R, Li T, Wang Y, Li D, et al. Zwitterionic surface modification of forward osmosis membranes using N-aminoethyl piperazine propane sulfonate for grey water treatment. Process Saf Environ Prot. 2018 May;116:632–9.
130. Li J, Yin L, Qiu G, Li X, Liu Q, Xie J. A photo-bactericidal thin film composite membrane for forward osmosis. J Mater Chem A Mater. 2015;3(13):6781–6.
131. Ni T, Ge Q. Highly hydrophilic thin-film composition forward osmosis (FO) membranes functionalized with aniline sulfonate/bisulfonate for desalination. J Memb Sci. 2018 Oct;564:732–41.
132. Chen G, Liu R, Shon HK, Wang Y, Song J, Li XM, et al. Open porous hydrophilic supported thin-film composite forward osmosis membrane via co-casting for treatment of high-salinity wastewater. Desalination. 2017 Mar;405:76–84.

133. Chen D, Werber JR, Zhao X, Elimelech M. A facile method to quantify the carboxyl group areal density in the active layer of polyamide thin-film composite membranes. J Memb Sci. 2017 Jul;534:100–8.
134. Zhang X, Tian J, Ren Z, Shi W, Zhang Z, Xu Y, et al. High performance thin-film composite (TFC) forward osmosis (FO) membrane fabricated on novel hydrophilic disulfonated poly(arylene ether sulfone) multiblock copolymer/polysulfone substrate. J Memb Sci. 2016 Dec;520:529–39
135. Wu X, Shaibani M, Smith SJD, Konstas K, Hill MR, Wang H, et al. Microporous carbon from fullerene impregnated porous aromatic frameworks for improving the desalination performance of thin film composite forward osmosis membranes. J Mater Chem A Mater. 2018;6(24):11327–36.
136. Yang E, Kim CM, Song J, Ki H, Ham MH, Kim IS. Enhanced desalination performance of forward osmosis membranes based on reduced graphene oxide laminates coated with hydrophilic polydopamine. Carbon N Y. 2017 Jun;117:293–300.
137. Li M, Karanikola V, Zhang X, Wang L, Elimelech M. A self-standing, support-free membrane for forward osmosis with no internal concentration polarization. Environ Sci Technol Lett. 2018 May 8;5(5):266–71.
138. Ghaemi N, Khodakarami Z. Nano-biopolymer effect on forward osmosis performance of cellulosic membrane: High water flux and low reverse salt. Carbohydr Polym. 2019 Jan;204:78–88.
139. Shakeri A, Salehi H, Rastgar M. Chitosan-based thin active layer membrane for forward osmosis desalination. Carbohydr Polym. 2017 Oct;174:658–68.
140. Salehi H, Rastgar M, Shakeri A. Anti-fouling and high water permeable forward osmosis membrane fabricated via layer by layer assembly of chitosan/graphene oxide. Appl Surf Sci. 2017 Aug;413:99–108.
141. Ma N, Wei J, Qi S, Zhao Y, Gao Y, Tang CY. Nanocomposite substrates for controlling internal concentration polarization in forward osmosis membranes. J Memb Sci. 2013 Aug;441:54–62.
142. Emadzadeh D, Ghanbari M, Lau WJ, Rahbari-Sisakht M, Rana D, Matsuura T, et al. Surface modification of thin film composite membrane by nanoporous titanate nanoparticles for improving combined organic and inorganic antifouling properties. Materi Sci Eng: C. 2017 Jun;75:463–70.
143. Shen L, Xiong S, Wang Y. Graphene oxide incorporated thin-film composite membranes for forward osmosis applications. Chem Eng Sci. 2016 Apr;143:194–205.
144. Amini M, Jahanshahi M, Rahimpour A. Synthesis of novel thin film nanocomposite (TFN) forward osmosis membranes using functionalized multi-walled carbon nanotubes. J Memb Sci. 2013 May;435:233–41.
145. Kim T, Lee Y, Kim E, Kim K. Fouling-resistant surface modification of forward osmosis membranes using MoS2-Ag nanofillers. Surf Interf. 2023 Jun;38:102844.
146. Cath TY, Elimelech M, McCutcheon JR, McGinnis RL, Achilli A, Anastasio D, et al. Standard methodology for evaluating membrane performance in osmotically driven membrane processes. Desalination. 2013 Mar;312:31–8.
147. Akhtar A, Singh M, Subbiah S, Mohanty K. Modelling, experimental validation and process design of forward osmosis process for sugarcane juice concentration. LWT. 2021 Apr 1;141:110852.
148. Wang J, Liu X. Forward osmosis technology for water treatment: Recent advances and future perspectives. J Clean Prod. 2021 Jan;280:124354
149. Heo J, Boateng LK, Flora JRV, Lee H, Her N, Park YG, et al. Comparison of flux behavior and synthetic organic compound removal by forward osmosis and reverse osmosis membranes. J Memb Sci. 2013 Sep;443:69–82.

150. Bai D, Kruczek B. Effect of membrane orientation and concentration of draw solution on the behavior of commercial osmotic membrane in a novel dynamic forward osmosis tests. Membranes (Basel). 2022 Mar 31;12(4):385.
151. Dsilva Winfred Rufuss D, Kapoor V, Arulvel S, Davies PA. Advances in forward osmosis (FO) technology for enhanced efficiency and output: A critical review. J Clean Prod. 2022 Jul;356:131769.
152. Ali SM, Kim JE, Phuntsho S, Jang A, Choi JY, Shon HK. Forward osmosis system analysis for optimum design and operating conditions. Water Res. 2018 Nov;145:429–41.
153. Seo J, Kim YM, Chae SH, Lim SJ, Park H, Kim JH. An optimization strategy for a forward osmosis-reverse osmosis hybrid process for wastewater reuse and seawater desalination: A modeling study. Desalination. 2019 Aug;463:40–9.
154. Kim C, Lee S, Shon HK, Elimelech M, Hong S. Boron transport in forward osmosis: Measurements, mechanisms, and comparison with reverse osmosis. J Memb Sci. 2012 Nov;419–420:42–8.
155. Cui Y, Ge Q, Liu XY, Chung TS. Novel forward osmosis process to effectively remove heavy metal ions. J Memb Sci. 2014 Oct;467:188–94.
156. Cornelissen E, Harmsen D, Dekorte K, Ruiken C, Qin J, Oo H, et al. Membrane fouling and process performance of forward osmosis membranes on activated sludge. J Memb Sci. 2008 Jul 1;319(1–2):158–68
157. Hau NT, Chen SS, Nguyen NC, Huang KZ, Ngo HH, Guo W. Exploration of EDTA sodium salt as novel draw solution in forward osmosis process for dewatering of high nutrient sludge. J Memb Sci. 2014 Apr;455:305–11.
158. Phuntsho S, Sahebi S, Majeed T, Lotfi F, Kim JE, Shon HK. Assessing the major factors affecting the performances of forward osmosis and its implications on the desalination process. Chem Eng J. 2013 Sep;231:484–96.
159. Li D, Zhang X, Simon GP, Wang H. Forward osmosis desalination using polymer hydrogels as a draw agent: Influence of draw agent, feed solution and membrane on process performance. Water Res [Internet]. 2013;47(1):209–15. Available from: http://dx.doi.org/10.1016/j.watres.2012.09.049
160. Zhao P, Gao B, Xu S, Kong J, Ma D, Shon HK, et al. Polyelectrolyte-promoted forward osmosis process for dye wastewater treatment—Exploring the feasibility of using polyacrylamide as draw solute. Chem Eng J. 2015 Mar;264:32–8.
161. Zhao S, Zou L. Effects of working temperature on separation performance, membrane scaling and cleaning in forward osmosis desalination. Desalination. 2011 Sep;278(1–3):157–64.
162. Wang W, Zhang Y, Esparra-Alvarado M, Wang X, Yang H, Xie Y. Effects of pH and temperature on forward osmosis membrane flux using rainwater as the makeup for cooling water dilution. Desalination. 2014 Oct;351:70–6.
163. McCutcheon JR, Elimelech M. Influence of concentrative and dilutive internal concentration polarization on flux behavior in forward osmosis. J Memb Sci. 2006 Nov 1;284(1–2):237–47.
164. Chanukya BS, Patil S, Rastogi NK. Influence of concentration polarization on flux behavior in forward osmosis during desalination using ammonium bicarbonate. Desalination. 2013 Mar;312:39–44.
165. Xie M, Price WE, Nghiem LD. Rejection of pharmaceutically active compounds by forward osmosis: Role of solution pH and membrane orientation. Sep Purif Technol. 2012 Jun;93:107–14.
166. Hizam SM, Bilad MR, Nordin NAH, Sambudi NS, Wirzal MDH, Yusof N, et al. Inclined forward osmosis module system for fouling control in sustainable produced water treatment using seawater as draw solution. J Water Process Eng. 2021 Apr;40:101752.

167. Lee WJ, Ng ZC, Hubadillah SK, Goh PS, Lau WJ, Othman MHD, et al. Fouling mitigation in forward osmosis and membrane distillation for desalination. Desalination. 2020 Apr;480:114338.
168. Firouzjaei MD, Seyedpour SF, Aktij SA, Giagnorio M, Bazrafshan N, Mollahosseini A, et al. Recent advances in functionalized polymer membranes for biofouling control and mitigation in forward osmosis. J Memb Sci. 2020 Feb;596:117604.
169. Xu D, Sun X, Wu S. A comprehensive review of the forward osmosis membrane modification strategies for membrane fouling mitigation in wastewater treatment. J Environ Chem Eng. 2025 Jun;13(3):116992
170. Ghorbani F, Shakeri A, Ali Vafaei M, Salehi H. Polyoxometalate-cored supramolecular star polymers as a novel crosslinker for graphene oxide-based forward osmosis membranes: Anti-fouling, super hydrophilic and high water permeable. Sep Purif Technol. 2021 Jul;267:118578.
171. Xie M, Price WE, Nghiem LD, Elimelech M. Effects of feed and draw solution temperature and transmembrane temperature difference on the rejection of trace organic contaminants by forward osmosis. J Memb Sci. 2013 Jul;438:57–64.
172. Zou S, Wang YN, Wicaksana F, Aung T, Wong PCY, Fane AG, et al. Direct microscopic observation of forward osmosis membrane fouling by microalgae: Critical flux and the role of operational conditions. J Memb Sci. 2013 Jun;436:174–85.
173. Boo C, Lee S, Elimelech M, Meng Z, Hong S. Colloidal fouling in forward osmosis: Role of reverse salt diffusion. J Memb Sci. 2012 Feb;390–391:277–84.
174. Zhang J, Loong WLC, Chou S, Tang C, Wang R, Fane AG. Membrane biofouling and scaling in forward osmosis membrane bioreactor. J Memb Sci. 2012 Jun;403–404:8–14
175. Ibrar I, Naji O, Sharif A, Malekizadeh A, Alhawari A, Alanezi AA, et al. A review of fouling mechanisms, control strategies and real-time fouling monitoring techniques in forward osmosis. Water (Basel). 2019 Apr 4;11(4):695.
176. Li Z, Wu C, Huang J, Zhou R, Jin Y. Membrane fouling behavior of forward osmosis for fruit juice concentration. Membranes (Basel). 2021 Aug 11;11(8):611.
177. Choi H, Son M, Yoon S, Celik E, Kang S, Park H, et al. Alginate fouling reduction of functionalized carbon nanotube blended cellulose acetate membrane in forward osmosis. Chemosphere. 2015 Oct;136:204–10.
178. Wang Y, Fang Z, Zhao S, Ng D, Zhang J, Xie Z. Dopamine incorporating forward osmosis membranes with enhanced selectivity and antifouling properties. RSC Adv. 2018;8(40):22469–81
179. Xu X, Zhang H, Yu M, Wang Y, Gao T, Yang F. Conductive thin film nanocomposite forward osmosis membrane (TFN-FO) blended with carbon nanoparticles for membrane fouling control. Sci Total Environ. 2019 Dec;697:134050
180. Fan X, Liu Y, Quan X, Chen S. Highly permeable thin-film composite forward osmosis membrane based on carbon nanotube hollow fiber scaffold with electrically enhanced fouling resistance. Environ Sci Technol. 2018 Feb 6;52(3):1444–52.
181. Lakra R, Balakrishnan M, Basu S. Development of cellulose acetate-chitosan-metal organic framework forward osmosis membrane for recovery of water and nutrients from wastewater. J Environ Chem Eng. 2021 Oct;9(5):105882.
182. Alfahel R, Azzam RS, Hafiz M, Hawari AH, Pandey RP, Mahmoud KA, et al. Fabrication of fouling resistant Ti3C2Tx (MXene)/cellulose acetate nanocomposite membrane for forward osmosis application. J Water Process Eng. 2020 Dec;38:
183. Deng L, Wang Q, An X, Li Z, Hu Y. Towards enhanced antifouling and flux performances of thin-film composite forward osmosis membrane via constructing a sandwich-like carbon nanotubes-coated support. Desalination. 2020 Apr;479:114311.

184. Khanafer D, Ibrahim I, Yadav S, Altaee A, Hawari A, Zhou J. Brine reject dilution with treated wastewater for indirect desalination. J Clean Prod. 2021 Nov;322:129129
185. Ibrar I, Yadav S, Altaee A, Safaei J, Samal AK, Subbiah S, et al. Sodium docusate as a cleaning agent for forward osmosis membranes fouled by landfill leachate wastewater. Chemosphere. 2022 Dec;308:136237.
186. Ibrar I, Yadav S, Altaee A, Samal AK, Zhou JL, Nguyen TV, et al. Treatment of biologically treated landfill leachate with forward osmosis: Investigating membrane performance and cleaning protocols. Sci Total Environ. 2020 Nov;744:140901.
187. Jung J, Ryu J, Yu Y, Kweon J. Characteristics of organic fouling, reversibility by physical cleaning and concentrates in forward osmosis membrane processes for wastewater reclamation. Chemosphere. 2020 Apr;245:125787.
188. Lee S, Boo C, Elimelech M, Hong S. Comparison of fouling behavior in forward osmosis (FO) and reverse osmosis (RO). J Memb Sci. 2010 Dec;365(1–2):34–9.
189. Ansari AJ, Hai FI, He T, Price WE, Nghiem LD. Physical cleaning techniques to control fouling during the pre-concentration of high suspended-solid content solutions for resource recovery by forward osmosis. Desalination. 2018 Mar;429:134–41.
190. Trishitman D. Ultrasonication assisted forward osmosis process: A comparative study on beetroot juice concentration. Ultrasonics. 2025 Sep;153:107676.
191. Song J, Yan M, Ye J, Zheng S, Ee LY, Wang Z, et al. Research progress in external field intensification of forward osmosis process for water treatment: A critical review. Water Res. 2022 Aug;222:118943.
192. Cao X, Zhang P, Xian Y, Zhang Y, Muratkhan M, Youravong W, et al. Nonthermal concentration of skimmed goat milk by forward osmosis: The insights into process performance and product quality. Chem Eng Res Des. 2024 Apr;204:450–60.
193. Zhou M, Cao X, Yang L, Wang W, Wang H, Li Z. A green approach for non-thermal concentration of skim milk by forward osmosis combined with membrane distillation for draw solution regeneration. Chem Eng Res Des. 2024 Oct;210:469–80.
194. Daly S, Casey E, Semião AJC. Osmotic backwashing of forward osmosis membranes to detach adhered bacteria and mitigate biofouling. J Memb Sci. 2021 Feb;620:118838.
195. Madaeni SS, Mansourpanah Y. Chemical cleaning of reverse osmosis membranes fouled by whey. Desalination. 2004 Feb;161(1):13–24.
196. Li L, Shi W, Yu S. Research on forward osmosis membrane technology still needs improvement in water recovery and wastewater treatment. Water (Basel). 2019 Dec 29;12(1):107.
197. Liu X, Wu J, Hou L an, Wang J. Fouling and cleaning protocols for forward osmosis membrane used for radioactive wastewater treatment. Nucl Eng Technol. 2020 Mar;52(3):581–8.
198. Chun Y, Mulcahy D, Zou L, Kim I. A short review of membrane fouling in forward osmosis processes. Membranes (Basel). 2017 Jun 12;7(2):30.
199. She Q, Wang R, Fane AG, Tang CY. Membrane fouling in osmotically driven membrane processes: A review. J Memb Sci. 2016 Feb;499:201–33.
200. Hancock NT, Xu P, Roby MJ, Gomez JD, Cath TY. Towards direct potable reuse with forward osmosis: Technical assessment of long-term process performance at the pilot scale. J Memb Sci. 2013 Oct;445:34–46.
201. Yadav S, Ibrar I, Bakly S, Khanafer D, Altaee A, Padmanaban VC, et al. Organic fouling in forward osmosis: A comprehensive review. Water (Basel). 2020 May 25;12(5):1505.
202. Wang Z, Tang J, Zhu C, Dong Y, Wang Q, Wu Z. Chemical cleaning protocols for thin film composite (TFC) polyamide forward osmosis membranes used for municipal wastewater treatment. J Memb Sci. 2015 Feb;475:184–92.
203. Xu H, Liu Y. Control and cleaning of membrane biofouling by energy uncoupling and cellular communication. Environ Sci Technol. 2011 Jan 15;45(2):595–601.

204. Zhang M, She Q, Yan X, Tang CY. Effect of reverse solute diffusion on scaling in forward osmosis: A new control strategy by tailoring draw solution chemistry. Desalination. 2017 Jan;401:230–7.
205. Gwak G, Hong S. New approach for scaling control in forward osmosis (FO) by using an antiscalant-blended draw solution. J Memb Sci. 2017 May;530:95–103.
206. Nguyen HT, Nguyen NC, Chen SS, Ngo HH, Guo W, Li CW. A new class of draw solutions for minimizing reverse salt flux to improve forward osmosis desalination. Sci Total Environ. 2015 Dec;538:129–36.
207. Ding C, Zhang X, Xiong S, Shen L, Yi M, Liu B, et al. Organophosphonate draw solution for produced water treatment with effectively mitigated membrane fouling via forward osmosis. J Memb Sci. 2020 Jan;593:117429.
208. Lu X, Romero-Vargas Castrillón S, Shaffer DL, Ma J, Elimelech M. In situ surface chemical modification of thin-film composite forward osmosis membranes for enhanced organic fouling resistance. Environ Sci Technol. 2013 Nov 5;47(21):12219–28.
209. Yu HY, Kang Y, Liu Y, Mi B. Grafting polyzwitterions onto polyamide by click chemistry and nucleophilic substitution on nitrogen: A novel approach to enhance membrane fouling resistance. J Memb Sci. 2014 Jan;449:50–7.
210. Shen L, Zuo J, Wang Y. Tris(2-aminoethyl)amine in-situ modified thin-film composite membranes for forward osmosis applications. J Memb Sci. 2017 Sep;537:186–201.
211. Huang L, Bui NN, Meyering MT, Hamlin TJ, McCutcheon JR. Novel hydrophilic nylon 6,6 microfiltration membrane supported thin film composite membranes for engineered osmosis. J Memb Sci. 2013 Jun;437:141–9.
212. Zhang X, Shen L, Lang WZ, Wang Y. Improved performance of thin-film composite membrane with PVDF/PFSA substrate for forward osmosis process. J Memb Sci. 2017 Aug;535:188–99.
213. Dabaghian Z, Rahimpour A. Carboxylated carbon nanofibers as hydrophilic porous material to modification of cellulosic membranes for forward osmosis desalination. Chem Eng Res Des. 2015 Dec;104:647–57.
214. Heikkinen J, Kyllönen H, Järvelä E, Grönroos A, Tang CY. Ultrasound-assisted forward osmosis for mitigating internal concentration polarization. J Memb Sci. 2017 Apr;528:147–54.
215. Chanukya BS, Rastogi NK. Ultrasound assisted forward osmosis concentration of fruit juice and natural colorant. Ultrason Sonochem. 2017 Jan;34:426–35.
216. Qasim M, Khudhur FW, Aidan A, Darwish NA. Ultrasound-assisted forward osmosis desalination using inorganic draw solutes. Ultrason Sonochem. 2020 Mar;61:104810.
217. Chen JC, Li Q, Elimelech M. In situ monitoring techniques for concentration polarization and fouling phenomena in membrane filtration. Adv Colloid Interface Sci. 2004 Mar;107(2–3):83–108.
218. Sun C, Zhang N, Li F, Ke G, Song L, Liu X, et al. Quantitative analysis of membrane fouling mechanisms involved in microfiltration of humic acid–protein mixtures at different solution conditions. Water (Basel). 2018 Sep 22;10(10):1306.
219. Bucs S, Jung O, Wagner M, Saravia F, Horn H, Picioreanu C. Raman micro-spectroscopy for the study of concentration polarization in the presence of biofouling in pressure driven membrane systems. J Memb Sci. 2025 Jan;713:123219.
220. Jung O, Saravia F, Wagner M, Heißler S, Horn H. Quantifying concentration polarization—Raman microspectroscopy for in-situ measurement in a flat sheet cross-flow nanofiltration membrane unit. Sci Rep. 2019 Nov 4;9(1):15885.
221. Rey C, Hengl N, Baup S, Karrouch M, Dufresne A, Djeridi H, et al. Velocity, stress and concentration fields revealed by micro-PIV and SAXS within concentration polarization layers during cross-flow ultrafiltration of colloidal Laponite clay suspensions. J Memb Sci. 2019 May;578:69–84.

222. Meng B, Li X. In situ visualization of concentration polarization during membrane ultrafiltration using microscopic laser-induced fluorescence. Environ Sci Technol. 2019 Mar 5;53(5):2660–9.
223. Shang W, Wang R, Dong W, Li X, Li M, Zhao L, et al. Laser induced fluorescence (LIF) technique visualizes and characterizes concentration polarization and fouling layer in the cross-flow nanofiltration. Sep Purif Technol. 2019 Apr;212:520–7.

# 3 Forward Osmosis System Modelling, Design, and Optimization

## 3.1 INTRODUCTION

Mathematical modelling is a powerful tool for understanding, designing, and optimizing complex processes across diverse industries. In FO applications—ranging from liquid food concentration to desalination, wastewater treatment, and other industrial separations—modelling provides critical insights into fluid dynamics, solute transport, membrane performance, and overall system behaviour under varying operating conditions.

To streamline analysis, mathematical models in this domain may be broadly categorized into three major types:

1. **Data-Driven Models:** These models are based on empirical observations and statistical methods. They include the following:
   - Empirical models, which describe the observed relationship between input and output variables;
   - Statistical models, useful for analysing trends and conducting regression;
   - Stochastic models, which account for randomness and uncertainty in system parameters.

   These models are particularly useful when the underlying physical mechanisms are complex or partially understood. They are often used for performance prediction, experimental design, and optimization when detailed mechanistic insights are unavailable.
2. **Mechanistic or Physics-Based Models:** These models are grounded in fundamental physical laws, such as mass, momentum, and energy balances. They describe transport phenomena like water flux, solute rejection, and CP using established theories (e.g. SD and irreversible thermodynamics). Mechanistic models offer deeper insights into system behaviour but often require simplifications due to the complexity of real-world food matrices.
3. **Hybrid Models:** Given the limitations of purely empirical or mechanistic approaches, hybrid models have emerged as a practical compromise. These models combine physics-based frameworks with experimentally estimated parameters or data-driven elements. Typically, parameters such as water permeability or solute diffusion coefficients are estimated via well-structured experiments and parameter estimation techniques. Hybrid models strike a balance between physical interpretability and practical applicability, making them highly relevant in FO food applications.

DOI: 10.1201/9781003657873-3

With recent advancements, stand-alone ML models have increasingly been adopted in FO membrane research (1). ML techniques, including neural networks, support vector machines (SVMs), and ensemble methods, offer robust capabilities for predicting membrane performance, identifying optimal membrane materials, and selecting suitable raw materials for FO membrane fabrication. These models leverage large datasets from experimental studies or simulations to uncover complex relationships and patterns that conventional modelling approaches might overlook. Consequently, ML-based models hold significant promise for enhancing predictive accuracy, streamlining experimental processes, and accelerating innovation in membrane materials and FO system optimization.

In this chapter, we first focus on building a practical understanding of FO process modelling in food applications. We introduce fundamentals of membrane transport and CP, governing equations for fluid and solute flow, and methods for estimating model parameters using experimental set-ups like membrane test cells and HF modules.

This foundation will be essential for model calibration, validation, and scaling up FO systems for real-world food processing operations.

## 3.2 FUNDAMENTALS OF FO MODELLING

Accurate modelling of FO systems is essential for understanding membrane performance, optimizing operating conditions, and facilitating the scale-up from laboratory to industrial applications, particularly in food processing, where precise control over temperature, concentration, and flux dynamics is critical.

The FO modelling approach involves developing a multilayered, hierarchical structure. This structure begins at the fundamental membrane transport, progressively incorporating subsystem dynamics, and cumulates in the integrated system model, as illustrated by the flow chart (*Figure 3.1*).

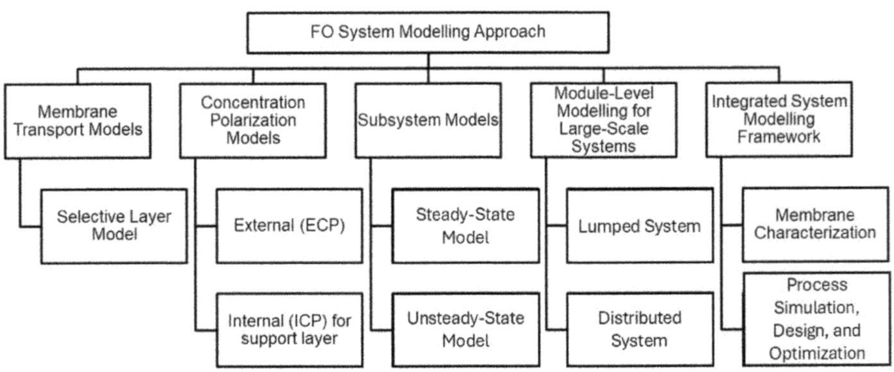


**FIGURE 3.1** Hierarchical framework of the FO system modelling approach

### 3.2.1 Membrane Transport Modelling

The fundamental building blocks of FO modelling lie within the membrane layers themselves. FO membranes are typically composed of two essential layers:

- **Selective (Active) Separation Layer**: This layer primarily governs the transport of water and solutes. The selective layer's transport phenomena are described by established theories such as the SD model or the Spiegler–Kedem (SK) irreversible thermodynamic model. Water flux in FO is driven by osmotic pressure gradients, while solute flux—referred to as RSF—is determined by solute permeability coefficients and concentration gradients across the membrane.
- **Support Layer:** The support layer offers mechanical strength for the separation layer and is critical in FO membrane design. Its thickness, porosity, and tortuosity must be optimized to minimize ICP, which can reduce water flux. Unlike reverse osmosis membranes, FO support layers should be highly porous and enable easy mass transport. Effective structural optimization of these parameters enhances membrane performance and efficiency, especially important in food processing applications.

### 3.2.2 Concentration Polarization Modelling

CP describes the localized build-up of solute concentration at membrane interfaces and within the porous structure. This phenomenon reduces the effective osmotic pressure gradient, thus diminishing membrane flux. CP is broadly categorized into:

- **ECP** occurs at the external surface of the selective layer due to solute accumulation, influenced primarily by FS and DS hydrodynamics.
- **ICP** occurs within the porous SL of the membrane due to the accumulation of solutes inside the pore structure. Unlike ECP, which develops at the membrane–solution interface, ICP takes place beneath the selective layer and is governed primarily by the structural characteristics of the support. The severity of ICP depends on factors such as porosity, tortuosity, and structural thickness. A dense or highly tortuous SL restricts mass transport, allowing solutes to accumulate internally and significantly diminishing the effective osmotic driving force for water transport. Conversely, a well-designed SL with optimized porosity and reduced thickness can mitigate ICP by facilitating more efficient solute and water transport. Compared to ECP, ICP exerts a much stronger influence on membrane performance, making its accurate representation critical in both experimental studies and process modelling.

The orientation of the membrane whether the SL faces the DS or FS determines whether CP is dilutive or concentrative. Therefore, a comprehensive FO model must capture all four CP scenarios (external and internal, dilutive and concentrative) to be applicable across configurations. Accurate CP modelling is vital for realistic performance prediction, reliable scale-up, and robust operational strategies.

In general, membrane separation processes are modelled under steady-state conditions, particularly when working with test cells or small membrane modules. In such cases, the primary objective is often to characterize the membrane performance using flat-sheet configurations. For processes like RO, this approach is valid because the feed is maintained at a sufficiently high volume. Even when a small fraction of water permeates through the membrane, the overall feed concentration remains nearly unchanged. Thus, the assumption of steady-state operation is reasonable.

However, FO processes, especially in food applications, present a fundamentally different scenario. Here, the FS is often extremely expensive. For example, aroma concentrates may cost over $1,000 per litre, making it impractical to operate with large feed volumes. Consequently, FO experiments must be conducted with low volumes of both FS and DS.

This low-volume constraint introduces a key challenge: The system no longer behaves as a steady state. Instead, the process inherently exhibits unsteady-state characteristics due to two coupled dynamics: (i) reduction in FS and DS volumes and (ii) simultaneous changes in their solute concentrations. As a result, the conventional experimental protocols and steady-state modelling approaches commonly used for RO cannot be directly applied to FO, particularly in food-related separations.

To address this, the FO lab-scale process may be modelled by incorporating subsystems that explicitly account for the storage tanks of both FS and DS. These storage tanks play a critical role in capturing the dynamic variations in concentration and solution mass (or volume) over time. By integrating these dynamics into the system model, a more realistic and predictive framework is obtained, enabling accurate description and optimization of FO processes in food applications.

### 3.2.3 Subsystem and Storage Tank Modelling

In practical FO set-ups, particularly laboratory-scale systems, the subsystem that includes storage tanks for FS and DS plays a significant role. These tanks act simultaneously as sources and sinks of solutes and solvent, dynamically influencing both volumes and concentrations due to the bidirectional flux across the membrane. Key considerations for subsystem modelling include the following:

- **Dynamic Volume Changes**: Flux between DS and FS leads to continuous variations in tank volumes, impacting process stability and operation time.
- **Dynamic Concentration Changes**: RSF and water flux dynamically alter the concentrations in feed and draw tanks, directly affecting osmotic pressure gradients and membrane performance.

Consequently, when FO experiments are carried out at laboratory scales—where FS and DS availability may be limited, unsteady-state models are essential. These models capture short-term dynamic changes accurately, ensuring precise characterization. In contrast, large-scale or pilot plant systems, often operated continuously, can be efficiently described using simplified steady-state models.

### 3.2.4 Module-Level Modelling for Large-Scale Systems

As FO systems scale up from bench-scale experiments to pilot or industrial applications, modelling becomes more complex due to increased module size and flow path complexity. Large-scale FO modules, such as HF or spiral-wound modules, require transitioning from lumped-parameter models to distributed-parameter models that incorporate spatially resolved transport phenomena. The distributed-parameter models include the following:

- Mass Balances: Capture solute and solvent flux variations along the entire module length, allowing precise predictions of concentration profiles.
- Momentum Balances: Account for pressure drop, flow velocity changes, and hydrodynamics along the flow path, critical for module and system design.

### 3.2.5 Integrated System Modelling Framework

Finally, a fully integrated FO system model combines all the previously described aspects into a coherent mathematical framework. The integrated model explicitly incorporates:

- Membrane transport equations (selective layer and SL),
- Detailed CP phenomena (ECP and ICP),
- Storage tank dynamics (volume and concentration variations), and
- Module-level distributed mass and momentum balances.

This integrated approach provides a robust foundation for:

- Membrane Characterization: Precisely interpreting laboratory-scale experiments and generating reliable membrane performance metrics.
- Process Design and Optimization: Informing decisions about operating conditions, system dimensions, and scale-up methodologies.
- Performance Simulation: Accurately simulating various operational scenarios including batch, semi-batch, and continuous operating modes.

In summary, the multi-level, integrated FO modelling framework supports both rigorous membrane characterization in research settings and robust, scalable design strategies in commercial food processing operations.

### 3.2.6 Membrane Transport Models

In FO systems using a single-solute DS and pure water feed, as depicted in *Figure 3.2*, the concentration profile of the solute across the membrane is inherently asymmetric, depending on whether the dense AL faces the DS (AL-DS) or the AL faces FS (AL-FS). In the AL-DS configuration, a sharp concentration gradient develops at the interface between the DS and the AL, resulting in pronounced ECP on the draw side, while ICP occurs within the porous SL. Conversely, in the AL-FS orientation, the

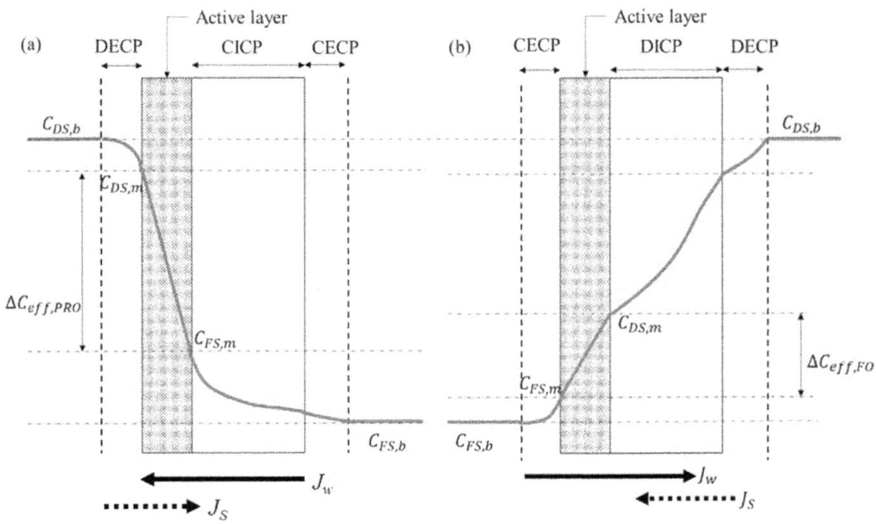


**FIGURE 3.2** Schematic representation of concentration polarization in forward osmosis with a single-solute draw solution and pure water feed: (a) asymmetric membrane with dense active layer facing draw solution (AL-DS) and (b) asymmetric membrane with porous support layer facing feed solution (AL-FS)

concentration gradient is more distributed within the SL adjacent to the FS, leading to different patterns of polarization and flux attenuation.

The solute flux direction is always from the region of higher concentration (DS) towards the lower concentration (FS), crossing the selective separation layer of the membrane. This movement is governed by diffusion and, to a lesser extent, convective phenomena, with the membrane's AL serving as the principal barrier modulating both magnitude and selectivity of flux. Understanding these gradients and flux directions is essential for formulating accurate mathematical models of mass transport in the membrane separation layer, which form the basis for the subsequent derivation of the FO separation layer model.

The SD model assumes that the solute and solvent are first absorbed on the dense membrane AL and then diffused from high to low chemical potential. The solvent and solute flux through the membrane in the SD model are defined by water permeability (Lp) and the solute permeability (B) coefficient. The SD model is used to describe the mass transfer across the FO membrane at the local point of the membrane. However, the SD does not consider the convective phenomenon or the coupling effects in the mass transport of the AL of the membrane. The local membrane mass transport in the FO process may also be described by models based on the theory of irreversible thermodynamics. These models are based on linear relations between fluxes and driving forces when the system is near equilibrium. The membrane acts as a "black box" between the two solutions, and the models considered that all forces on one side of the membrane cause observable fluxes through the membrane. However, these models are not very effective for optimizing the membrane performance based on the membrane structure and properties. The Kedem–Katchalsky model is one of the IT models, which may be used for the process if the applied forces are small (such as in a biological process). However, the phenomenological parameters of this model may show dependency on solute concentration when the driving forces are sufficiently large. On the other hand, the SK model is the modified version of this model, which has been shown to be more accurate and has been successfully applied to describe the mass transport in RO, UF, and NF processes.

In this section, SD and SK models are discussed, describing the local mass transport in the AL of the membrane and combining it with local CP models for obtaining models for both spiral-wound and HF membrane modules.

### 3.2.6.1 Solution–Diffusion Model

The SD model is developed here for a simplified case of a pure water feed with a low solute concentration and a draw solution containing a single solute. It is a foundational two-parameter model commonly used to describe the transport of solvent and solute across a homogeneous dense membrane, specifically within the selective (separation) layer. In this model, water flux $J_w$ is described by modified Fick's law with respect to chemical potential,

$$J_w = -c_w \frac{D_w}{RT}\frac{d\mu_w}{dx} \tag{3.1}$$

where $J_w$ is the water flux (mol m$^{-2}$ s$^{-1}$), $D_w$ is the water diffusion coefficient within the membrane (m$^2$ s$^{-1}$), $c_w$ is the water concentration across the membrane (mol m$^{-3}$), and $x$ is the coordinate across the membrane thickness (m).

For solute flux,

$$J_s = -c_s \frac{D_s}{RT}\frac{d\mu_s}{dx} \tag{3.2}$$

where $J_s$ is the solute flux, expressed in mol m$^{-2}$ s$^{-1}$, $D_s$ denotes the solute's diffusion coefficient within the membrane, in m$^2$ s$^{-1}$, $c_s$ represents the solute concentration inside the membrane, in mol m$^{-3}$, and $x$ is the spatial coordinate measured across the membrane's thickness. The term R represents the universal gas constant, in J mol$^{-1}$ K$^{-1}$, and T is the absolute temperature, in kelvins (K),

$$\text{where } d\mu_i = RTd\ln\hat{a}_i + \overline{v}_i dP \tag{3.3}$$

By differentiating along the membrane thickness,

$$\frac{d\mu_i}{dx} = RT\frac{d\ln\hat{a}_i}{dx} + \overline{v}_i\frac{dP}{dx} \tag{3.4}$$

By substituting Equation (3.4) in (3.1) and integrating across the membrane thickness:

$$\int_0^{\Delta x} J_w dx = -\int_{DS}^{FS} c_w \frac{D_w}{RT}\left(RTd\ln\hat{a}_w + \overline{v}_w dP\right)dx \tag{3.5}$$

$$J_w = \frac{c_w D_w}{\Delta x RT}\left[-RT\ln\left(\frac{\hat{a}_w|_{x=\Delta x}}{\hat{a}_w|_{x=0}}\right) + \overline{v}_w\left(P_{DS} - P_{FS}\right)\right] \tag{3.6}$$

For the FO process,

$$\left(\Delta P\right)_{critical}|_{J_w=0} = \frac{RT}{\overline{v}_w}\ln\left(\frac{\hat{a}_w|_{x=\Delta x}}{\hat{a}_w|_{x=0}}\right) = \Delta\pi = \pi_{DS,m} - \pi_{FS.m} \tag{3.7}$$

$$\therefore RT\ln\left(\frac{\hat{a}_w|_{x=\Delta x}}{\hat{a}_w|_{x=0}}\right) = \overline{v}_w\Delta\pi \tag{3.8}$$

$$J_w = \frac{c_w D_w}{\Delta x RT}\left(-\overline{v}_w\Delta\pi + \overline{v}_w\Delta P\right) \tag{3.9}$$

Water flux for the FO membrane separation layer (2,3):

$$J_w = A\left(\Delta P - \Delta\pi\right) = A\left(\left(P_{DS} - P_{FS}\right) - \left(\pi_{DS,m} - \pi_{FS.m}\right)\right) \tag{3.10}$$

where $A = \frac{c_w \overline{v}_w D_w}{\Delta x RT}$

General chemical potential differential (for solute “s”)

$$d\mu_s = -RTd\ln\hat{a}_s + \overline{\upsilon}dP \tag{3.11}$$

Differentiate across the membrane thickness “$x$”

$$\frac{d\mu_s}{dx} = RT\frac{d}{dx}\left(\ln\hat{a}_s\right) + \overline{\upsilon}_s\frac{dP}{dx} \tag{3.12}$$

Activity term is written as $\hat{a}_s = \gamma_s x_s$ ; so,

$$\frac{d}{dx}\left(\ln\hat{a}_s\right) = \frac{d}{dx}\left(\ln\gamma_s + \ln x_s\right) = \frac{1}{\gamma_s}\frac{d\gamma_s}{dx} + \frac{1}{x_s}\frac{dx_s}{dx} \tag{3.13}$$

Substituting into Equation (3.12):

$$\frac{d\mu_s}{dx} = RT\left[\frac{1}{\gamma_s}\frac{d\gamma_s}{dx} + \frac{1}{x_s}\frac{dx_s}{dx}\right] + \overline{\upsilon}_s\frac{dP}{dx} \tag{3.14}$$

Simplifications

- Ideal Solution: $\gamma_s \approx 1 \Rightarrow d\gamma_s/dx \approx 0$
- Dilute Solution: Since $x_s << 1$, the factor $1/x_s$ is large $\rightarrow$ the composition term dominates $dP/dx$ term; therefore, $dP/dx \approx 0$

The chemical potential difference across the membrane

$$\frac{d\mu_s}{dx} = \frac{RT}{x_s}\frac{dx_s}{dx} \tag{3.15}$$

$$J_s = -\frac{c_s D_s}{RT}\frac{d\mu_s}{dx} \tag{3.16}$$

$$J_s = -\frac{c_s D_s}{RT}\frac{RT}{x_s}\frac{d\mu_s}{dx} = -cx_s\frac{D_s}{x_s}\frac{dx_s}{dx} \tag{3.17}$$

$$J_s = -D_s\frac{d\,c\,x_s}{dx} = -D_s\frac{d\,c_s}{dx} \tag{3.18}$$

$$J_s = -D_s\frac{(c_{FS})_m - (c_{DS})_m}{\Delta x} = D_s\frac{(c_{DS})_m - (c_{FS})_m}{\Delta x} \tag{3.19}$$

$$\frac{(c_{FS})_m}{c_{FS.m}} = K_1 \,\&\, \frac{(c_{DS})_m}{c_{DS,m}} = K_2\,;\; K_1 = K_2 = K$$

$$J_s = D_s\frac{K\,c_{DS} - K\,c_{FS}}{\Delta x} = \frac{D_s K}{\Delta x}\left(c_{DS,m} - c_{FS,m}\right) \tag{3.20}$$

$$J_s = B\Delta c \tag{3.21}$$

where $B = \frac{D_s K}{\Delta x}$

In FO, water and solute transport occur in opposite directions across the membrane. With the coordinate defined from the draw side to the feed side, water permeation from the feed to the draw solution is therefore represented by a negative water flux in Eq. (3.10). Conversely, solute transport proceeds from the draw to the feed side, opposite to the direction of water flow, and is thus identified as reverse solute flux, as expressed in Eq. (3.21).

The key assumptions of the model include (4):

- The AL of the asymmetric membrane is functionally responsible for species selectivity, and the SL is a porous structure with significantly larger pore sizes than the solutes' Stokes radii.
- Solute transport across the membrane is one-dimensional, normal to the area of the AL.
- Electroneutrality is maintained in the membrane matrix and the bulk solution under steady-state conditions.
- Transport is driven by diffusion in the absence of convection and electromigration.
- There are continuous chemical potentials in the solution and the membrane, due to chemical equilibrium across the solution–membrane interface.

- There is pressure uniformity across the AL of the membrane, manifesting an attenuation of the species concentration gradient across the AL.

In practical FO applications, both the FS and DS often contain more than one solute, leading to a multicomponent transport problem. A realistic model must therefore extend beyond the idealized binary SD approach, which assumes a pure water FS and a single solute DS.

*Figure 3.3* illustrates the conceptual transport mechanisms and concentration profiles for a three-solute FO system using a decoupled SD framework. In practical FO applications, both the FS and DS often contain multiple solutes, necessitating a model that accounts for the simultaneous transport of more than one solute species. In this representative system:

- The FS contains a single solute (solute 3).
- The DS contains two distinct solutes (solutes 1 and 2).
- Transport occurs across a thin, dense AL supported by a porous structure, with flux governed by osmotic pressure gradients.

The SD model given below provides a simplified, decoupled approach to understanding individual water and solute fluxes in a multicomponent setting, aiding in the analysis and design of FO systems for complex solutions, such as those encountered in food concentration, wastewater treatment, or pharmaceutical separations.

The net water flux across the dense AL is driven by the combined osmotic pressure differences contributed by all solutes:

$$J_w = -A\left(\Delta\pi_1 + \Delta\pi_2 - \Delta\pi_3\right) \tag{3.22}$$

where A represents the membrane's water permeability coefficient, and $\Delta\pi_i$ denotes the osmotic pressure difference for solute i across the membrane. The draw solutes

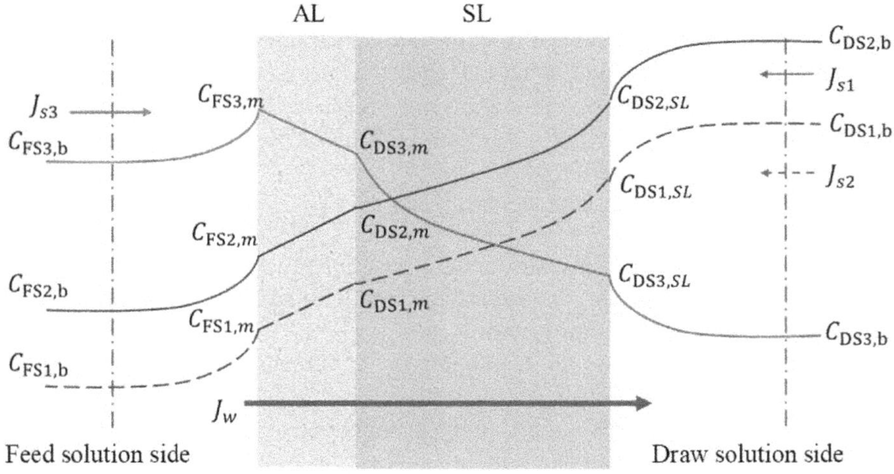


**FIGURE 3.3** Three-solute transport in a forward osmosis system: multicomponent solution–diffusion model

1 and 2 enhance the osmotic driving force, while solute 3 in the feed opposes it by generating back pressure.

The RSF of solute from the draw side to the feed side:

$$J_{s1} = B_1(C_{DS1,m} - C_{FS1,m}) \tag{3.23}$$

$$J_{s2} = B_2(C_{DS2,m} - C_{FS2,m}) \tag{3.24}$$

where $B_i$ represents the permeability coefficient of solute i through the membrane. $C_{im1}$ and $C_{im2}$ denote the concentrations of solute i at the membrane interface on the draw side and feed side, respectively. RSF is undesirable because it lowers the osmotic driving force and introduces contaminants into the feed.

The forward solute flux from the feed side to the draw side is

$$J_{s3} = B_3(C_{DS3,m} - C_{FS3,m}) \tag{3.25}$$

Both water flux ($J_w$) and forward solute flux ($J_{s3}$) are expected to be negative and aligned in the same direction, opposite to the reverse solute fluxes ($J_{s1}$, $J_{s2}$). When interactions between solutes are considered in the SD model, the system shifts from a decoupled to a coupled or multicomponent framework. In this case, solute transport is no longer independent, and the flux of each solute depends on the concentration gradients of all solutes present. This becomes particularly significant in FO systems containing multiple ionic or organic solutes, where electrostatic, osmotic, and steric interactions can take place within the membrane matrix.

Water flux:

$$Jw = A_{w,eff} \Delta\pi \tag{3.26}$$

$$A_{i,eff} = A_{i,ss} + A_{i,ex}(C_{DSi,m}, C_{FSi,m}) \tag{3.27}$$

However, the decoupling of solvent flux is justified when the orders of magnitude of the coefficients are considered (5). For a wide range of solutes commonly investigated in water purification, the mole fractions of solutes at saturation are approximately one order of magnitude smaller than the solvent. The solvent-to-solute mole fraction is much greater than that for concentrations used in actual applications. In this limit, the solvent's cross-diffusion ($D_{w,j}$) tends to zero, while the pseudo-binary diffusion coefficient ($D_{w,eff}$) tends towards the solvent's binary diffusion coefficient ($D_w$) (6). As a result, the "$A_{w,ex}$" is assumed as zero and decouples the solute flux.

Solute flux

$$J_i = B_{i,eff}\left(\left|C_{DSi,m} - C_{FSi,m}\right|\right) \tag{3.28}$$

Similar to the treatment of excess properties in the thermodynamics of solutions, the effective permeability coefficient can be modelled as an arithmetic sum of the binary solute permeability coefficient ($B_{i,ss}$) and an excess solute permeability coefficient ($B_{i,ex}$). The excess permeability captures the deviations arising from nonidealities due to solute coupling (mixing effects) within the membrane matrix and is a function of the concentrations of the interacting solutes. $B_{i,ex}$ regress to zero with the concentrations of the counter solutes, ensuring convergence with binary mixtures in that limit.

$$B_{i,eff} = B_{i,ss} + B_{i,ex}\left(C_{DSi,m}, C_{FSi,m}\right) \tag{3.29}$$

Further to accurately model species chemical potentials, multicomponent activity coefficient models are used to capture the nonidealities arising from the solute–solvent interactions. In this study, the author employs the Pitzer-Kim (7) and the eUNIQUAC (8) equations to model multi-electrolyte and organic electrolyte solutions.

#### 3.2.6.2 Spiegler–Kedem Model

The SK model is derived from the irreversible thermodynamic principle. Here, both $J_w$ and $J_s$ are calculated by mapping fluxes with their corresponding forces, such as pressure, osmotic pressure difference, and the interaction effect between them. The final flux of the SK model consists of the reflection coefficient in addition to the other two parameters of the SD model. The flux equation of the SK models is expressed as follows (9):

$$J_w = -A\left[\frac{iRT}{M_w}\sigma\left(C_{DS,m} - C_{FS,m}\right)\right] \tag{3.30}$$

$$J_s = \frac{J_w\left(1-\sigma\right)\left[C_{DS,m} - C_{FS,m}\exp\frac{J_w\left(1-\sigma\right)}{B}\right]}{\left[\exp\frac{J_w\left(1-\sigma\right)}{B} - 1\right]} \tag{3.31}$$

where $J_w$, $J_s$, $L_p$, $B$, and $\sigma$ represent water flux ($Lm^{-2}h^{-1}$), RSF ($gm^{-2}h^{-1}$), pure water permeability ($Lm^{-2}h^{-1}bar^{-1}$), solute permeability ($Lm^{-2}h^{-1}$), and reflection coefficient, respectively. $C_{DS,m}$ and $C_{FS,m}$ are the DS and FS concentrations at the AL of the membrane surface, respectively. Furthermore, the terms $i$, $R$, $T$, and $M_w$ represent the van't Hoff factor, universal gas factor, temperature of the solution, and molecular weight, respectively.

The osmotic pressure plays a significant role in determining the overall FO performance. Therefore, the determination of appropriate osmotic pressure is crucial. The van't Hoff equation assumes a linear correlation between the osmotic pressure and the solution concentration. The linear assumptions are considered accurate for low-concentration (or ideal) solutions. However, the deviation from the ideal behaviour is more prominent for high concentrations.

The SK model is a phenomenological framework from irreversible thermodynamics that adds a reflection coefficient (σ) to the SD picture, capturing coupling between water and solute transport through the dense AL. In FO, it is commonly written with three membrane parameters: water permeability A, solute permeability B, and reflection coefficient σ. For a single solute, the familiar SK expressions for water/solute flux use the osmotic pressure difference at the membrane surface (not the bulk) and include an exponential convective–diffusive term for solute transport across the AL.

The SK model stems from irreversible thermodynamics and links fluxes to their conjugate driving forces: hydraulic pressure, osmotic pressure, and the coupling between solvent and solute transport. In contrast to the two-parameter SD model, SK introduces a reflection coefficient (σ) that quantifies solute–solvent coupling (solute "drag" with water) in addition to the pure-water permeability (Lp) and the solute permeability (B). For FO, where hydraulic pressure differences are small, SK has proven effective at capturing partial rejection and RSF.

As described in Figure 3.2, we consider three solutes: two draw-side solutes (solutes 1 and 2) and one feed-side solute (solute 3). This mirrors the three-solute SD schematic used earlier (two species in DS that enhance the driving force and one in FS that generates "back-pressure"). The membrane is represented by a dense AL and a porous SL, and all interfacial concentrations below refer to the membrane-surface values (m), not bulk. CP relations used to obtain these interfacial values are exactly those given in the CP subsection (external CP film model and internal CP via the structural parameter S).

Analogous to the SD model, a multi-solute SK model can be formulated; it requires additional parameters to capture interactions among the solutes.

The multi-solute SK expression for water flux is the linear Onsager form with solute-specific reflection coefficients:

$$J_w = -A(\sigma_1 \Delta\pi_1 + \sigma_2 \Delta\pi_2 + \sigma_3 \Delta\pi_3) \tag{3.32}$$

where each osmotic pressure difference is $\pi_i = \pi_i(C_{i,DS,m}) - \pi_i(C_{i,FS,m})$.

$$\Delta\pi_i = \pi_i(C_{i,DS,m}) - \pi_i(C_{i,FS,m}). \tag{3.33}$$

For each solute i ∈ {1, 2, 3}, the SK solute flux relation across the AL (including convective coupling) can be expressed in the closed form commonly used in FO analyses:

$$J_i = \frac{J_w(1-\sigma_i)\left[C_{i,DS,m} - C_{i,DS,m}\exp(s_i)\right]}{\exp(s_i)-1}; \quad s_i \equiv \frac{J_w(1-\sigma_i)}{B_i} \tag{3.34}$$

$$\text{Equivalently,}\ J_i = \frac{J_w(1-\sigma_i)}{1-\exp(-s_i)}\left[C_{i,DS,m}\exp(s_i) - C_{i,FS,m}\right] \tag{3.35}$$

These expressions reduce to the expected limits:

1 when $s_i \to 0$, $J_i \approx B_i(C_{i,DS,m} - C_{i,FS,m}) - (1-\sigma_i)J_w\overline{C}_i$ (diffusion minus solvent drag), and
2 when $\sigma_i \to 1$ (nearly perfect rejection), the convective term disappears.

Table 3.1 summarizes the relative strengths and limitations of the SD and SK models in FO systems relevant to liquid food concentration. While both models are widely used in FO research and engineering, their applicability depends heavily on the solute system, concentration range, and desired predictive accuracy. The SD model is particularly advantageous for dilute systems or when solute–solvent coupling is negligible, offering a compact, mechanistic framework that extends naturally to multicomponent mixtures. In contrast, the SK model provides a more empirically flexible structure by incorporating the reflection coefficient ($\sigma$), which allows it to capture partial rejection, solvent drag, and RSF features critical in food applications involving high-Brix draws or preservative-laden systems.

As shown in Table 3.1, the SD model may suffice in early-stage concentration processes or design studies, while the SK model becomes essential in later-stage, high-concentration scenarios where convective diffusive interactions and nonideal osmotic behaviour dominate. In multi-solute systems, such as those with two draw solutes enhancing osmotic pressure and one feed solute generating back-pressure, the SK model's ability to differentially weight osmotic contributions via $\sigma_1$, $\sigma_2$, and $\sigma_3$ makes it particularly useful.

In conclusion, both SD and SK frameworks play complementary roles in FO modelling. The SD model provides mechanistic simplicity and structural insight, whereas the SK model offers empirical precision, especially when extended to account for multi-solute transport and nonideal osmotic behaviour. Selection between the two should be guided by the specific application requirements, membrane characteristics, and concentration regime, particularly when modelling FO systems for liquid food processing and value-added nutrient recovery.

### 3.2.7 Concentration Polarization

CP is inevitable in any membrane process and significantly influences solvent movement through the membrane. Due to CP, the real osmotic pressure difference between the FS and DS is much lower, thus resulting in the actual permeate flux being much lower than expected. As shown in Figure 3.3, FO membranes consist of an AL and a porous SL.

The FO process can be operated either in AL facing FS (AL-FS) or AL facing DS (AL-DS).

As shown in Figure 3.3a, in AL-DS, the permeate flows through SL without hydraulic resistance. However, since the solutes of FS may come along with water and get trapped in the porous SL, the accumulation of the FS solutes within the SL may create concentrative ICP (CICP). Similarly, on membrane AL, the accumulation of transported molecules dilutes DS, thereby generating dilutive ECP (DECP). As shown in Figure 3.3b, in AL-DS, in the AL-FS orientation, concentrative ECP (CECP) forms, leading to feed-solute accumulation on the AL surface. Additionally, dilution of the DS within the SL causes dilutive ICP (DICP), which lowers both the water flux and the RSF from the bulk DS to the membrane interface.

As shown in Figure 3.3, the dilute CP occurs in the DS side, and the concentrated CP occurs in the FS side of the membrane when the membrane AL is in contact with DS. The concentration of DS at the membrane surface ($C_{DS,m}$) is lower than that of the

**TABLE 3.1**
**Comparison of solution–diffusion (SD) and Spiegler–Kedem (SK) models for forward osmosis in food applications**

| Model | Advantages | Limitations | Remarks in the Context of Food Applications |
|---|---|---|---|
| Solution–diffusion (SD) | Simple, mechanistic basis based on Fick's law<br>Requires only water (A) and solute ($B_i$) permeability<br>Easy to calibrate from RO/FO data—naturally extends to multi-solute systems via diffusion matrix or effective permeability | Does not account for the convective interaction between water and solute transport<br>Unable to represent partial solute rejection or the reflection coefficient ($\sigma$)<br>- Predictions of RSF can become inaccurate at high fluxes, and the model has difficulty accounting for nonideal osmotic pressures at elevated Brix levels. | Effective for dilute food streams (e.g. juice pre-concentration)<br>Suitable when convection and RSF are low<br>Excellent for model-based design of mixtures when membrane convection is negligible |
| Spiegler–Kedem (SK) | Incorporates a distinct reflection coefficient ($\sigma_i$) for each solute<br>Accounts for the coupling between water and solute fluxes<br>Provides a better model prediction capability to experimental FO/RO measurements<br>Enables accurate flux predictions under high-Brix or high-salt draw conditions<br>Accommodates opposing flux directions in systems with multiple solutes | Requires estimation of additional parameters ($\sigma_i$), which may correlate with $B_i$<br>Parameters are often concentration-dependent—less mechanistic—and may need experimental regressions<br>Nonideal behaviours reduce physical interpretability | Preferred for concentrated food solutions (e.g. final concentration of tea, tomato juice, and whey)<br>Supports accurate RSF prediction with preservative solutes (e.g. NaCl and citric acid) and nutrient retention modelling |

bulk ($C_{DS,b}$) due to $J_w$ from FS to DS. Therefore, the effective osmotic pressure will be low in the presence of a highly dilute CP.

#### 3.2.7.1 CP Model for a Single-Solute System

The DECP model equation is defined according to the film theory model, that is solute balance across a DS boundary layer near the membrane AL, including diffusive and convective components:

$$J_s = D_s \frac{dC(x)}{dx} - J_w C(x) \tag{3.36}$$

For the steady-state case, the integration of the resulting equation across the following ECP boundary layer:

$C(x) = C_{DS,m}$ *at* $x = 0; C(x) = C_{DS,b}$ *at* $x = \delta$ , yields

$$C_{DS,m} = C_{DS,b} \exp\left(\frac{-J_w}{k_{DS}}\right) + \frac{(1-\sigma)\left[C_{DS,m} - C_{FS,m} \exp\left(\frac{J_w(1-\sigma)}{B}\right)\right]}{\exp\left(\frac{J_w(1-\sigma)}{B}\right)}\left[\exp\left(-\frac{J_w}{k_{DS}}\right) - 1\right] \tag{3.37}$$

where the term $k_{DS}$ represents the mass transfer coefficient on the DS side.

Similarly, the boundary condition for the SL is as follows:

$C(x) = C_{FS,p}$ *at* $x = 0; C(x) = C_{FS,m}$ *at* $x = \delta$ , yields

At steady state, integrating the governing equation across the SL yields the CICP boundary-layer equation used to determine the concentration of FS at the interface between the porous SL and the selective layer. The equation is as follows:

$$C_{FS,m} = \left[1 - \exp\left(\frac{J_w(1-\sigma)}{B}\right)\right] \exp(J_w K) \frac{C_{FS,p}}{C_{DS,m}} + (1-\sigma)\left[1 - \exp(J_w K)\right] \tag{3.38}$$

The following relation may be obtained for the CICP in the SL by integrating the SK and CP models:

$$\left(\frac{C_{FS,m}}{C_{DS,m}}\right) = \frac{\left[1 - \exp\left(\frac{J_w(1-\sigma)}{B}\right)\right] \exp(J_w K) \frac{C_{FS,p}}{C_{DS,m}} + (1-\sigma)\left[1 - \exp(J_w K)\right]}{\left[1 - \exp\left(\frac{J_w(1-\sigma)}{B}\right)\right] + (1-\sigma)\exp\left(\frac{J_w(1-\sigma)}{B}\right)\left[1 - \exp(J_w K)\right]} \tag{3.39}$$

The solute resistivity "*K*" is defined as the ratio of the structural parameter "*S*" and the diffusion coefficient "$D_s$" of the solute at the SL.

$$K = \left(\frac{S}{D_{diff}}\right) = \left(\frac{t_s \tau}{\varepsilon D_{diff}}\right) \tag{3.40}$$

where K, S, $D_{diff}$, $t_s$, $\tau$, and $\varepsilon$ represent solute resistivity, structural parameter, diffusion coefficient of the solute at porous SL, thickness, tortuosity, and porosity of the SL, respectively.

Furthermore, the following correlations can be used for estimating the FO test cell mass transfer coefficient under laminar flow conditions:

$$Sh = a\,\mathrm{Re}^{k_b}\, Sc^{1/3} \tag{3.41}$$

$$\mathrm{Re} = \frac{\rho u d_h}{\mu} \tag{3.42}$$

$$Sc = \frac{\mu}{\rho D_{diff}} \tag{3.43}$$

$$k = \frac{D_{diff}\, Sh}{d_h} \tag{3.44}$$

where ρ, u, $d_h$, μ, and $D_{diff}$ are density, velocity, hydraulic diameter, viscosity, and diffusion coefficient, respectively. The hydraulic diameter might be estimated using:

$$d_h = \frac{4(\mathit{volume\ of\ flow\ channel})}{\mathit{wetted\ surface}} \tag{3.45}$$

The mass transfer coefficient can also be expressed as a power function of the FS and DS velocity after simplification of the equation:

$$k = k_a u^{k_b} \tag{3.46}$$

where $k_a$ and $k_b$ are mass transfer coefficient correlation constants.

Similarly, the boundary condition for the FS boundary layer near the membrane SL is as follows:

$C(x) = C_{FS,b}\ at\ x = 0; C(x) = C_{FS,p}\ at\ x = \delta$, yields

At steady state, integrating the governing equation across the FS boundary layer yields the CECP boundary-layer equation used to determine the concentration of FS at the interface between the porous SL and FS boundary layer. The equation is as follows:

$$C_{FS,p} = C_{FS,b}\exp\left(\frac{J_w}{k_{FS}}\right) + \frac{(1-\sigma)\left[C_{DS,m} - C_{FS,m}\exp\left(\frac{J_w(1-\sigma)}{B}\right)\right]}{\exp\left(\frac{J_w(1-\sigma)}{B}\right)}\left[\exp\left(\frac{J_w}{k_{FS}}\right) - 1\right] \tag{3.47}$$

where $C_{FS,p}$ and $C_{FS,b}$ are the concentrations of FS at the surface of the porous SL and the feed bulk phase. The term $k_{FS}$ represents the mass transfer coefficient in the FS side.

Similarly, the CP equation for the AL facing the FS solution is summarized in *Table 3.2*.

**TABLE 3.2**
**Summary of the CP equation for the FO process for a single solute**

| | | |
|---|---|---|
| AL-DS | DECP | $C_{DS,m} = \frac{J_s}{J_w}\exp\left(\frac{-J_w}{k_d} - 1\right) + C_{DS,b}\exp\left(\frac{-J_w}{k_d}\right)$ |
| | CICP | $C_{FS,m} = C_{FS,b}\exp\left[J_w\left(\frac{1}{k_f} + \frac{S}{D_d}\right)\right] + \frac{J_s}{J_w}\exp\left[J_w\left(\frac{1}{k_f} + \frac{S}{D_d}\right) - 1\right]$ |
| | CECP | $C_{FS,p} = C_{FS,b}\exp\left(\frac{J_w}{k_{FS}}\right) + \frac{(1-\sigma)\left[C_{DS,m} - C_{FS,m}\exp\left(\frac{J_w(1-\sigma)}{B}\right)\right]}{\exp\left(\frac{J_w(1-\sigma)}{B}\right)}\left[\exp\left(\frac{J_w}{k_{FS}}\right) - 1\right]$ |
| AL-FS | CECP | $C_{FS,m} = \frac{J_S}{J_w}\exp\left[\frac{J_w}{k_f} - 1\right] + J_w C_{FS,b}\exp\left[\frac{J_w}{k_f}\right]$ |
| | DICP | $C_{DS,m} = C_{DS,b}\exp\left[J_w\left(\frac{1}{k_d} + \frac{S}{D_d}\right)\right] + \frac{J_S}{J_w}\exp\left[-J_w\left(\frac{1}{k_d} + \frac{S}{D_d}\right) - 1\right]$ |
| | DECP | $C_{DS,p} = C_{DS,m} - \frac{J_s}{J_w}\left[1 - \exp\left(-\frac{J_w}{k_d}\right)\right]\exp\left(\frac{J_w}{k_d}\right)$ |

### 3.2.7.2 CP Model for a Two-Solute System

In the previous section, the film theory of CP was described for a single-solute system under two orientations of the membrane: AL-DS and AL-FS. The present section extends this analysis to a two-solute system, where one solute (DS1) is present in the DS and the other solute (DS2) is present in the FS, as illustrated in *Figure 3.4*.

As in the single-solute case, multiple interfaces are formed due to the presence of boundary layers. For the AL-DS configuration, these interfaces occur at:

- the feed boundary layer and the SL,
- the SL and the AL, and
- the AL and the DS boundary layer.

For the AL-FS configuration, the interfaces occur at:

- the DS boundary layer and the SL,
- the SL and the AL, and
- the AL and the feed boundary layer.

The concentration profiles of both solutes (S1 and S2) across these boundary layers are obtained by solving material balance equations that incorporate convective

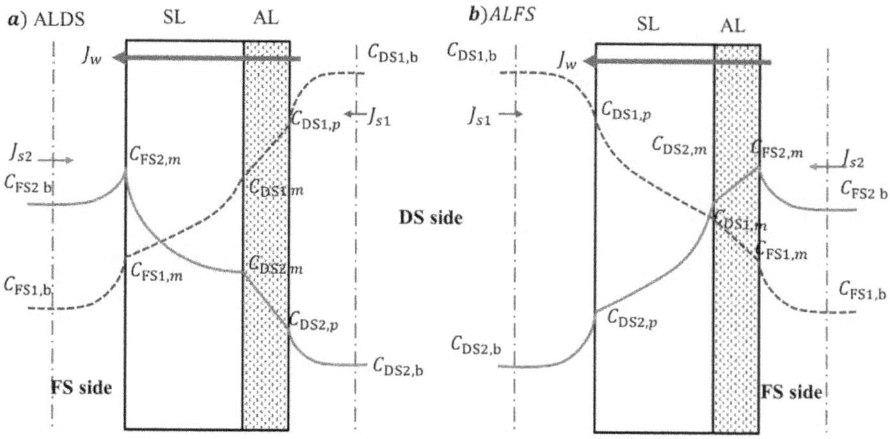


**FIGURE 3.4** Schematic representation of concentration polarization in forward osmosis with a single-solute draw solution and single-solute feed solution: (a) asymmetric membrane with dense active layer facing draw solution (AL-DS) and (b) asymmetric membrane with porous support layer facing feed solution (AL-FS)

and diffusive transport within each layer. These balances, together with the specified boundary conditions described in the previous section, provide the basis for calculating solute concentrations at the different interfaces.

The solution of these equations yields a CP model for the two-solute system, enabling prediction of solute concentrations at all membrane–boundary layer interfaces. The calculated interface concentrations are summarized in *Table 3.3*, which presents the results derived from the film theory model for this configuration.

The CP modelling framework discussed earlier can be systematically extended to multicomponent systems involving more than two solutes. Whether dealing with multiple draw solutes or multiple feed solutes, the forward water flux and RSF become functions of the respective CP effects. These, in turn, are governed by solute balance equations. By following the principles and derivations outlined for the binary system, it is feasible to derive corresponding equations for multicomponent systems. This approach offers a scalable and adaptable framework for simulating and analysing FO systems involving complex solute compositions.

### 3.2.8 Storage Tank and Subsystem Dynamics

To determine the parameters of membrane transport and CP models, carefully designed laboratory-scale experiments are conducted. These experiments typically measure water and solute fluxes under varying feed conditions, such as solute concentration, flow rate, fluid velocity, and temperature. In addition to the membrane module or test cell, the FS and DS tanks play a crucial role in accurately determining experimental fluxes during FO testing. When the FS and DS are continuously recycled back to their respective tanks, the need for explicit tank modelling

**TABLE 3.3**
**Overview of the CP equation for the FO process with two solutes**

| | | |
|---|---|---|
| AL-DS | DECP | $C_{DS1,m} = \frac{J_{s,1}}{J_w}\left[1-\exp\left(\frac{-J_w}{k_{d,1}D_d}\right)\right] + C_{DS1,b}\exp\left(-\frac{J_w}{k_{d,1}D_d}\right)$ |
| | | $C_{DS2,m} = \frac{J_{s,2}}{J_w}\left[1-\exp\left(\frac{-J_w}{k_{d,2}D_d}\right)\right] + C_{DS2,b}\exp\left(-\frac{J_w}{k_{d,2}D_d}\right)$ |
| | CICP | $C_{FS,1,m} = C_{FS,1,b}\exp\left[J_w\left(\frac{1}{k_{f,1}}+\frac{S}{D_{1,eff}}\right)\right] + \frac{J_{s,1}}{J_w}\left[\exp\left\{J_w\left(\frac{1}{k_{f,1}}+\frac{S}{D_{1,eff}}\right)\right\}-1\right]$ |
| | | $C_{FS,1,m} = C_{FS,2,b}\exp\left[J_w\left(\frac{1}{k_{f,2}}+\frac{S}{D_{2,eff}}\right)\right] + \frac{J_{s,2}}{J_w}\left[\exp\left\{J_w\left(\frac{1}{k_{f,2}}+\frac{S}{D_{2,eff}}\right)\right\}-1\right]$ |
| | CECP | $C_{FS1,p} = C_{FS1,b}\exp\left(\frac{J_w}{k_{f,1}}\right) - \frac{(1-\sigma_1)J_wC_{FS1,m} - B_1(C_{FS1,m}-C_{DS1,m})}{k_{f,1}}\left[\exp\left(\frac{J_w}{k_{1,f}}\right)-1\right]$ |
| | | $C_{FS2,p} = C_{FS2,b}\exp\left(\frac{J_w}{k_{f,2}}\right) - \frac{(1-\sigma_2)J_wC_{FS2,m} - B_2(C_{FS2,m}-C_{DS2,m})}{k_{f,2}}\left[\exp\left(\frac{J_w}{k_{2,f}}\right)-1\right]$ |
| AL-FS | CECP | $C_{FS1,m} = \frac{J_{s1}}{J_w}\exp\left(\frac{J_w}{k_{f,1}}-1\right) + J_wC_{FS1,b}\exp\left(\frac{J_w}{k_{f,1}}\right)$ |
| | | $C_{FS2,m} = \frac{J_{s2}}{J_w}\exp\left(\frac{J_w}{k_{f,2}}-1\right) + J_wC_{FS2,b}\exp\left(\frac{J_w}{k_{f,2}}\right)$ |
| | DICP | $C_{DS1,m} = C_{DS1,b}\exp\left[J_w\left(\frac{1}{k_d}+\frac{S}{D_d}\right)\right] + \frac{J_{s,1}}{J_w}\exp\left[-J_w\left(\frac{1}{k_{d1}}+\frac{S}{D_d}\right)-1\right]$ |
| | | $C_{DS2,m} = C_{DS2,b}\exp\left[J_w\left(\frac{1}{k_d}+\frac{S}{D_d}\right)\right] + \frac{J_{s,2}}{J_w}\exp\left[-J_w\left(\frac{1}{k_{d2}}+\frac{S}{D_d}\right)-1\right]$ |
| | DECP | $C_{DS1,p} = C_{DS1,m} - \frac{J_{s,1}}{J_w}\left[1-\exp\left(-\frac{J_w}{k_{d,1}}\right)\right]\exp\left(\frac{J_w}{k_{d,1}}\right)$ |
| | | $C_{DS2,p} = C_{DS2,m} - \frac{J_{s,2}}{J_w}\left[1-\exp\left(-\frac{J_w}{k_{d,2}}\right)\right]\exp\left(\frac{J_w}{k_{d,2}}\right)$ |

becomes significant, and its importance depends largely on the membrane surface area employed. This leads to two distinct scenarios:

Small Membrane Area (Quasi Steady-State System): When a very small membrane coupon is employed, the overall changes in feed and draw concentrations are negligible due to the low magnitudes of water permeation and solute reflux (both forward and reverse). Under such conditions, the feed and draw concentrations may be reasonably assumed constant throughout the duration of the experiment. This allows the system to be modelled as a lumped steady state, in which water and solute fluxes are evaluated at nearly constant bulk concentrations. The membrane transport parameters, water permeability (A), solute permeability (B), reflection coefficient ($\sigma$), and CP mass transfer coefficients, can be estimated directly by fitting measured fluxes to the SD or SK model formulations.

Large Membrane Area (Dynamic System): When the effective membrane surface area is larger, the fluxes of water and solutes cause appreciable temporal changes in the bulk concentrations of the feed and draw tanks. In this case, assuming constant concentrations is no longer valid, and the system must be treated as dynamic. Two approaches can then be adopted:

- Two-Tank Batch Configuration: A single feed tank and a single draw tank are employed. The solute and solvent balances vary continuously with time and must be described by unsteady-state mass balance equations along with membrane transport equations (Figure 3.5). The estimated transport parameters therefore emerge from fitting the dynamic concentration–time profiles to the coupled SD/SK and CP model equations.
- Four-Tank Continuous Configuration: Both the feed and draw streams use separate supply and receiving tanks. This configuration enables continuous replenishment of fresh FS and DS, keeping inlet concentrations at the module nearly constant. As a result, you can track inlet and outlet concentrations and flow rates to determine steady-state fluxes more accurately. The trade-off is the need for larger solution volumes and uninterrupted pumping, which is feasible only when there is ample membrane surface area.

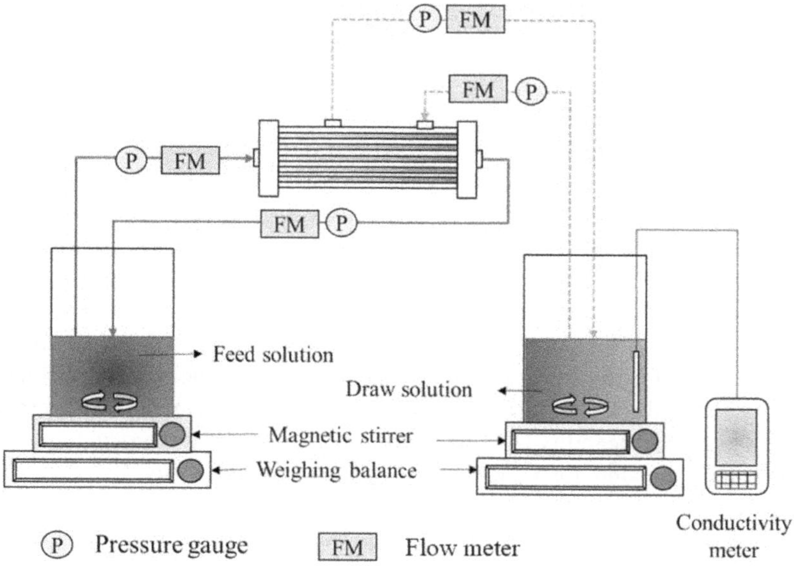


**FIGURE 3.5** Schematic representation of bench-scale experimental set-up for forward osmosis process (10)

#### 3.2.8.1 Flux Measurement Strategies

Single Solute Systems: Conductivity measurements of the FS and DS can be used to track concentration changes and solute flux, since ionic conductivity correlates linearly with solute concentration.

Multi-Solute Systems: Separate analytical methods (e.g. HPLC, ion chromatography, or spectrophotometry) are required to independently quantify the concentrations of each solute. This enables estimation of both forward and reverse fluxes for each solute component.

#### 3.2.8.2 Summary of Experimental Levels

To align the parameter estimation protocol with the system overview described earlier, experiments can be organized into the following levels:

- Level I: lumped steady-state experiments with negligible concentration change (small membrane area).
- Level II: semi-lumped or batch experiments with observable but moderate concentration variations.
- Level III: fully dynamic experiments where time-dependent concentration profiles in both feed and draw tanks must be modelled, and fluxes are quantified from weight change and solute concentration measurements.

This classification ensures that both steady-state and transient models can be validated under controlled laboratory conditions. Ultimately, the chosen configuration depends on the membrane area, expected fluxes, and available instrumentation. The extracted parameters water permeability (A), solute permeability (B), reflection coefficient ($\sigma$), and CP mass transfer coefficients provide the basis for predictive modelling of FO performance in realistic applications.

#### 3.2.8.3 Steady-State Modelling Approach

The steady-state modelling approach for FO offers a systematic framework for predicting water and solute transport under constant operating conditions. In this approach, the water flux is driven by the effective osmotic pressure difference across the membrane, which is corrected for the impacts of ICP and ECP. Experimental data such as solute concentration, flow rate, pressure, and weight changes in both the FS and DS are used to estimate membrane transport parameters. These parameters are obtained by minimizing the error between model predictions and experimental measurements, typically using an error function defined as the squared absolute difference between the two. The model incorporates essential membrane properties, including water permeability (A), solute permeability (B), and the structural parameter (S), which accounts for SL thickness, porosity, and tortuosity. By capturing the balance between water and solute fluxes, steady-state models enable reliable estimation of system performance indicators such as water recovery, DS dilution, and feed concentration. This approach is particularly valuable for the design and optimization of FO systems in applications ranging from desalination and wastewater treatment to resource recovery.

### 3.2.8.4 Dynamic Modelling Approach

In situations involving a single solute, where both the DS and FS are composed of demineralized water, the mass balance equation for the DS and FS tanks can be described as follows. Assuming there is no evaporation of water from the tank, the equation accounts for all relevant variables within the system and provides a foundation for further analytical investigation.

$$\frac{dM_{DS}}{dt} = J_w A_m \rho_w - J_s A_m \tag{3.48}$$

$$\frac{dM_{FS}}{dt} = J_s A_m - J_w A_m \rho_w \tag{3.49}$$

where $J_w A_m \rho_w = Q_P$, permeate flow from feed to draw side, $J_s A_m = Q_S$, solute flow from draw to feed side, and $J_s$ is RSF from draw to feed side.

In dynamic modelling of FO systems, mass balance equations for FS and DS tanks are applied to capture temporal changes in volume and concentration. A critical consideration in such experiments is water evaporation, particularly when open tanks are used without cooling jackets.

If evaporation is neglected, the calculated water flux ($J_w$) may be significantly overestimated or under-estimated. This is because any loss of water mass due to evaporation would be mistakenly attributed to transmembrane transport. Similarly, solute concentrations may appear to change faster than expected, not solely due to osmotic transport but also due to water loss, leading to erroneous estimates of solute permeability and CP effects.

In practical laboratory set-ups:

- Open FS and DS tanks with continuous recirculation are especially prone to evaporation, as surface exposure and elevated operating temperatures accelerate water loss.
- Absence of temperature control (e.g. cooling jackets) can further amplify this effect, since solution heating from pumps or ambient conditions increases evaporation rates.
- These combined effects can introduce systematic errors in measured fluxes, misguiding parameter estimation and subsequent model validation.

For this reason, the dynamic model must explicitly account for evaporation terms in both solvent and solute balance equations. Alternatively, experimental set-ups should minimize evaporation by:

- Using closed tanks or lids,
- Employing cooling jackets around FS and DS tanks, and
- Maintaining controlled laboratory conditions (temperature and humidity).

By incorporating evaporation effects into the model—or by designing experiments that suppress them—researchers can avoid misinterpretation of flux data and ensure accurate estimation of membrane transport parameters.

Similarly, the DS and FS tank total mass balance equations in the presence of water evaporation are given by:

$$\frac{dM_{DS}}{dt} = J_w A_m \rho_w - J_s A_m - E_{DS} A_{DS-air} \tag{3.50}$$

$$\frac{dM_{FS}}{dt} = J_s A_m - J_w A_m \rho_w - E_{FS} A_{FS-air} \tag{3.51}$$

The solute mass balance in the DS and FS tanks is given by:

$$\frac{dC_{DS} V_{DS}}{dt} = -J_s A_m \tag{3.52}$$

$$\frac{dC_{FS} V_{FS}}{dt} = J_s A_m \tag{3.53}$$

where the terms V, M, C, and E represent volume, mass, concentration, and water evaporation rate of superscript DS and FS, respectively. The terms $A_m$ and $\rho_w$ represent the membrane surface area and density of water, respectively. Furthermore, $A_{FS-air}$ and $A_{DS-air}$ represent the surface area of FS and DS which are in contact with air, respectively.

The DS and FS tank volumes and densities are given by:

$$V_{DS} = \frac{M_{DS}}{\rho_{DS}} \tag{3.54}$$

$$V_{FS} = \frac{M_{FS}}{\rho_{FS}} \tag{3.55}$$

In Equations (3.43) and (3.44), the terms $\rho_{DS} = \rho_w + C_{DS}$ and $\rho_{FS} = \rho_w + C_{FS}$ represent the density of DS and FS, respectively.

In the given study, $E_{FS}$ and $E_{DS}$ are expressed as:

$$E_{FS} = \frac{1.97}{1.2 \times 10^6 A_{FS-air}} \tag{3.56}$$

$$E_{DS} = \frac{1.5233 - 9 \times 10^{-3} C_{DS}}{1.2 \times 10^6 A_{DS-air}} \tag{3.57}$$

The steady-state water and solute flux was experimentally evaluated using the following equations:

$$J_{w,\exp} = \frac{V_{DS,final} - V_{DS,initial}}{A_m t_{total}} \tag{3.58}$$

$$J_{s,\exp} = \frac{\left(V_{FS,final} \times C_{FS,final}\right) - \left(V_{FS,initial} \times C_{FS,initial}\right)}{A_m \times t_{total}} \tag{3.59}$$

where the terms $A_m$ and $t_{total}$ represent the membrane area and total duration of the experiment, respectively. The superscript final and initial denote the end and start of the experiment, respectively.

### 3.2.9 Membrane Module-Level Modelling for Large-Scale Systems

In scaling FO systems from laboratory test cells to full membrane modules, the choice between lumped and distributed modelling approaches becomes critical. When the module length is large, treating the system as lumped can introduce significant errors. This is because spatial variations in concentration, flux, and polarization effects along the flow path are neglected, leading to inaccurate parameter estimation. As a result, model predictions may deviate under different operating conditions, limiting the model's usefulness for system design and optimization.

While lumped models may be acceptable for small- or laboratory-scale modules with relatively short flow paths and uniform conditions, large-scale systems require distributed models that capture axial variations. Distributed modelling enables more reliable prediction of performance across a range of operating conditions, making it essential for the design, scale-up, and optimization of FO processes.

#### 3.2.9.1 Lumped System Approach

The lumped system approach is a modelling technique used to simplify complex systems by treating them as a single "lump" or uniform mixed system. In this approach, it is assumed that all properties (such as temperature, pressure, or concentration) of the system are uniform throughout its volume. This simplification allows for the application of balance equations without the need to account for spatial variations in steady-state models. In dynamic modelling, the lumped-parameter approach assumes that all properties evolve uniformly over time, using ordinary differential equations to describe transient behaviour and non-linear equations to represent steady-state conditions. The lumped system approach is particularly useful for modelling systems where spatial gradients are negligible or when a simplified analysis is sufficient. Therefore, both steady-state and dynamic models can be effectively adapted to the lumped system approach with minimal modifications. This makes it a powerful and efficient tool for analysing systems under these conditions.

#### 3.2.9.2 Distributed System Approach

Designing and optimizing FO systems for liquid food processing requires understanding how transport phenomena, membrane properties, solution chemistry, and module configuration interact. Mathematical models help predict system behaviour under different operating conditions. This is especially important for liquid food concentration, where FSs are complex and sensitive to temperature, pressure, and concentration.

In practice, HF and spiral-wound membrane modules are common, with HFs being more widely used. To model these systems, a distributed system approach is often required. Unlike the lumped approach, which assumes uniform conditions throughout, the distributed approach recognizes that variables like concentration, pressure, and flux change along the length of the module.

In FO modules, especially with long HF or spiral-wound designs, water travels across extended paths, creating strong spatial variations in osmotic pressure, solute concentration, and flux. A distributed model divides the module into small regions, capturing these local variations.

By accounting for spatial gradients, distributed system models provide more accurate predictions of FO performance and support better system design and optimization.

#### *3.2.9.2.1 Modelling of HFFO Membrane Module*

This section outlines the critical steps in developing a mathematical model for an HFFO membrane module.

*3.2.9.2.1.1 Overview of Hollow-Fibre Forward Osmosis Membranes* HF modules may be configured in either radial-flow or axial-flow geometries. In FO or osmosis-like applications (such as PRO), radial-flow configurations are typically less efficient than axial-flow configurations. This inefficiency arises because in radial flow, the DS becomes diluted and the FS becomes concentrated in a spatially non-uniform manner (whether in co-current or countercurrent flow), thereby reducing the osmotic driving force along many flow paths (3,11–14).

In contrast, an axial-flow HFFO module allows more systematic variation of DS dilution and FS concentration along the length of fibres. Empirical and modelling studies (3,11–14) show that both DS and FS concentrations vary significantly with axial position and that neglecting this (i.e. using lumped or radial-flow assumptions) leads to overestimation of performance. Thus, distributed-parameter models (which account for concentration gradients along the module axis) are essential for predicting flux, RSF, and overall module efficiency.

This section presents a batch-scale FO study employing an axial-flow AQP-integrated HFFO membrane module. The flow configuration of the HFFO module is depicted in Figure 3.6. Depending upon the membrane orientation, the FS and DS are allowed to pass through either the shell lumen side of the membrane module. Due to the continuous dilution and concentration of DS and FS, a non-uniform osmotic driving force is present along the length of the module. Unlike the radial-flow HFFO mathematical model, FS and DS concentrations are expected to vary along the axial direction. Thereby, a precise mathematical model for the axial flow HFFO model is vital to predict and optimize the overall HFFO membrane performance (2,15). The mathematical model discussed in this section is primarily intended to be used for the experimental validation of liquid food concentration. Furthermore, the validated model can be potentially used to optimize the HFFO membrane module design and operating conditions.

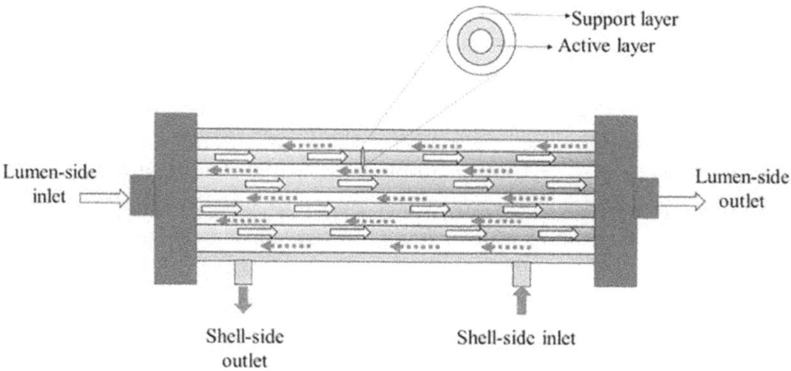


**FIGURE 3.6** Hollow fibre for osmosis membrane module in countercurrent flow configuration

*3.2.9.2.1.2 Flux Equation for FO Membrane* At the module level, two contributions determine the effective transmembrane driving force:

1 The driving forces are the transmembrane osmotic pressure difference, $\Delta\pi = \pi dm - \pi fm$, and
2 The pressure difference across the membrane arises from hydraulic losses along the module, caused by fluid friction on either the lumen or the shell side.

When pressure drop is neglected, model predictions tend to overestimate water flux and under-estimate RSF, particularly in long HF bundles. Thus, incorporation of hydraulic effects is essential for accurate large-scale FO modelling.

The SD model is used to describe solute and solvent transport through the dense selective layer of the membrane. The model assumes that solute and solvent molecules are first adsorbed on the membrane surface and then diffuse down their respective chemical potential gradients. The governing equations for solvent flux (L m$^{-2}$ h$^{-1}$) and RSF (g m$^{-2}$ h$^{-1}$) are given below:

$$J_w = A[(\Delta\pi_{dm,NaCl} - \Delta\pi_{fm,NaCl} - \sum_{i=1}^{n} \pi_{fm,tea_i}) - (P_d - P_f)] \quad (3.60)$$

$$J_{s,NaCl} = B(C_{dm,NaCl} - C_{fm,NaCl}) \quad (3.61)$$

where $A$, $B$, $P$, and $C$ are solvent permeability coefficient (L m$^{-2}$ h$^{-1}$ bar$^{-1}$), solute permeability coefficient (L m$^{-2}$ h$^{-1}$), transmembrane pressure (bar), and concentration (g L$^{-1}$) of the draw and feed sides, respectively. The $\pi_{dm}$ and $\pi_{fm}$ represent the osmotic pressure at the AL of the membrane surface on the draw and feed sides, respectively.

In liquid food concentration, the feed typically contains non-permeating solutes such as alcohols, aldehydes, ketones, esters, and various heterocyclic compounds. Within the SD model formulation, the osmotic pressure contribution of these solutes must be included when calculating the effective driving force across the membrane. Experimental results, however, consistently show that the forward flux of tea solutes is negligible (i.e. $J_{s,tea} = 0$). This implies that while tea solutes contribute to the osmotic pressure gradient governing water transport, they themselves do not permeate through the membrane. Accordingly, in the subsequent model development, the osmotic pressure due to tea components is retained in the flux equations, but their solute flux is assumed to be zero. Accurate determination of osmotic pressure remains essential, as it strongly influences the overall FO performance.

The van't Hoff equation assumes a linear correlation between the osmotic pressure and the solution concentration. The linear assumptions are considered accurate for low-concentration (or ideal) solutions. However, the deviation from the ideal behaviour is more prominent for high concentrations. In this study, the bulk osmotic pressure ($\pi$) for FS and DS is calculated considering water activity ($a_w$):

$$\pi = -\left(\frac{RT}{V}\right)\ln(a_w) \tag{3.62}$$

where $R$, $T$, $V$, and $a_w$ represent the gas constant, temperature, molar volume, and water activity, respectively.

The water activity ($a_w$) is estimated using the Pitzer equation:

$$a_w = \exp\left(-0.01802\phi\sum_i M_i\right) \tag{3.63}$$

*3.2.9.2.1.3 Concentration Polarization* Figure 2.7 (a) demonstrates the concentration gradient across a TFC membrane operating in AL-DS mode (the porous SL facing the FS). In AL-DS mode, due to water permeation from the FS to the DS side, the DS becomes diluted in the vicinity of AL; this phenomenon is commonly termed a "DECP". Whereas the membrane is permeable to water only, and the solute in the membrane accumulates inside the SL of the membrane, this phenomenon is called ICP.

$$C_{dm} = \frac{J_S}{J_w}\exp\left[\frac{-J_w}{k_d} - 1\right] + C_{db}\exp\left[\frac{-J_w}{k_d}\right] \tag{3.64}$$

where $C_{dm}$ and $C_{db}$ represent the concentration of DS on the membrane surface and bulk, respectively. The terms $k_d$, $J_w$, and $J_s$ denote the mass transfer coefficient of DS, water flux, and RSF, respectively.

$$C_{fm} = C_{fb}\exp\left[J_w\left(\frac{1}{k_f} + \frac{S}{D_d}\right)\right] + \frac{J_S}{J_w}\exp\left[J_w\left(\frac{1}{k_f} + \frac{S}{D_d}\right) - 1\right] \tag{3.65}$$

The term $C_{fm}$ denotes the FS concentration at the selective SL interface and can be expressed as the sum of the two components. The first component is expressed in terms of concentrative ICP and ECP, whereas the second component is associated with RSF.

The water and RSF were expressed in terms of the structural parameter (S), pure water permeability ( $L_p$), solute permeability coefficient (B), and mass transfer coefficient (*k*).

The mass transfer coefficient "*k*" for the DS and FS sides of the membrane module was calculated using Equations (3.66 and 3.67), where the terms *D*, *Sh*, and $d_h$ represent the diffusion coefficient, Sherwood number, and hydraulic diameter, respectively:

$$k_d = \frac{DSh_d}{d_{i,fiber}} \tag{3.66}$$

$$k_f = \frac{DSh_f}{d_{o,fiber}} \tag{3.67}$$

Sherwood number (Sh) is a function of two other dimensionless numbers, Reynolds number (Re) and Schmidt number (Sc):

$$Sh_d = \alpha_d \, \mathrm{Re}_d^{\beta_d} \, Sc^{\gamma_d} \tag{3.68}$$

$$Sh_f = \alpha_f \, \mathrm{Re}_f^{\beta_f} \, Sc^{\gamma_f} \tag{3.69}$$

where the terms $\alpha_f$, $\alpha_d$, $\beta_f$, $\beta_d$, $\gamma_f$, and $\gamma_d$ are considered as the tuning parameters which are determined during model validation. The Schmidt number (*Sc*) for FS and DS can be expressed as:

$$Sc_f = \frac{\mu_f}{\rho_f D_f} \tag{3.70}$$

$$Sc_d = \frac{\mu_d}{\rho_d D_d} \tag{3.71}$$

where the terms $\rho$, $\mu$, and *D* represent density, dynamic viscosity, and diffusion coefficient, respectively.

The mass transfer coefficient is expected to vary with time as the FS becomes concentrated due to the changes in FS density and viscosity.

*3.2.9.2.1.4 Mass Balance Equation* Total mass balance for the FS along the length of the HF module

$$\frac{dQ_f}{dx} = \rho_w J_w A_m \tag{3.72}$$

The distribution of solute 1 (i.e. NaCl) in the DS and its mass balance in the FS along the length of the HF module.

$$\frac{dQ_f C_{f,NaCl}}{dx} = J_{s,NaCl} A_m \tag{3.73}$$

Similarly, the mass balance of solutes in the FS (here, the tea components) is presented below.

$$\frac{dQ_f C_{f,tea_i}}{dx} = J_{s,tea_i} A_m \tag{3.74}$$

The mass balance of the DS along the module length is expressed as follows:

$$\frac{dQ_d}{dx} = -\rho J_w A_m \tag{3.75}$$

$$\frac{dQ_d C_{d,NaCl}}{dx} = -J_{s,NaCl} A_m \tag{3.76}$$

where the terms $C$, $A_m$, and $Q$ represent the concentration, active membrane area, and flow rate, respectively. Subscripts f and d stand for FS and DS, respectively.

*3.2.9.2.1.5 Tank Mass Balance Equation* The unsteady-state mass balance equation determines the change in volume and concentration in FS and DS tanks during the FO process. The equation depicts the mass balance for the FS tank:

$$\frac{dV_f}{dt} = Q_{f,in} - Q_{f,out} \tag{3.77}$$

$$\frac{dV_f C_{f,out,NaCl}}{dx} = C_{f,in} Q_{f,in,NaCl} - C_{f,out} Q_{f,out,NaCl} \tag{3.78}$$

$$\frac{dV_f C_{f,out,tea_i}}{dx} = C_{f,in} Q_{f,in,tea_i} - C_{f,out} Q_{f,out,tea_i} \tag{3.79}$$

$$M_f = \frac{V_f}{\rho_f} \tag{3.80}$$

For an ideal solution, $\rho_f = \rho_{water} + C_f$

Similarly, the mass balance for the DS tank can be represented using Equations (3.81 to 3.83):

$$\frac{dV_d}{dt} = Q_{d,in} - Q_{d,out} \tag{3.81}$$

$$\frac{dV_dC_{d,out,NaCl}}{dt} = Q_{d,in}C_{d,in,NaCl} - Q_{d,out}C_{d,out,NaCl} \tag{3.82}$$

$$M_d = \frac{V_d}{\rho_d} \tag{3.83}$$

For an ideal solution, $\rho_d = \rho_{water} + C_d$

*3.2.9.2.1.6 Pressure Drop Equation* Pressure drop in the system is calculated using established correlations for different flow regions in the HF module. In food applications, the DS is typically passed through the fibre lumen to reduce fouling; if feed enters the lumen and concentrates, it can clog the small fibres. Therefore, the diluted DS flows inside the lumen, while the feed passes outside, minimizing fouling and blockage. In this set-up, pressure drop (ΔPlumen) in the lumen is determined by the draw flow rate, using the Hagen–Poiseuille equation for laminar flow.

$$\frac{dP_{lumen}}{dx} = \frac{128\mu_d Q_d}{\pi d_i^4} \tag{3.84}$$

where:

- $\mu_d$ = dynamic viscosity of the DS (Pa s)
- x = length of the fibre (m)
- $Q_d$ = volumetric flow rate of draw through an HF module lumen side ($m^3/s$)
- di = inner diameter of the fibre (m)

This equation assumes fully developed, steady-state laminar flow and becomes especially relevant for dilute solutions and low Reynolds' numbers commonly encountered in HF lumen side flow configuration.

The FS is pumped into the shell side of the module, which acts like a packed bed. For this shell side, where flow moves through a packed fibre arrangement, the Ergun equation estimates the pressure drop ($\Delta P_{shell}$).

$$\frac{dP_{shell}}{dx} = \left[\frac{150\mu_f(1-\varepsilon)^2 v_{s,f}}{d_o^2\varepsilon^3}\right] + \left[\frac{1.75\rho_f(1-\varepsilon)v_{s,f}^2}{d_o\varepsilon^3}\right] \tag{3.85}$$

where:

- P= shell side pressure
- $\mu_f$ = dynamic viscosity of the solution (Pa·s)
- $\rho_f$ = density of the fluid ($kg/m^3$)
- x = length of the HF module (m)
- $V_{s,f}$ = superficial velocity (m/s)
- dp = equivalent fibre diameter or characteristic particle diameter (m)
- ε = bed porosity (dimensionless)

The first term accounts for viscous losses, while the second term represents kinetic (inertial) losses. The Ergun equation is suitable for both laminar and turbulent regimes and captures the complex flow behaviour in the shell side packed with fibres.

These established equations allow for comprehensive modelling of pressure profiles in both the fibre lumen and shell regions, providing vital information for optimal design and scale-up of the HFFO module.

*3.2.9.2.1.7 Methods for Solving the Model Equation and Membrane Parameter Estimation* In our one-dimensional representation of the HFFO module, we treat "n" finite-volume elements in series as a single plug-flow unit. First, we derived a model for an individual continuous stirred tank by solving and integrating the membrane transport, CP, and material balance equations. This yielded a dynamic lumped-parameter description of both the FS and DS tanks. Next, we assembled the full FO flowsheet by linking the FS tank, DS tank, and HFFO membrane equations.

Furthermore, the developed model was used to estimate the membrane transport parameters by minimizing the error between the model and experimental data:

$$Error = \sum_{t=0}^{t} \left[ \begin{aligned} & \left( \frac{M_{d,\exp} - M_{d,\text{mod}}}{M_{d,\exp}} \right)^2 + \left( \frac{M_{f,\exp} - M_{f,\text{mod}}}{M_{f,\exp}} \right)^2 + \left( \frac{C_{d,\exp} - C_{d,\text{mod}}}{C_{d,\exp}} \right)^2 \\ & + \left( \frac{C_{f,\exp} - C_{f,\text{mod}}}{C_{f,\exp}} \right)^2 + \sum_{i=1}^{8} \left( \frac{C_{f,tea_i,\exp} - C_{f,tea_i,\text{mod}}}{C_{f,tea_i,\exp}} \right)^2 \end{aligned} \right] \tag{3.86}$$

where $M$ and $C$ denote the mass and concentration, respectively, whereas the subscripts exp and mod represent the experimental and model outputs, respectively.

The following model membrane parameters can be estimated by using appropriate optimization techniques:

A = water permeability coefficient
B = solute permeability coefficient
S = structural parameter
k = mass transfer coefficient
$\sigma$ = reflection coefficient
$\varepsilon$ = porosity

The number of finite-volume elements in the HFFO module is a critical parameter to minimize the overall simulation time. To estimate the optimal number of finite-volume elements, the HFFO module was simulated by varying the number of finite-volume elements. The simulation results were analysed by comparing the improvement in the model prediction against the number of finite-volume elements. The simulation results were analysed by comparing them in terms of permeate flow rate. For instance, if the flow rate reaches an asymptomatic value beyond optimal finite-volume elements, then the optimal CST units for the countercurrent AQP HFFO model were fixed at 6 (16).

*3.2.9.2.1.8 Model Simulation and Validation* The model equation for the FO process can be solved using the Modelica language in the Dymola software tool. The estimated model parameter for AQP HFFO membrane module, for single-component inorganic DS, the model parameters ($L_p$ = 0.889 ± 0.004 L $m^{-2}$ $h^{-1}$ $bar^{-1}$; B = 0.079 ± 0.010 L $m^{-2}$ $h^{-1}$; S= 210±0.000 μm; $\alpha_f$ = 0.046 ± 0.003; $\alpha_d$ = 0.075 ± 0.006; $\beta_f$ =1.816 ± 0.123; $\beta_d$=0.796 ± 0.07; $\gamma_f$ =0.320 ± 0.007; $\gamma_d$ = 0.329 ± 0.000) were able to predict the performance of the AQP HFFO membrane module for the preparation of concentrated tea extract within the maximum error of 12.12% for change in mass and concentration of DS and FS tanks, respectively (10).

#### *3.2.9.2.2 Modelling of Spiral-Wound FO Membrane Module*

This section presents the development of a distributed model for the spiral-wound FO (SWFO) module. To ensure consistency and avoid redundancy, all non-geometric sub-models and methodologies established earlier in Chapter 3, Section 3.2.4.2.1, and more detailed models are reported in the literature (9). These include the following:

- Membrane transport relations (including SD and SK models) for local water and solute fluxes, $J_w$ and $J_{s,i}$
- Closures for CP, specifically the film-theory ECP and ICP in porous media, characterized by the structural parameter "S"
- Osmotic thermodynamics, using activity-based evaluations of osmotic pressure
- Tank and subsystem dynamics, which include unsteady mass and volume balance equations and allow for the inclusion of evaporation terms as needed
- Mass transfer correlations and workflows for parameter estimation
- A numerical strategy that utilizes finite-volume discretization of the module and time-integration schemes

All notation, constitutive equations, calibration constants, and procedural steps for the solution align exactly with those defined in previous sections, removing the need for rederivation.

The primary distinction in this model lies in the module-level distributed balances within the spacer-filled planar leaf channels, in contrast to the HF configuration (refer to Figure 3.7). Differences stem from variations in geometry, flow direction, and pressure drop behaviour. Specifically, for the SWFO module, the following adaptations are implemented:

- Planar material (component) balances are applied on the unwound x–y leaf for both feed and draw channels. These balances capture in-plane convection and transmembrane coupling via water and solute fluxes (Jw and Js,i) and utilize interfacial concentrations provided by the CP sub-model.

- Momentum and pressure drop relationships for the spacer-filled channels are incorporated using a friction-factor approach with module-specific exponents and parameters, ensuring accurate calculation of local superficial velocities.

This modelling framework adheres to the standard unwound-leaf formulation validated in previous SWFO and spiral-wound NF (SWNF) module-scale studies. The implementation mirrors these approaches by applying mass and pressure balances across the x–y plane, with all relevant inlet and outlet boundary conditions.

The osmotic pressure driving force across the FO membrane transfers solvent (water) from the feed side to the draw side, leading to variations in concentrations, velocity, and pressure drop on both sides of the membranes. The present study discusses the mass balance and pressure drop equations for both FO and NF membranes to understand overall transport phenomena in the spiral-wound modules.

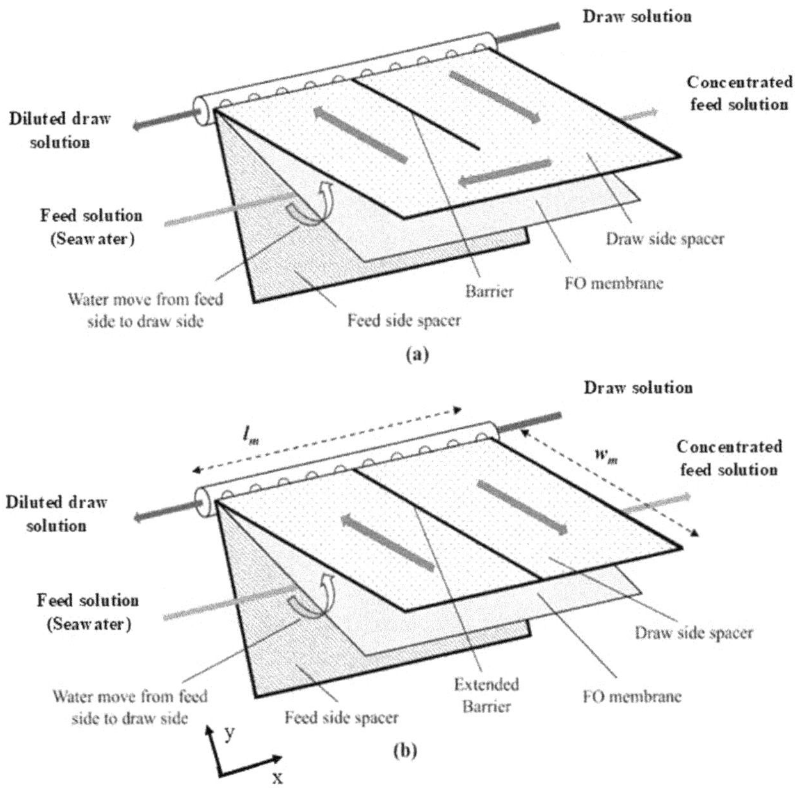


**FIGURE 3.7** Comparison of forward osmosis membrane modules with different barrier configurations. A standard forward osmosis membrane module with a (a) localized and (b) extended barrier component within the draw side spacer, showing the flow of feed solution and draw solution (17)

The mass balance and pressure drop equations on the XY plane for the feed-side channel in the FO membrane, where feed flows towards the x-axis and DS along the y-axis, may be expressed as:

$$\frac{dv_f}{dx} = \frac{-2J_w}{H_F} \tag{3.87}$$

$$\frac{dC_f}{dx} = \frac{2\left[J_w + C_F J_w\right]}{H_F V_f} \tag{3.88}$$

$$\frac{dP_f}{dx} = -f_f \mu_f v_f \tag{3.89}$$

where $H_F$ is the height of the bulk side channel. Similarly, for the draw-side channel, the mass balance and pressure drop equations on the XY plane can be expressed as:

$$\frac{dv_D}{dy} = \frac{2J_w}{H_P} \tag{3.90}$$

$$\frac{dC_d}{dy} = \frac{-2\left[J_s + C_d J_w\right]}{H_P v_D} \tag{3.91}$$

$$\frac{dP_d}{dy} = -f_P \mu_d v_D \tag{3.92}$$

where $H_P$ is the height of the permeate side channel, and $f_F$ and $f_P$ are the friction parameters. Similar equations of mass and pressure balances were used by Dinesh et al. for module-scale FO modelling (18). The initial boundary conditions for the velocity, concentration, and pressure equations (i.e. Equations 11 to 16) at the feed and draw sides of the FO module are as follows:

$$v_f = \frac{Q_{fin}}{Ac_F} at\, x = 0; 0 \leq \ y \leq \mathrm{W}/2 \tag{3.93}$$

$$C_f = C_{f,bulk}\, at\, x = 0; 0 \leq \ y \leq \mathrm{W}/2 \tag{3.94}$$

$$P_f = P_{fin}\, at\, x = 0; 0 \leq \ y \leq \mathrm{W}/2 \tag{3.95}$$

$$V_d = \frac{Q_{din}}{Ac_D} at\, y = 0; 0 \leq \ x \leq \mathrm{L} \tag{3.96}$$

$$C_d = C_{d,bulk}\, at\, y = 0; 0 \leq \ x \leq \mathrm{L} \tag{3.97}$$

$$P_d = P_{din}\, at\, y = 0; 0 \leq \ x \leq \mathrm{L} \tag{3.98}$$

where $A_{CF}$ and $A_{CP}$ are the cross-sectional areas of feed and permeate channels; $C_{f,buk}$, $Q_{fin}$, and $P_{fin}$ are the FS concentration inside the bulk, flow rate, and pressure, respectively. Similarly, $C_{d,\text{ bulk}}$, $Q_{din}$, and $P_{din}$ are the DS concentration inside the bulk, flow rate, and pressure, respectively.

## 3.2.10 Integrated System Modelling Framework

### 3.2.10.1 Membrane Characterization Without Fouling

In membrane-based separation processes, understanding the intrinsic properties and performance of the membrane itself is essential for optimizing its efficiency and longevity. Membrane characterization without fouling refers to the process of evaluating the membrane's behaviour under ideal conditions, where the impact of fouling (the accumulation of particles, microorganisms, or solutes on the membrane surface) is minimized or absent.

As discussed in Section 3.2.3, the steady-state experimental approach is most relevant to Level I and Level II, which involve lumped steady-state experiments with negligible concentration changes (typically using small membrane coupons) and semi-lumped or batch in nature (with observable but moderate concentration variations), respectively. While the dynamic experimental approach is specifically associated with Level III, which involves fully dynamic experiments, where time-dependent concentration profiles in both the feed and draw tanks must be modelled.

Model validation and calibration are essential steps that determine the credibility and practical applicability of the developed FO model. Validation is the process of confirming that the model accurately predicts system behaviour under a wide range of operating conditions, while calibration involves fine-tuning model parameters to align with experimental observations. In FO applications for liquid food processing, this process is complicated by the complexity of the feed matrix, nonideal solution behaviour, and variability in membrane performance.

Calibration typically begins with laboratory-scale experiments in which controlled FO tests are conducted using known FS and DS compositions. Key performance indicators (KPIs) such as water flux, RSF, osmotic pressure difference, and solute concentration profiles are measured across various operating conditions—such as different flow rates, DS concentrations, membrane orientations, and module geometries. These experimental datasets serve as a baseline for parameter estimation.

The calibration process uses optimization algorithms, often in MATLAB or Python environments, to minimize the objective function—commonly the sum of squared errors (SSE) between the predicted and measured values of fluxes and concentrations. Parameters such as the membrane water permeability ($L_p$), solute permeability ($B$), and structural parameter ($S$) are adjusted iteratively. Constraints are applied to ensure physically meaningful values, and sensitivity analysis is used to determine which parameters exert the greatest influence on model outcomes.

Once the model is calibrated, it must be validated using independent datasets not employed during calibration. This may include experiments at different feed concentrations, alternative draw solutes, or long-duration runs that simulate batch operation. The predictive accuracy of the model is assessed using statistical indicators such as the coefficient of determination ($R^2$), root mean square error (RMSE), and normalized mean absolute error (NMAE). A high $R^2$ and low RMSE across test scenarios suggest a well-calibrated and generalizable model.

Importantly, for FO in food systems, validation must also include qualitative comparisons—such as the model's ability to predict the trend of component concentration or retention of thermosensitive compounds (e.g. catechins in tea extract). Models that can account for component-specific behaviours, fouling trends, and nonideal osmotic interactions offer greater confidence when used in scale-up or optimization studies.

In conclusion, a model is only as useful as its ability to reflect real-world behaviour. Rigorous validation and calibration ensure that the FO model becomes a reliable tool not only for academic study but also for industrial process development, techno-economic analysis (TEA), and digital twin applications in smart food processing systems.

###### *3.2.10.1.1 Dynamic Experiment Approach*

The dynamic experiment approach plays a critical role in capturing the transient nature of system performance under non-steady-state conditions. This method involves the collection of time-resolved data, reflecting variations in KPIs such as water flux, solute concentration, and osmotic pressure throughout the operational cycle. It is particularly significant when high-flux membranes are employed, as these exhibit pronounced temporal fluctuations due to evolving concentration gradients and ICP. Unlike steady-state models, dynamic experimentation enables the FO model to account for short-term phenomena such as initial flux surges, the onset and progression of membrane fouling, and real-time dilution effects in the DS. Such transient responses are crucial for simulating practical scenarios like batch processing and feed variability in liquid food applications. Calibrating a dynamic FO model typically involves solving time-dependent mass transfer and osmotic balance equations using numerical techniques, including finite difference or Runge–Kutta methods. This calibration ensures that the model faithfully replicates the inertia and lag observed in real systems, thereby enhancing its applicability for scale-up, process optimization, and control strategies in FO-based concentration processes.

###### *3.2.10.1.2 Steady-State Experiment Approach*

The steady-state experimental approach, often derived from test cell data, is based on measurements taken once the FO system has reached equilibrium. In these conditions, variables such as water flux and solute concentration remain constant over time, simplifying the calibration process. Steady-state data are ideal for initial model development and provide a stable reference for tuning key membrane parameters like A, B, and S. Using simplified algebraic formulations of the mass transfer equations, steady-state models offer analytical clarity and are easier to calibrate with optimization tools. These experiments are typically conducted under controlled lab conditions with consistent FS and DS compositions, allowing reproducible performance benchmarking. However, they may not capture time-dependent effects like DS dilution or fouling build-up, which limits their scope for dynamic operational forecasting.

Nonetheless, steady-state approaches are widely adopted in membrane technology due to their simplicity and suitability for comparative studies across different membranes, configurations, and solutes.

#### 3.2.10.2 Membrane Characterization with Fouling

The accumulation of foulants at the membrane interface presents distinct challenges in FO systems. Several key membrane characteristics, including the absence of high hydraulic pressure, enhanced membrane hydrophilicity, and larger pore dimensions, contribute to a fundamentally different foulant cake structure in FO compared to pressure-driven processes such as RO. These properties typically yield more loosely arranged deposits on the FO membrane surface compared to the densely compacted foulant cake of RO systems. Despite their structural openness, these cake layers still create significant mass transfer barriers by restricting solute movement near the membrane surface, leading to localized concentration imbalances, a phenomenon termed "cake-enhanced concentration polarization" (CEOP).

CEOP is primarily a mechanism contributing to flux decline in the cake foulant layer. Understanding and modelling CEOP is essential for predicting membrane performance. However, developing a comprehensive mass transfer model for CEOP is challenging due to the complex flow patterns and diverse membrane geometric characterization of FO systems. To simplify the modelling process, a quasi-steady, one-dimensional approach focusing on solute and water transport perpendicular to the membrane is often adopted. Within this framework, the cake filtration model provides a foundation for quantifying CEOP. This model incorporates both membrane intrinsic resistance ($R_m$) and cake resistance ($R_c$) in the estimation of water flux, given as:

$$J_w = \frac{\Delta P - \Delta \pi_m^*}{\mu(R_m + R_c)} \tag{3.99}$$

where $\Delta P$ and $\Delta \pi_m^*$ represent the hydraulic pressure difference and the transient osmotic pressure difference, respectively. The magnitude of osmotic and hydraulic resistance exerted by the cake layer, represented by the cake resistance ($R_c$), can be described by the Kozeny–Carman equation, as given in the equation below. The mass of the foulant cake can be expressed as a function of time to account for the dynamic nature of foulant deposition and its impact on water over time.

$$R_c = \left[\frac{180(1-\varepsilon_c)}{\rho_p d_p^2 \varepsilon_c^3}\right] m_c \tag{3.100}$$

One of the limitations of the cake filtration model is that it does not consider ECP effects. This can cause significant inaccuracy when modelling high salinity conditions, where ECP effects are significant. Furthermore, the model's inability to provide detailed concentration profiles at different CO interfaces limits its application in optimizing FO processes where severe CP often occurs.

The modified layered transport model addresses these limitations by dividing the perpendicular flow into multiple discrete layers, allowing for the quantification of individual CP layers and CEOP under varying solute concentrations. This model extends the unfouled FO model by incorporating a distinct foulant cake layer between the CECP layer and the AL, as illustrated in Figure 3.8.

The porous convection–diffusion model, originally developed for the SL, was adapted to describe mass transfer within the foulant cake layer. Given the low-velocity and high-porosity characteristics within the cake layer, a cake structural parameter ($S_c$), analogous to the SL structural parameter ($S$), was introduced to access how the cake layer affects solute and water transport.

$$J_s = \frac{\left[J_w\left(C_c e^{-J_w S_c/D_c} - C_m\right)\right]}{\left(e^{-J_w S_c/D_c} - 1\right)} \quad (3.101)$$

where $C_c$ and $C_m$ represent the solute concentration at the cake surface and at the cake–AL interface, respectively. Table 3.4 summarizes the applicability and limitations of existing fouling model studies.

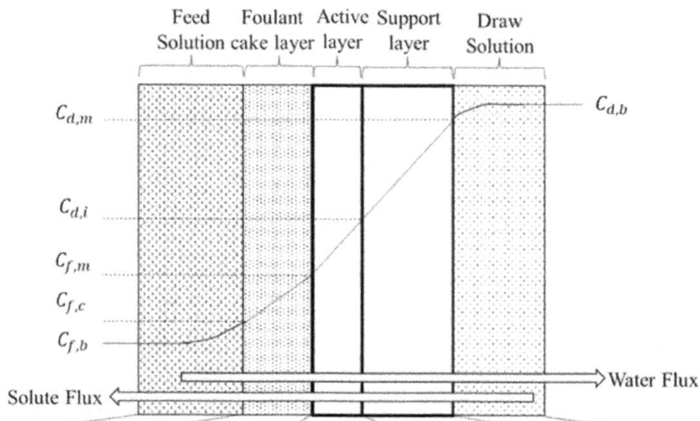


| CECP | CEOP | Active Layer Resistance | ICP | DECP |
|---|---|---|---|---|
| Film theory | Salt: porous convection–diffusion; Water: pore-flow | Solution–diffusion | Porous convection and diffusion and dispersion | Film theory |
| $k_f = \frac{Sh_f D}{d_h}$ | $K_c = \frac{S_c}{D_c}$ | $J_w = A\pi_{eff}$ | $K = \frac{S}{D}$ | $k_d = \frac{Sh_d D}{d_h}$ |

**FIGURE 3.8** Summary of layers and their respective models

**TABLE 3.4**

**Summary of cake-fouling model studies and their key findings**

| Model | Applicability | Limitation | Ref. |
|---|---|---|---|
| Cake filtration model<br>$J_w = \frac{\Delta P - \Delta \pi_m^*}{\mu(R_m + R_c)}$<br>$R_c = \left[\frac{180(1-\varepsilon_c)}{\rho_p d_p^2 \varepsilon_c^3}\right] m_c$ | • $R_c$ accounts for both hydraulic resistance and foulant accumulation in terms of the osmotic pressure differential across the cake.<br>• Applicable to a wide range of processes, including RO, FO, and NF | • The model cannot calculate the concentration profiles within individual CP layers.<br>• The model may not accurately predict performance for processes with severe ECP.<br>• The model does not account for the time-dependent deposition of the foulant layer. | Hoek and Elimelech (2003) (19) |
| Simplified cake filtration model<br>$C_{f,m} = C_{f,b}\left[1 - R + R.\exp\left(\frac{-J_w}{k^*}\right)\right]^{-1}$<br>$k^* = \left[\delta_c\left(\frac{1}{D_c} - \frac{1}{D_b}\right) + \frac{1}{k}\right]^{-1}$ | • Derived from the Hoek and Elimelech model and the film theory model for the FO process by assuming no hydraulic pressure difference across the membrane | • The model does not account for the time-dependent deposition of the foulant layer | Park et al. (2013) (20) |
| Modified layered transport model<br>$J_s = \frac{J_w\left(C_{f,m} e^{-J_w S_c/D_c} - C_{f,c}\right)}{\left(e^{-J_w S_c/D_c} - 1\right)}$<br>$S_c \approx \frac{m_c}{\rho_c(1-\varepsilon_c)}$<br>$\Delta P = \frac{32\mu\tau_c S_c J_w}{d_p^2}$ | • Capable of calculating concentration profiles within individual concentration polarization layers | • The given model does not account for the time-dependent deposition of the foulant layer | Tow and Lienhard (2016) (21) |

The above equation does not take the hydraulic resistance of the foulant cake layer into account. This resistance can be expressed as the pressure drop across the cake layer:

$$\Delta P_c = \frac{32\mu\tau_c S_c J_w}{d_p^2} \tag{3.102}$$

While both the cake filtration model and the modified layer transport model are valuable tools for predicting fluxes under fouling conditions, they have limitations in capturing the time-dependent development of the foulant layer. To accurately model the gradual deposition of foulant molecules convectively transported to the membrane surface, $(dM_c/dt)$ is proportional to the water flux $(J_w)$ and foulant concentration $(C_c)$:

$$\left(\frac{dM_c}{dt}\right) \propto J_w C_c \tag{3.103}$$

However, the deposition of foulant molecules onto the membrane surface is not solely determined by convective transport. Only a fraction of foulant molecules colliding with the membrane surface adhere. To account for this, an attachment coefficient ($\alpha$) is introduced to represent the probability of foulant adhesion. The attachment coefficient is influenced by various factors, including water flux, membrane properties, foulant characteristics, and hydrodynamic conditions. Under fixed membrane and foulant conditions, the attachment coefficient is determined by the balance between permeation drag and the electrostatic repulsion between foulant particles and the membrane surface (22). At high permeate flux, the effect of permeation drag dominates, leading to a larger $\alpha$. However, when the permeate flux is reduced to a level where the effect of repulsion exceeds permeation drag, the attachment coefficient approaches zero, leading to a negligible fouling rate. This threshold permeate flux is known as the critical flux.

Determining the attachment coefficient is challenging due to its complex dependence on various factors. Experimental studies are necessary to quantify the attachment coefficient, yet such data remain limited. Critical flux, which can be determined experimentally and correlated with cross-flow velocity, offers a more practical approach for modelling foulant deposition.

The empirical function of critical flux ($J_c$) is often given as:

$$J_c = av^b + c \tag{3.104}$$

where $v$ is the cross-flow velocity, and a, b, and c are coefficients estimated experimentally. However, it is essential to acknowledge that the empirical relationship between critical flux and velocity may be specified for certain foulant species and operating conditions.

By combining and rearranging the above two equations, the rate of foulant deposited $(dm_c/dt)$ can be calculated in terms of the critical flux expressed as:

$$\frac{dm_c}{dt} = C_c\left(J_w - J_c\right) \tag{3.105}$$

The term $C_cJ_w$ represents the rate at which foulant particles are convectively transported towards the membrane surface, while $C_cJ_c$ represents the rate at which these particles are diffused back into the bulk solution. This approach offers an alternative for modelling foulant deposition, particularly when the attachment coefficient is challenging to quantify.

#### 3.2.10.3 Process Simulation, Design, and Optimization

Process simulation plays a pivotal role in the design and optimization of FO systems. It serves as a powerful tool for predicting the behaviour of a membrane separation system under various operating conditions, helping to guide decision-making throughout the design and operational phases. By utilizing mathematical models and computational tools, process simulation enables a deeper understanding of how the system behaves, interacts, and performs under different conditions.

During the design phase, process simulation is integral to optimizing system configurations and selecting the most appropriate components. It allows engineers to explore different membrane designs, module configurations, and system geometries to determine which set-up provides the best performance in terms of water flux, solute rejection, and energy efficiency. Additionally, simulation helps in the selection of key operational parameters such as membrane material, FS and DS concentrations, flow rates, and pressure conditions. These simulations enable the designer to anticipate how different factors will affect system performance before physical prototypes are built, thus saving both time and resources.

Another critical role of process simulation in the design stage is the estimation of energy consumption and operational costs. By modelling the energy requirements for pumping, pressurizing, and maintaining the FO system, simulations offer valuable insights into the most cost-effective system design. Furthermore, they provide the ability to evaluate various membrane configurations and operational strategies for minimizing energy use while maintaining high levels of performance. In systems prone to fouling, simulations can also predict how fouling might affect performance over time, allowing for the incorporation of strategies to mitigate or delay its impact, such as choosing more fouling-resistant membrane materials or designing cleaning protocols.

Once the FO system has been designed and is operational, process simulation continues to be an essential tool for optimizing performance. It can be used to monitor real-time performance and make adjustments as needed to improve efficiency, reduce energy consumption, and extend the lifespan of the system. By comparing real-time operational data with the predicted performance from simulations, operators can identify discrepancies and quickly make corrections to bring the system back to optimal performance levels. This could involve adjusting parameters like flow rates, pressures, or concentrations of the FS and DS.

Simulations also help operators adjust operational parameters dynamically, ensuring that the system runs efficiently over time. By simulating the effects of changing conditions, operators can make informed decisions on how to fine-tune the process to achieve better flux rates, lower energy consumption, or minimize fouling. This ability to continuously optimize the system in real time is especially important for systems that experience transient behaviours or fluctuations in feed quality, as it allows for immediate adjustments rather than waiting for degradation to occur.

Moreover, dynamic simulations provide a deeper understanding of the system's response to time-dependent factors such as initial flux surges, the onset of fouling, and the impact of dilution in the DS. These transient phenomena are often difficult to capture with steady-state models, but process simulations can model them accurately, enabling operators to anticipate short-term challenges and take pre-emptive measures. By simulating these real-world conditions, operators can develop strategies to manage the system's behaviour effectively throughout its operational cycle.

Long-term performance predictions are another significant benefit of process simulation. As membranes naturally degrade over time due to fouling, scaling, or general wear, simulations allow for the prediction of when maintenance or membrane replacement may be needed. This foresight helps operators optimize cleaning cycles, reduce downtime, and ensure that the system continues to perform optimally without unexpected failures. Simulations also help in modelling the impact of fouling on water flux and solute rejection, providing insights into how to mitigate the effects of fouling through more effective cleaning or system adjustments.

Finally, process simulation plays a critical role in energy and resource efficiency. By simulating different operational scenarios, such as varying DS concentrations or adjusting flow paths, it becomes possible to identify the most energy-efficient operating conditions. This ensures that the FO system not only performs effectively but also minimizes operating costs, which is crucial in large-scale industrial applications where energy consumption is a significant cost factor.

Through scenario analysis, process simulation also allows for the exploration of alternative configurations, feed water qualities, or membrane types to identify potential improvements in system performance. Whether investigating changes in the FS characteristics or evaluating new membrane materials, simulations offer the flexibility to test various "what-if" scenarios without the need for expensive physical testing.

In conclusion, process simulation is indispensable during both the design and operational phases of an FO system. In the design phase, it aids in selecting the best configuration, materials, and operational parameters, while also predicting performance under a range of conditions. Once the system is operational, simulation continues to support optimization by enabling real-time adjustments, predictive maintenance, and scenario-based analysis. By leveraging process simulation throughout the life cycle of an FO system, operators can ensure optimal performance, reduce costs, and extend the lifespan of the membrane system, leading to more efficient and sustainable membrane-based separation processes.

## 3.3 DESIGN AND SCALE-UP PRINCIPLES

To successfully design and scale up an FO process plant for food applications, engineers follow a systematic, stepwise approach. Figure 3.9 illustrates this model-informed design workflow from lab experiments to industrial deployment. Each step focuses on conceptual and procedural principles (rather than detailed equations), ensuring the FO system meets performance targets while maintaining product quality and process efficiency. Below, we outline ten key steps in FO plant design and scale-up for food applications, with recent literature insights (2020–2025) highlighted for each step.

### 3.3.1 Step 1: Define Application and Performance Targets

Any FO system design begins by clearly defining the application and objectives. This involves identifying the FS (e.g. type of juice, extract, or dairy stream) and the desired outcomes such as concentration factor, final solids content, and quality constraints. Key product quality targets include preserving aroma, flavour compounds, and nutrients that are heat-sensitive. FO's advantage is operation at ambient or mild temperatures, which minimizes loss of volatiles and nutritional components compared to evaporation (23). The design must also specify the throughput capacity (volume to process) and any regulatory or hygiene requirements (for food safety). Clearly defining these targets frames all subsequent design decisions. FO could effectively concentrate dairy liquids to high solids without thermal damage (24). Such application-specific goals (e.g. concentrating milk or tea extract while retaining essential bioactive components (16)) guide the selection of suitable membranes, DSs, and process conditions. In summary, Step 1 establishes what the FO process must achieve and why FO is being chosen, providing the foundation for design.

### 3.3.2 Step 2: Characterize Feed and Candidate Draw Solutions

With the application defined, the next step is to characterize the FS properties and evaluate potential DSs. Important feed properties include osmotic pressure (or water activity), viscosity, fouling propensity (e.g. suspended solids or organics that could deposit on the membrane), and sensitivity to shear. Measuring these informs the design: For instance, a very high feed osmotic pressure or viscosity might limit

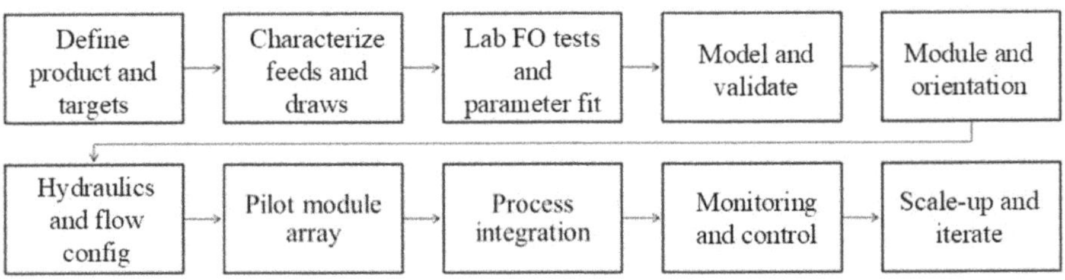


**FIGURE 3.9** Forward osmosis design and scale-up—model-guided flowsheet

achievable flux or necessitate pretreatment. Meanwhile, DS selection is critical in FO, especially for food applications where draw solutes must be food-safe or easily removed. Common food-grade draws include inorganic salts (e.g. NaCl), sugars (glucose and fructose), and food additives like glycerol or lactose, chosen for high osmotic strength and low toxicity (25). Recent advances have expanded DS options to stimuli-responsive or easily recoverable solutes—for example, $CO_2$-responsive salts, MnPs, or hydrogels that can be separated with minimal energy (26,27). An ideal draw solute provides high water flux and minimal RSF, is cost-effective, and can be regenerated efficiently. In this step, engineers shortlist draw solutes by considering these criteria in light of the feed characteristics. For instance, a sugar draw might be suitable for concentrating fruit juice (food-compatible), whereas a fertilizer salt draw could be used if product carryover is acceptable. Economic factors are also weighed: Even if a draw solute is cheap, its regeneration energy or complexity could make it unattractive (28). Ultimately, Step 2 produces a well-characterized feed profile and a choice of one or two optimal DSs to carry forward.

### 3.3.3 Step 3: Lab-Scale FO Testing and Parameter Fitting

Before designing a full-scale plant, lab-scale FO experiments are conducted to assess feasibility and gather data for modelling. In this step, small bench-scale FO units or membrane test cells (using flat-sheet coupons or HFs) are used with the real feed and candidate draw under various conditions. The goals are to measure baseline performance metrics—water flux, RSF, salt rejection, and any fouling tendencies—and to determine key membrane transport parameters. Typically, experiments vary factors like flow rates, draw concentration, temperature, and membrane orientation (i.e. AL-DS or AL-FS) to observe their effects (29). From these tests, parameters such as the membrane's pure water permeability "Lp", solute permeability "B", structural parameter "S" (which influences ICP), and mass transfer coefficients are obtained via a parameter estimation process defined in section 2.2.3 (30). For example, one might fit the FO water flux data to the classical SD model with ECP/ICP to extract A, B, and S (31). These experimentally determined parameters anchor the design by quantifying the membrane's performance. Additionally, lab tests help flag potential issues early: for example, rapid flux decline indicating severe fouling with the chosen feed or high RSF with a particular draw solute. Such findings would prompt reconsideration of pretreatment or draw choice before proceeding. In summary, Step 3 uses lab experiments to validate that the FO process can meet targets at a small scale and provides the quantitative parameters needed for design calculations.

### 3.3.4 Step 4: Develop and Validate an FO Process Model

Using the lab data, engineers then develop a process model for the FO system. This is often a physics-based model (augmented by empirical elements) that can predict FO performance under different conditions. The model typically incorporates the membrane transport parameters from Step 3, including water flux equations (e.g. using the van't Hoff osmotic pressure difference), salt flux and rejection (e.g. using SK or SD equations), and CP effects. It also represents the FO module configuration (e.g. co-current vs. countercurrent flow and axial variation in concentration)—for instance, using a distributed model for long FO SWRO/HF modules to account for concentration changes along the flow path (2,3). Validation of the model is crucial: The model's predictions are compared against independent lab experiments or small pilot tests to ensure it can accurately predict water flux, solute build-up, etc., within a reasonable error. A well-calibrated model effectively becomes a digital twin of the FO process, allowing designers to explore "what-if" scenarios and optimize design choices without extensive physical trials. For example, Ryu et al. (2023) developed a dynamic FO model for multistage operation and validated it, enabling them to predict flux decline and optimize staging before building the system (32). By the end of Step 4, the engineers have a validated FO process model that will guide scale-up decisions—instilling confidence that the design will perform as expected when scaled.

### 3.3.5 Step 5: Select Membrane Module Type and Orientation

With a reliable model and data in hand, the design now focuses on the membrane module configuration best suited for the application. Two primary decisions are made in this step: (i) what type of module (and membrane material) to use and (ii) which orientation to operate it in (AL-FS vs. AL-DS). For FO in food processing, common module formats are plate-and-frame (spiral wound) or HF modules. Plate-and-frame modules (often spiral-wound cassettes using flat-sheet membranes) are advantageous for viscous or high-fouling feeds because they can be disassembled and cleaned easily (33). They have lower packing density, but the ease of cleaning is critical in food applications (e.g. juice concentrates prone to fouling). HF modules offer much higher surface area per volume (compactness) and can achieve high throughput in a small footprint. However, they tend to be more difficult to clean and may be more susceptible to fouling in the fibre lumens. The choice is informed by the model's projections: If simulations (and experiments) indicate severe fouling for the feed, a plate-and-frame design is favoured to allow frequent cleaning. If fouling is manageable and space is at a premium, HFs might be selected for their efficiency. Alongside module type, the membrane orientation is chosen for optimal performance. In FO, the membrane can be oriented with the AL-FS or AL-DS. AL-FS orientation places the dense AL against the feed, which tends to minimize fouling on the membrane's support (since the fouling layer forms on the easily cleanable active surface) (34). AL-FS also reduces ECP on the AL. However, AL-FS suffers more from ICP within the SL, potentially lowering flux (35). AL-DS orientation, in contrast, often yields higher initial flux because the SL is on the feed side (reducing ICP) (36). It can also enhance solute rejection, as the SL provides an additional barrier to feed solutes reaching the draw. The downside is that fouling occurs on the SL (facing the feed), which is harder to clean and can cause long-term flux decline. The model can compare these orientations by simulating flux and fouling rates for each. Designers select the orientation that maximizes net performance for the specific case—for example, if the feed is highly fouling (like viscous fruit pulp), AL-FS might be chosen to facilitate cleaning, whereas for a cleaner feed where maximizing flux is key, AL-DS could be employed. In a case study of tea extract concentration, an FO system was run in AL-DS mode to boost water flux, with the understanding that pretreatment would mitigate fouling on the SL (10). By the end of Step 5, the design specifies the membrane material (often a TFC PA FO membrane for its high flux and rejection or a CTA membrane if higher fouling resistance is needed), the module type, and the orientation—all aligned with the feed's characteristics and the process goals.

### 3.3.6 Step 6: Optimize Hydrodynamics and Flow Configuration

To achieve the predicted performance at scale, the FO system's flow conditions and hydrodynamic design must be optimized. A key principle in FO design is mitigating CP through intelligent flow management. One aspect is setting the appropriate cross-flow velocity on the feed and draw sides. Sufficient cross-flow helps sweep away boundary layers on the membrane surface, reducing external CP and keeping the osmotic driving force high. However, excessive velocity can damage delicate food products or waste energy. McCutcheon and Elimelech (2006) showed that increasing cross-flow in FO significantly improves flux up to a point, beyond which energy costs outweigh gains (37). Thus, in the design, pumps and channel dimensions are chosen to deliver velocities in the optimal range (see Step 7 for pump sizing). Another hydrodynamic element is the feed spacer in plate-and-frame modules. Spacers create turbulence and circulation at the membrane surface, disrupting boundary layers. The spacer mesh size, thickness, and orientation are selected (or even custom-designed) based on model simulations that capture their effect on mass transfer coefficients. Recent FO studies use 3D modelling of flow in spacer-filled channels to minimize CP "dead zones" while avoiding too much pressure drop.

In addition to local hydrodynamics, the overall flow configuration (flow pattern through the module array) is decided. The two primary configurations are co-current flow and countercurrent flow between the feed and draw streams. In co-current operation, feed and draw enter at the same end of the module and flow in the same direction. This is simpler to implement but causes the driving osmotic gradient to continuously decrease along the module length (as the feed becomes more concentrated and the draw becomes more diluted in the same direction). Countercurrent flow, where feed and draw flow in opposite directions, maintains a higher average gradient because fresh (less concentrated) feed continuously meets the more dilute draw exiting the module, and the most concentrated feed at the exit meets the freshest draw entering—effectively preserving driving force across the module. Modelling and experiments consistently show that countercurrent arrangements yield higher water flux and recovery compared to co-current. For example, an FO sugar concentration study found ~7% higher water flux with countercurrent flow versus co-current in the same module (confirmed by both model and experimental data) (38). Figure 3.9 versus Figure 3.10 illustrates this effect: The countercurrent mode (Figure 3.9) sustains a higher flux throughout the module, whereas the co-current mode (Figure 3.10) sees flux drop off towards the outlet. Designers therefore tend to favour countercurrent operation for performance, especially in high-solids food concentration, where every bit of driving force matters. However, the trade-off is hydraulic complexity: Countercurrent systems require more careful manifold design to distribute flow evenly and may have a higher pressure drop. In contrast, co-current systems are hydraulically simpler and may suffice for small-scale or less demanding applications.

In practice, many commercial FO units (e.g. spiral-wound modules) approximate countercurrent via staging or recirculation. In Step 6, the choice between co-current and countercurrent is made based on the performance gains (quantified by the model) versus the complexity and cost. By optimizing cross-flow velocities, spacer design, and flow pattern, the design ensures the membrane can operate near its ideal performance (minimizing CP) in the real system.

### 3.3.7 Step 7: Design the Module Array and Scale-Up Staging

When scaling up from a single module to an industrial system, multiple FO modules must be arranged to handle the required throughput and achieve the desired concentration. In Step 7, engineers determine the number of modules needed and how to arrange them—in series, parallel, or a combination (multi-stage)—to meet capacity and concentration targets. This involves principles of staging familiar from RO design, but with FO-specific considerations. Series configuration (multistage FO) means that the feed passes through one module, is partially concentrated, then goes into a second FO module with a fresh DS, and so on. Staging in series can incrementally increase concentration without hitting a flux plateau in a single pass. Modelling can predict diminishing returns for additional stages—often there are an optimal number of stages beyond which the gain is marginal. Parallel configuration means multiple modules operate with the same feed in parallel to increase flow capacity (throughput) without further concentration per pass. A parallel set-up is useful if a single module is too small to handle the volumetric feed flow required. In practice, large FO plants might use racks of modules where feeds are split among modules (parallel) and/or a train of modules in series for higher concentration. The design also accounts for modular increments: FO is typically run in batch or semi-batch mode for food (feed tank and draw tank), so a continuous multistage design may include intermediate DS adjustments or recirculation loops. The scale-up modelling performed in Step 4 is invaluable here—it can simulate multi-module arrangements to find the best configuration. Akhtar et al. (2021) used modelling to design a two-stage FO system for sugarcane juice, finding that a single stage could not reach the target solids without losing flux, whereas two stages maintained a higher average driving force (38). The model indicated that splitting the process into two steps improved overall flux and efficiency. Similarly, models help ensure that parallel modules will each get the appropriate flow distribution and that the DS dilution in each is manageable. Another aspect of module array design is planning for module replacement and maintenance—for example allocating spare modules or bypass lines so that one module can be cleaned or serviced without shutting down the whole plant. By the end of Step 7, a full process flow diagram of the FO module array is defined (how many modules, how connected, and where pumps and valves are placed), and often a small pilot system is built to verify that the scaled arrangement performs as predicted. Pilot trials at this stage are used to validate scale-dependent factors like pressure drop, flow maldistribution, or fouling behaviour in a multi-module system. Insights from the pilot may lead to iterative refinement of the design (e.g. adding another stage and changing module length) before finalizing the industrial-scale design.

### 3.3.8 Step 8: Integrate with Pretreatment and Draw Solution Recovery

FO systems in food applications do not operate in isolation—they are part of a process chain. Step 8 addresses the upstream and downstream integration needed for a viable operation. Upstream, feed pretreatment is often essential to protect the FO membrane from fouling or damage. Given the feed characterization from Step 2 and any fouling observed in lab tests, designers specify unit operations to condition the feed. For example, fruit juice or tea extract might be filtered (microfiltration or fine screening) to remove pulp or tea fines before FO. If modelling predicts rapid flux decline due to particulates or high turbidity, a clarification step (centrifuge, sedimentation, or enzymatic treatment in juice) will be included. Bardhan et al. (2020) found that a simple prefilter greatly reduced fouling when concentrating tea, as anticipated by their FO model (16). The design will also consider feed temperature control (if FO performance is temperature-sensitive) and possibly bioburden reduction (UV or pasteurization) if needed for food safety, since FO is often run at ambient temperature. The goal of pretreatment is to ensure that the feed entering the FO meets the assumptions of the design model (e.g. free of large particles and within a certain viscosity range) so that the FO unit can achieve the expected flux.

Downstream, after FO, the diluted DS must be processed to recover the product water and reconcentrate the draw solute for reuse. This is a distinctive aspect of FO—unlike RO, FO requires a regeneration step for the draw. The design must include a DS recovery unit, which could be an RO system, an MD unit, an evaporator, or another separation process, depending on the draw solute chosen. The mass balance from the FO model (how much water is transferred to draw, final draw dilution) informs the sizing of this unit. Efficiency is key: Because the draw recycle is a major energy consumer, the design looks for synergy—for example using waste heat for an MD step if available or selecting a draw that can be regenerated with minimal energy (such as a volatile draw that can be removed by low-grade heat). Figure 3.12 shows a typical integrated FO process with a draw regeneration loop: The FO module concentrates the feed and dilutes the draw; then, an RO unit reconcentrates the draw and produces clean water. Recent studies have explored FO–RO hybrids (FO for primary concentration and RO for draw recovery) and FO–MD systems for juice concentration (39). The integration is planned such that the draw loop operates continuously with FO, and any make-up or purge streams are accounted for (to remove impurities that might build up in a recycled draw). Additionally, Step 8 considers integration with other process steps in the facility. For example, if FO is deployed upstream of an existing evaporator, the evaporator can be smaller or operate at a lower duty since FO pre-concentrated the product. Similarly, in dairy processing, FO could be integrated before spray drying to reduce dryer load (as Chen et al. noted, FO can achieve high concentration factors for whey, easing the drying step (42)). Integration also means ensuring that the FO unit can be cleaned in place (CIP) easily alongside other equipment, that its materials are food-grade and compatible with cleaning chemicals, and that the overall process flow is logical (e.g. recovered water from draw RO might be reused for cleaning or other purposes). By completing Step 8, the FO system is embedded into a holistic process: It has appropriate feed pretreatment to run reliably and a downstream draw recovery (and possibly product polishing) so that the outputs (concentrated product and clean water) are delivered as needed in a closed loop. This integration is crucial for the practical, continuous operation of an FO plant in the food industry.

### 3.3.9 Step 9: Implement Monitoring and Control Strategies

Designing a robust FO plant also involves planning how it will be monitored and controlled to maintain optimal performance. In Step 9, the insights from modelling (Step 4) are used to define the KPIs and control loops for the process. Common KPIs for an FO system include the permeate (water) flux, feed and draw concentrations (which can be tracked via conductivity or refractive index), temperature, flow rates, and tank levels. The model sensitivity analysis can identify which variables most strongly affect performance—for example, if the model shows that feed solids content has a large impact on flux, the design would include an online sensor for feed concentration or viscosity. Operators would monitor this KPI and take action if it deviates from the expected range. Modern FO systems may employ a form of model-predictive control (MPC), wherein the calibrated model runs in parallel with the plant to predict trends and recommend adjustments. For instance, if the feed concentration starts rising (perhaps due to upstream variability) and the model predicts a flux decline, the control system could proactively increase the draw flow rate or concentration to maintain flux. Even simpler control rules can be derived: for example maintain draw concentration such that the feed-to-draw osmotic pressure difference stays above a set point, or adjust pump speeds to keep flux at a target value. Ensuring steady-state operation is particularly important for FO to avoid performance degradation—unlike pressure-driven membranes, FO is sensitive to osmotic equilibrium and thus needs a careful balance of flows and concentrations. The response time of the system (e.g. tank volumes causing time lags) might be modelled to tune controllers (pumps and valves) so they act neither too slowly nor too aggressively. Additionally, alarms and safeguards are defined: For example, if flux drops below a threshold or if an unexpected rise in RSF is detected (indicating a membrane issue), the system should alert operators or initiate corrective action. In essence, Step 9 ensures the FO plant will run as intended in real-world conditions, despite disturbances. It leverages the model as a "digital twin" to inform control strategy, as seen in recent work using MPC for membrane systems. With appropriate sensors (conductivity, flow, pressure, turbidity for fouling, etc.) and a control logic that might include automated DS dosing or periodic flow reversals, the FO process can be kept in its optimum regime continuously.

### 3.3.10 Step 10: Plan Fouling Management and Cleaning Schedules

Finally, long-term operation requires a strategy for fouling management and membrane cleaning. FO is less prone to fouling than high-pressure processes, but in food applications, the feed often contains organics, proteins, or particulates that can foul membranes over time. In Step 10, designers establish how fouling will be controlled and when cleaning will be performed, using both model predictions and industry best practices. Fouling models (if developed in Step 4) can predict the rate of flux decline or increase in resistance over time for a given feed and flow condition. In the absence of detailed fouling models, empirical data from pilot runs or analogous processes inform the schedule (some food FO systems might do a daily water flush and a weekly chemical clean, for instance). The cleaning protocol itself is designed to be effective yet gentle: Common approaches include flushing with reverse osmotic flow, using food-safe cleaning agents (enzymatic cleaners for organic fouling, acid for scale, etc.), and sanitization with agents like peracetic acid or chlorine (if the membrane tolerates it) to control microbial growth. The design of the FO unit facilitates cleaning—for example valves to allow flow reversal (running draw on the feed side to dislodge cake), the ability to soak membranes in cleaning solution, and smooth module surfaces that don't trap debris. Real-time monitoring (Step 9) aids fouling management too: By comparing the actual flux to the model-predicted clean flux, operators can detect early signs of fouling if the actual falls below the expected. This could trigger an earlier cleaning before flux is severely lost. Regular cleaning and sanitization are especially important in food processing to maintain product quality and safety (41). Industry guidelines often require periodic sanitization; the FO design ensures this can be done (material compatibility with hot water or chemicals, etc.). The cleaning schedule is balanced to minimize downtime—overly frequent cleaning reduces productivity, but too infrequent leads to irreversible fouling. By integrating such recommendations, the design achieves a sustainable operation. In summary, Step 10 provides a fouling mitigation plan (comprising pretreatment from Step 8, along with potential anti-fouling strategies such as periodic backwashing or feed pulsing) and a CIP schedule that ensures the FO plant operates efficiently throughout its lifespan.

Overall, through these ten steps, the FO process is systematically designed and scaled from concept to industrial reality. Each step is informed by modelling and empirical data, ensuring that when the plant is built, it operates as intended—achieving high product concentrations at low energy cost, with maintained quality and hygienic standards. By 2025, numerous case studies (tea extract, dairy whey, sugar solutions, etc.) will have validated these principles (42,43), making FO an increasingly viable and optimized technology for the food and beverage industry. The marriage of robust modelling with sound engineering design principles is key to realizing FO's potential in sustainable food processing.

## 3.4 CASE STUDIES: APPLICATION OF FORWARD OSMOSIS FOR CONCENTRATION OF SPECIFIC LIQUID FOODS

In the liquid food and dairy processing industry demands, FO systems that handle high fouling ensure product recovery and optimize energy consumption. Membranes with exceptional fouling resistance and solute rejection are necessary. Food-grade DS with high osmotic pressure and low cost is preferred. HF or plate-and-frame modules facilitate cleaning. Energy recovery options should align with specific process conditions. Regular cleaning and sanitization are essential to maintain product quality.

In this section, FO process/system design procedure is explained in the given section. The process/system design is the step-by-step procedure explained. The following outline provides a comprehensive overview of the design process for an FO system using examples of liquid food concentration. It describes the critical steps involved, from defining the application of the process to the final design step. While this example focuses on liquid food concentration, these underlying principles and methodology can be adapted for other FO applications. It is essential to note that successful FO system design necessitates a deep understanding of chemistry, engineering, and specific application requirements. This outline serves as a foundational framework and should be complemented by expert guidance and rigorous analysis for optimal results. The step-by-step procedure for the process/system design.

### 3.4.1 Liquid Food Concentration

Figure 3.10 illustrates the process of concentrating liquid food using the FO method, which is an energy-efficient, membrane-based technology driven by natural osmotic pressure. In this set-up, the FS tank holds the diluted liquid food, such as fruit juice or dairy serum. A DS tank consisting of a highly concentrated osmotic agent (like salts or sugars) creates the osmotic pressure gradient required for water movement. In the FO membrane module, water from the FS naturally diffuses through a semi-permeable membrane into the DS, leaving behind a more concentrated food product. The concentrated FS is collected for further use, while the diluted DS, now containing extra water, is sent to a regeneration unit. This unit separates water from the draw solutes using techniques such as RO or MD, enabling the reuse of the DS in a closed-loop system.

For a deeper understanding, the given section discusses the design of the FO system for the concentration of liquid food extract (freshly brewed tea extract) using the FO process. In the given study, the author demonstrated the feasibility of concentrating liquid food extract (freshly brewed tea extract) using the FO process.

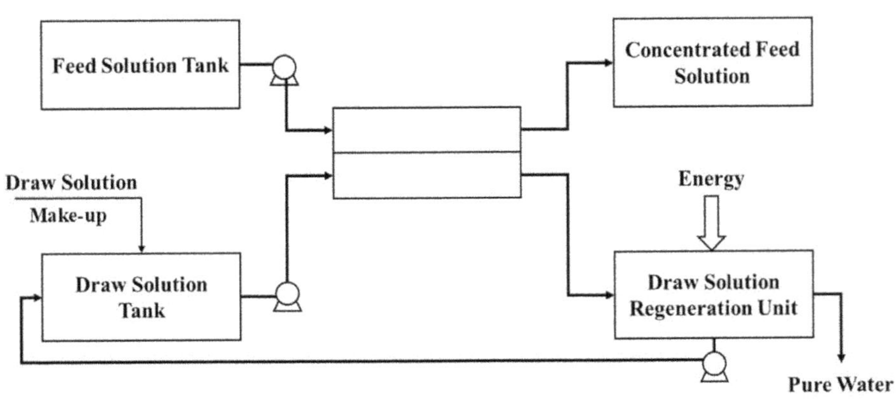


**FIGURE 3.10** Process flow diagram of the forward osmosis process for liquid food concentration

#### 3.4.1.1 Application of FO System and Target Volume

The application of the FO system targets the concentration of tea (*Camellia sinensis var. assamica*) extract, aiming to reduce water content while preserving sensory and nutritional qualities. FO offers significant advantages over conventional thermal methods by operating at ambient temperatures, thereby minimizing the loss of essential thermosensitive aromatic tea components (such as catechin, epicatechin, epigallocatechin, and L-theanine). The target volume reduction was set to 20–30% to enhance storage stability and ease of transportation.

#### 3.4.1.2 Selection of Draw Solution

Due to ease of handling, low cost, and non-toxic nature, the inorganic salt solution was chosen as DS. The concentration of the DS was maintained as such to provide an optimal osmotic gradient while allowing manageable post-treatment.

#### 3.4.1.3 Membrane Selection and Orientation

TFC membrane with PA AL is selected due to its high selectivity, permeability, and stability against FS components (such as tea polyphenols and organic acids). The membrane is oriented with AL-DS, optimizing the flux and minimizing fouling, which ensures consistent permeation rates and prolonged membrane lifespan.

#### 3.4.1.4 Model Development

Development of a mathematical model based on mass transfer equations and osmotic pressure gradient. This includes equations for osmotic pressure, water flux, CP, and fouling dynamics. Computational simulations guide optimization of system parameters, ensuring efficient operational control and accurate prediction of concentration outcomes. Detailed steps for model development for the given system are discussed in the Annexure.

#### 3.4.1.5 Membrane Array Design and Flow Configuration

- FS and DS Flow: For DS and FS, a countercurrent flow configuration was chosen to maintain a high osmotic driving force throughout the membrane module.
- Number of Modules and Operating Mode:
  For a lab-scale experiment, a single FO module was utilized. However, for large-scale applications, a modular configuration consisting of two or more FO modules arranged sequentially can be employed to achieve incremental concentration steps, ensuring optimal flux maintenance.
- Pretreatment: The tea leaves were brewed and filtered to remove suspended solids, thus reducing the propensity of membrane fouling.
- Post-treatment: The diluted DS can be regenerated utilizing the MD process, recovering concentrated NaCl solution and pure water for reuse.

#### 3.4.1.6 Pump and Flow Rate Selection

- Pressure Difference: The FO process operates under minimal hydrostatic pressure ($< 1.0$ bar), emphasizing osmotic pressure rather than hydraulic driving force.

- Pumping System: Centrifugal pumps are selected due to their gentle handling of fluid, preventing damage to sensitive tea components.
- Selection of Pumps: Pumps rated for low pressure and medium flow (10–20 L $min^{-1}$) with adjustable speed control are chosen to precisely manage flow rates.
- Piping and Valves: For lab scale, food-grade polymers or corrosion-resistant stainless steel (for large scale) and manual butterfly valves to ensure sanitary operation and straightforward maintenance.
- Energy Consumption and Efficiency Improvements: Energy-efficient pumps and optimized flow rates minimize energy consumption. Variable frequency drives (VFDs) can be employed to maintain precise flow control, significantly improving system efficiency and reducing operational costs.

### 3.4.2 Dairy Product Processing

FO offers a promising approach for concentrating dairy products such as milk, whey, and functional beverages. Compared to thermal methods, FO operates at low temperatures, helping preserve the nutritional and sensory qualities of dairy constituents. However, due to the complex nature of dairy fluids—rich in proteins, fats, and sugars—mathematical modelling becomes essential for understanding the process for optimizing FO system performance. As dairy fluids differ significantly from simple aqueous solutions, FO models need to account for:

- High viscosity and non-Newtonian behaviour (especially in concentrated milk or whey).
- Protein and fat interactions with the membrane, which can affect water flux and fouling.
- Colloidal and macromolecule effects, which influence mass transfer and RSF.

A mathematical model was developed to understand and optimize FO for dairy processing applications, where the product streams such as milk, whey, lactose solution, and WPCs are concentrated prior to thermal evaporation or spray drying (44). FO offers significant benefits in this context, including higher achievable solids concentrations, lower energy input, and better preservation of product quality compared to pressure-driven membrane technologies.

The studies employed FO for concentrating various dairy streams using AQP and CTA membranes with magnesium chloride as DS. The mathematical modelling efforts aimed to:

- Predict water flux accurately under varying feed concentrations and temperatures.
- Estimate structural parameters of commercial FO membranes.
- Quantify the effects of ECP and ICP.
- Incorporate non-linear relationships between solute concentration and osmotic pressure in dairy matrices and $MgCl_2$ DSs.

The FO water flux was modelled using a modified version of the classical SD approach:

$$J_w = L_p\left(\pi_{DS,i} - \pi_{FS,m}\right)$$

where $L_p$ is the water permeability coefficient, and the terms $\pi_{DS,i}$ and $\pi_{FS,m}$ represent the effective osmotic pressure at the draw and feed interfaces, respectively.

Unlike the traditional model, the given study accounted for:

- Non-linear osmotic pressure–concentration relationships of real dairy streams
- Temperature-dependent diffusion coefficient for $MgCl_2$
- Dynamic variations in mass transfer coefficients with increasing feed viscosity

These were expressed in the flux model:

$$J_w = L_p\left[f_\pi\left(C_{DS}^b\right).e^{J_w S/D} - f_\pi\left(C_{FS}^b\right).e^{J_w/k}\right] \quad (3.106)$$

where $f_\pi(C)$ denotes the experimentally fitted function for osmotic pressure versus concentration, and S refers to the membrane structure parameter. The terms $D$ and $k$ refer to the diffusivity of $MgCl_2$ and the mass transfer coefficient, respectively.

Membrane properties were determined through experiments using pure water and 1.6 M MgC12 at varying temperatures (10/20/50 °C). Structural parameters were estimated by fitting water flux data to the model. The models were validated against experimental data, showing close agreement for both flux and solute behaviour.

The modelling approach accurately captured trends in:

- Declining water flux with increasing feed total solids,
- The impact of viscosity and fouling on CP behaviour, and
- Temperature effects on water permeability and solute diffusivity

Lactose and whey exhibited minimal fouling and more stable flux decline due to low protein content. In contrast, WPC and non-fat milk, with higher protein levels, showed greater CP factors and sharper flux decline, all captured within the modelling framework.

CPF values estimated from the model ranged from:

- 1.2 to 1.4 for lactose (low fouling)
- 1.8 to 3.0 for WPC and skim milk (severe CP and fouling)

The modelling allowed the prediction of maximum achievable concentration for each dairy stream, solute retention confirming minimal loss of protein or lactose, and RSF impacts notably <100 mg $Mg^{2+}$ per 100 g powder for $MgCl_2$, ensuring product compliance.

The framework supports upscaling by mimicking industrial conditions where DS concentration is kept constant via regeneration. It also provides insights for cleaning strategies, indicating that enzyme cleaning is required for protein-rich feeds to restore flux, especially under the pH limitations of CTA membranes.

## 3.5 CHALLENGES AND LIMITATIONS OF MODELLING FO IN LIQUID FOODS

The development and implementation of mathematical models for FO in liquid food systems have provided critical insights into system optimization and scale-up. However, the modelling FO of such complex environments is accompanied by multiple technical and practical challenges. These arise from the inherent variability of liquid foods, the non-linear behaviour of membrane transport under high-solute conditions, and the lack of standardization across the experimental and modelling methodologies.

This section aims to explore these constraints in detail, identify the specific limitations of existing modelling approaches, and propose realistic strategies to enhance the model's robustness, accuracy, and applicability.

### 3.5.1 Common Challenges of Mathematical Modelling

Mathematical modelling of FO processes in liquid food systems presents several inherent challenges that stem largely from the complexity and variability of the feed materials, as well as from the limitations in current theoretical frameworks. One of the foremost challenges is the compositional complexity of liquid foods themselves. Unlike binary or simple aqueous systems used in desalination research, liquid food matrices are comprised of a wide variety of solutes, including sugars, polyphenols, amino acids, organic acids, proteins, and suspended particles. These components interact not only with one another but also with the membrane, leading to non-linear behaviour that is difficult to represent using conventional SD models. For instance, osmotic pressure in such systems cannot be accurately predicted by classical van't Hoff relations, particularly at high concentrations where nonideal interactions between solutes dominate.

Another major obstacle is the accurate representation of CP effects, both external and internal, which critically influence water flux and solute retention. ICP occurring within the membrane's porous SL significantly reduces the effective osmotic driving force and is highly sensitive to solute diffusivity and membrane structure. In food systems, increased viscosity and the presence of macromolecules further hinder diffusion, exacerbating ICP beyond levels observed in simpler systems. ECP, although theoretically easier to mitigate through hydrodynamic control, still poses modelling difficulties due to localized turbulence, channel geometry effects, and changes in viscosity as concentration increases during the process.

Modelling becomes even more complicated when accounting for membrane heterogeneity. Commercial FO membranes, especially HF modules, often exhibit non-uniformity in terms of AL thickness, porosity, and tortuosity across the membrane area. These variations, though subtle, can result in significant spatial disparities in water flux and solute transmission, particularly in long or densely packed modules. Moreover, as the membrane is subjected to prolonged operation, properties such as permeability and structural integrity may evolve due to compaction, fouling, or partial wetting—factors that are seldom incorporated into static modelling frameworks.

A further challenge lies in the scarcity and inconsistency of physicochemical data required for accurate model input. Key parameters such as water activity, solute-specific diffusion coefficients, viscosity as a function of concentration, and dynamic fouling characteristics not only are difficult to measure but also vary between different food types and processing conditions. As a result, modellers are often forced to use generalized assumptions or literature values obtained under dissimilar conditions, which compromises model fidelity. In many cases, essential data for characterizing real food solutes—particularly under high concentration regimes—are either unavailable or derived under diluted conditions, leading to unreliable extrapolation when applied to industrial scenarios.

Collectively, these challenges underscore the inherent complexity of modelling FO in liquid food applications. The interplay among diverse solutes, nonideal thermodynamics, evolving membrane behaviour, and a lack of comprehensive experimental data makes it difficult to develop universally applicable and accurate models. These obstacles highlight the need for continued efforts to refine existing models, incorporate more realistic system representations, and establish robust experimental frameworks to support future model development.

### 3.5.2 Limitations of Current Models

Despite the valuable insights offered by current mathematical models for FO, their application to liquid food systems remains significantly constrained. One of the primary limitations lies in the foundational assumptions these models rely upon—most notably, the assumption of ideal solution behaviour. In many cases, FO models are based on the SD framework, which simplifies the transport of water and solutes across the membrane by treating the FS and DS as ideal or near-ideal systems. This simplification does not hold true for concentrated liquid foods, which contain complex mixtures of sugars, polyphenols, proteins, and other bioactive compounds. These solutes interact in non-linear ways and significantly alter osmotic pressure, viscosity, and diffusivity, leading to deviations from the predicted behaviour in idealized models.

Additionally, many existing models fail to capture time-dependent behaviours that are inherent to real FO processes. In practical settings, parameters such as solute concentration, flux, and membrane properties do not remain constant. Over time, FSs become more concentrated, DSs are diluted, and membrane fouling progresses, all of which influence system performance dynamically. However, most models operate under steady-state assumptions or utilize overly simplistic batch tank representations that ignore changes in volume and concentration during operation. This lack of temporal resolution limits the ability of such models to simulate long-duration or high-recovery FO processes with any degree of accuracy.

A further limitation is the insufficient incorporation of fouling phenomena. Although membrane fouling—whether organic, biofouling, or particulate—is a well-documented issue in FO applications, its effects are typically treated in a generalized manner, such as through the use of fixed flux decline coefficients or simplified resistance terms. These approaches do not distinguish between different fouling mechanisms or account for their evolving nature during continuous processing. Consequently, they provide limited predictive value for long-term system behaviour or for evaluating the efficacy of cleaning protocols and pretreatment strategies.

Current models also tend to rely on simplified representations of membrane modules. Most use one-dimensional flow assumptions and do not account for geometric complexities such as radial gradients, fibre packing irregularities, or dead zones that are common in HF and spiral-wound modules. These oversimplifications overlook the impact of local variations in flow velocity, pressure, and concentration, which can significantly influence membrane performance, particularly in large-scale modules.

In sum, while existing FO models serve as useful starting points for understanding the system under ideal or semi-ideal conditions, their capacity to predict the behaviour of real-world FO processes in liquid food systems is hampered by a series of critical limitations. These include unrealistic solution assumptions, a lack of dynamic modelling capabilities, inadequate representation of fouling effects, and oversimplified geometrical configurations. Bridging these gaps requires a more holistic and nuanced approach to modelling—one that aligns closely with the complexities of food matrices, membrane module designs, and operational realities.

### 3.5.3 Strategies to Overcome These Challenges

To address the various limitations outlined in current mathematical models for FO in liquid food systems, a combination of theoretical refinement, empirical validation, and computational innovation is essential. One promising approach involves the integration of empirical and semi-empirical corrections into the existing theoretical models. Since food systems exhibit nonideal solution behaviour—particularly at high solute concentrations—experimental correlations that account for osmotic pressure deviations, effective solute diffusivity, and flux decline trends under realistic operating conditions can enhance model accuracy. These empirical formulations, grounded in experimental data, can bridge the gap between idealized transport equations and the nuanced behaviour observed in concentrated food matrices.

In tandem with empirical refinements, the adoption of a multiscale modelling approach has shown considerable promise. This strategy acknowledges the layered complexity of the FO process, integrating phenomena at molecular, membrane, and system scales. At the molecular level, modelling solute–solute and solute–membrane interactions helps capture the basis for fouling, osmotic effects, and component-specific transport resistance. At the membrane scale, dynamic models of flux behaviour, ICP and ECP, and structural parameter variations offer insight into spatial heterogeneity along the membrane surface. At the process scale, flow dynamics, tank-level mass balances, and module configurations are incorporated to simulate overall system performance. The synergy between these scales provides a more holistic view and enables better optimization of FO modules in practical applications.

Moreover, the incorporation of data-driven techniques, including ML and statistical modelling, offers a complementary solution to traditional mechanistic approaches. These methods can identify patterns and non-linear relationships between input variables and system responses without the need for fully specified transport equations. Coupling ML models with high-resolution experimental datasets allows for predictive modelling of FO performance under varying conditions, while also enabling real-time adjustments and process control. When integrated with sensor technologies that provide continuous feedback on parameters like conductivity, pH, viscosity, and flow rate, these models can be dynamically calibrated to reflect actual process behaviour, thus improving prediction fidelity.

Refinements in thermodynamic modelling also play a critical role in overcoming the limitations of current FO models. Rather than relying on van't Hoff's law, which assumes ideal behaviour and fails at high solute concentrations, more advanced formulations such as Pitzer models or virial expansions can be used to calculate osmotic pressure based on water activity and solute interactions. These models are particularly well-suited for multicomponent food systems where solute interactions can either amplify or suppress osmotic gradients. Such precision in osmotic pressure calculation enhances the accuracy of water flux predictions, particularly when applied to highly concentrated or viscous food extracts.

Addressing spatial variability in membrane modules requires more detailed modelling of flow fields and mass transfer, particularly in configurations like HF or spiral-wound modules. The application of CFD, even in simplified forms, enables visualization of flow maldistribution, dead zones, and concentration gradients across the module. While full-scale CFD simulations may be computationally intensive, simplified two-dimensional or three-dimensional models can still provide critical insights into hydrodynamic behaviour that one-dimensional models cannot capture. These insights inform better module design and help in identifying regions prone to fouling or under-utilization.

Finally, the development of sub-models that explicitly simulate membrane fouling and cleaning dynamics is crucial for long-term process modelling. Such sub-models should include mechanisms for the formation, growth, and structure of fouling layers, as well as their impact on membrane resistance and surface area availability. They should also account for flux recovery behaviour following various cleaning strategies, whether physical (e.g. hydraulic backwashing) or chemical (e.g. EDTA rinsing and alkaline washes). Incorporating these elements into a larger modelling framework ensures that predictions extend beyond initial performance metrics and encompass operational sustainability, maintenance cycles, and product quality over time.

In conclusion, advancing FO modelling for liquid food systems necessitates a shift from static, idealized models towards dynamic, data-rich, and system-aware frameworks. By incorporating empirical corrections, multiscale modelling, advanced thermodynamics, data-driven prediction tools, spatial flow modelling, and fouling dynamics, the next generation of FO models can achieve the precision and flexibility needed to support industrial-scale adoption. These strategies not only improve the scientific robustness of the models but also make them more applicable to the complex, variable, and quality-sensitive nature of liquid food processing.

## REFERENCES

1. Nurhayati M, Jeong K, Lee H, Park J, Hong BU, Kang HG, et al. Predicting and optimizing forward osmosis membrane operation using machine learning. Desalination. 2024 Dec;592:118154.
2. Munubarthi KK, Gautam DK, Reddy KA, Subbiah S. Distributed parameter system modeling approach for the characterization of a high flux hollow fiber forward osmosis (HFFO) membrane. Desalination [Internet]. 2020;496(September):114706. Available from: https://doi.org/10.1016/j.desal.2020.114706
3. Teklu H, Gautam DK, Subbiah S. Axial flow hollow fiber forward osmosis module analysis for optimum design and operating conditions in desalination applications. Chem Eng Sci [Internet]. 2020;216:115494. Available from: https://doi.org/10.1016/j.ces.2020.115494
4. Foo ZH, Rehman D, Coombs OZ, Deshmukh A, Lienhard JH. Multicomponent Fickian solution-diffusion model for osmotic transport through membranes. J Memb Sci. 2021 Dec;640:119819.
5. Roy Y, Lienhard JH. On the presence of solute-solvent transport coupling in reverse osmosis. J Memb Sci. 2020 Oct;611:118272.
6. Wambui Mutoru J, Firoozabadi A. Form of multicomponent Fickian diffusion coefficients matrix. J Chem Thermodyn. 2011 Aug;43(8):1192–203.
7. Pitzer KS, Kim JJ. Thermodynamics of electrolytes. IV. Activity and osmotic coefficients for mixed electrolytes. J Am Chem Soc. 1974 Sep 1;96(18):5701–7.
8. Thomsen K, Iliuta MC, Rasmussen P. Extended UNIQUAC model for correlation and prediction of vapor–liquid–liquid–solid equilibria in aqueous salt systems containing non-electrolytes. Part B. Alcohol (ethanol, propanols, butanols)–water–salt systems. Chem Eng Sci. 2004 Sep;59(17):3631–47.
9. Mahawer K, Mutto A, Khanna R, Kumar Gupta S. A modeling and experimental comparative study of a closed-loop forward osmosis-nanofiltration and standalone nanofiltration systems for fouling-prone wastewater treatment. Chem Eng J. 2024 Apr;486:150283.
10. Ananya B. Forward osmosis for the concentration of black tea extract: Insights into the development of draw solute and extraction of essential tea components [Internet]. [Guwahati]: Indian Institute of Technology Guwahati; 2023 [cited 2024 Jan 10]. Available from: https://gyan.iitg.ac.in/handle/123456789/2354
11. Sekino M. Mass transport analysis of a hollow fiber forward osmosis module via two-layer membrane model derived from the irreversible thermodynamics. Chem Eng Sci. 2022 Jan;247:116703.
12. Ali SM, Im SJ, Jang A, Phuntsho S, Shon HK. Forward osmosis system design and optimization using a commercial cellulose triacetate hollow fibre membrane module for energy efficient desalination. Desalination. 2021 Aug;510:115075.
13. Xiao D, Li W, Chou S, Wang R, Tang CY. A modeling investigation on optimizing the design of forward osmosis hollow fiber modules. J Memb Sci. 2012 Mar;392–393:76–87.
14. Sugawara Y, Amamiya S, Yamaguchi T. Comprehensive simulation to uncover the ideal properties of a hollow fiber forward osmosis membrane module for the seawater desalination process. Desalination. 2022 Sep;538:115923.
15. Kumano A, Matsuyama H. Analysis of hollow fiber reverse osmosis membrane module of axial flow type. J Appl Polym Sci [Internet]. 2012;123(1):463–71. Available from: https://doi.org/10.1002/app.34404
16. Bardhan A, Subbiah S, Mohanty K. Modelling and experimental validation of osmotic driven energy efficient process for tea solution concentration. Environ Technol Innov [Internet]. 2020;20:101065. Available from: https://doi.org/10.1016/j.eti.2020.101065

17. Goi YK, Liang YY. A general modeling framework for FO spiral-wound membrane and its fouling impact on FO-RO desalination system. Desalination. 2025 Jan;593:118236.
18. Attarde D, Jain M, Gupta SK. Modeling of a forward osmosis and a pressure-retarded osmosis spiral wound module using the Spiegler-Kedem model and experimental validation. Sep Purif Technol. 2016 May 30;164:182–97.
19. Hoek EMV, Elimelech M. Cake-enhanced concentration polarization: A new fouling mechanism for salt-rejecting membranes. Environ Sci Technol. 2003 Dec 1;37(24): 5581–8.
20. Park M, Lee J, Boo C, Hong S, Snyder SA, Kim JH. Modeling of colloidal fouling in forward osmosis membrane: Effects of reverse draw solution permeation. Desalination. 2013 Apr;314:115–23.
21. Tow EW, Lienhard VJH. Quantifying osmotic membrane fouling to enable comparisons across diverse processes. J Memb Sci. 2016 Aug;511:92–107.
22. Hong S, Elimelech M. Chemical and physical aspects of natural organic matter (NOM) fouling of nanofiltration membranes. J Memb Sci. 1997 Sep;132(2):159–81.
23. Blandin G, Ferrari F, Lesage G, Le-Clech P, Héran M, Martinez-Lladó X. Forward osmosis as concentration process: Review of opportunities and challenges. Membranes (Basel). 2020;10(10):1–40.
24. Zhou M, Cao X, Yang L, Wang W, Wang H, Li Z. A green approach for non-thermal concentration of skim milk by forward osmosis combined with membrane distillation for draw solution regeneration. Chem Eng Res Des. 2024 Oct;210:469–80.
25. Bardhan A, Senthilmurugan S, Mohanty K. Optimisation of multi-component inorganic salt composition as draw solute for preparation of concentrated tea extract using forward osmosis process. Food and Bioproducts Processing [Internet]. 2023;138:126–38. Available from: https://doi.org/10.1016/j.fbp.2023.01.007
26. Bardhan A, Subbiah S, Mohanty K, Altaee A, Au AA. Feasibility of poly (vinyl alcohol)/poly (diallyldimethylammonium chloride) polymeric network hydrogel as draw solute for forward osmosis process. Membranes 2022. 2022;12:1097.
27. Kim YC, Han S, Hong S. A feasibility study of magnetic separation of magnetic nanoparticle for forward osmosis. Water Sci Technol. 2011 Jul 1;64(2):469–76.
28. Ma Z, Inastiti Noor I, Liu L, Wang X, Wang Q, Wang Z, et al. Optimal synergy between FO membrane and draw solute for high-efficient fruit juice concentration and simultaneous fruiter fertigation. Chem Eng J. 2024 May;488:150959.
29. Asghar N, Lee H, Jang D, Jang A. Recovery of volatile fatty acids using forward osmosis: Influence of solution chemistry, temperature, and membrane orientation. Chemosphere. 2022 Sep;303:134814.
30. Cath TY, Elimelech M, McCutcheon JR, McGinnis RL, Achilli A, Anastasio D, et al. Standard methodology for evaluating membrane performance in osmotically driven membrane processes. Desalination. 2013 Mar;312:31–8.
31. Wijmans JG, Baker RW. The solution-diffusion model: A review. J Membr Sci. 1995; 107(1–2):1–21. https://doi.org/10.1016/0376-7388(95)00102-I.
32. Ryu H, Mushtaq A, Park E, Kim K, Chang YK, Han JI. Dynamical modeling of water flux in forward osmosis with multistage operation and sensitivity analysis of model parameters. Water (Basel). 2019 Dec 19;12(1):31.
33. Eyvaz M, Arslan S, Koseoglu-Imer D, Yuksel E, Koyuncu I. Forward osmosis membranes: A review: Part I. https://doi.org/10.5772/intechopen.72287
34. Suzaimi ND, Goh PS, Ismail AF, Mamah SC, Malek NANN, Lim JW, et al. Strategies in forward osmosis membrane substrate fabrication and modification: A review. Membranes (Basel). 2020 Nov 7;10(11):332.
35. Al-Mutaz IS, Alalawi A, Bin Darwish N. Explicit expression for water permeation flux in forward osmosis desalination process. Appl Water Sci. 2024 Jul 6;14(7):149.

36. Bai D, Kruczek B. Effect of membrane orientation and concentration of draw solution on the behavior of commercial osmotic membrane in a novel dynamic forward osmosis tests. Membranes (Basel). 2022 Mar 31;12(4):385.
37. McCutcheon JR, Elimelech M. Influence of concentrative and dilutive internal concentration polarization on flux behavior in forward osmosis. J Memb Sci. 2006 Nov 1; 284(1–2):237–47.
38. Akhtar A, Singh M, Subbiah S, Mohanty K. Modelling, experimental validation and process design of forward osmosis process for sugarcane juice concentration. LWT. 2021 Apr 1;141:110852.
39. Park HW, Baek J, Kim WJ. Forward osmosis and direct contact membrane distillation: Emerging membrane technologies in food and beverage processing. Innov Food Sci Emerg Technol. 2024 May;93:103626.
40. Aydiner C, Sen U, Topcu S, Sesli D, Ekinci D, De Altınay A, et al. Techno-economic investigation of water recovery and whey powder production from whey using UF/RO and FO/RO integrated membrane systems. Desalination Water Treat. 2014 Jan;52(1–3): 123–33.
41. Ryther R, Stokes R. Development of a comprehensive cleaning and sanitizing program for food production facilities. In: Food safety management. Elsevier; 2023. p. 679–98.
42. Chen GQ, Artemi A, Lee J, Gras SL, Kentish SE. A pilot scale study on the concentration of milk and whey by forward osmosis. Sep Purif Technol [Internet]. 2019;215:652–9. Available from: https://www.sciencedirect.com/science/article/pii/S1383586618338085
43. Bardhan A, Subbiah S, Mohanty K. Modeling and experimental validation for the preparation of concentrated tea extract using a forward osmosis process using a food-grade inorganic draw solute. Ind Eng Chem Res. 2022; 62(7), 3315–26.
44. Chen GQ, Gras SL, Kentish SE. The application of forward osmosis to dairy processing. Sep Purif Technol. 2020 Sep;246:116900.

# 4 Application of Forward Osmosis in Liquid Food and Dairy Processing Industries

## 4.1 APPLICATION OF FO PROCESS IN LIQUID FOOD AND BEVERAGE PROCESSING INDUSTRIES

FO has recently attracted interest in the liquid food and beverage processing industry since it is reported as a green and energy-saving process for product concentration, nutrient retention, waste retention, and waste reduction to cope with the growing trend of consumer demand for minimally processed foods free of preservatives (1). FO improves the sustainability of liquid food transport, and it does this by preconcentrating liquid foods (such as coffee extract or fruit puree), decreasing the volume to ship and refrigerate, and cutting logistics costs and emissions. FO pre-concentration of juice has the potential to reduce the carbon footprint of the transport of juices globally by 25% based on estimated reduced weight and spoilage (2). Unlike thermal concentration techniques, such as evaporation or spray drying, which inactivate the heat-sensitive ingredients such as colour, flavour, and preservatives, FO is capable of operating at ambient temperatures, preserving the sensory and nutritional characteristics of products. This trait of FO renders it especially preferred for not only the concentration of fruit and vegetable juice but also the concentration of dairy and plant extracts without damaging bioactive compounds. For instance, FO has been utilized for the concentration of orange juices, retaining 95% of vitamin C and volatile aromatic compounds, outperforming thermal performance, where up to 40% of these substances are depleted (3). The study revealed FO can potentially concentrate liquid food while efficiently preserving the essential thermosensitive compounds (Table 4.1).

In the FO process, the DS nature and characteristics are particularly important. For food applications, a DS must be food-grade and have low cost, high osmotic pressure, and low viscosity. In general, lower molecular solutes can provide higher osmotic pressures due to higher molarity at the same mass-based concentration. In addition, lower molecular weight solutes exhibit lower viscosity, which facilitates diffusion; while this can increase water flux, it may also lead to increased RSF due to faster diffusion across the membranes' selective layer. Based on literature and considering the specific needs for concentration of liquid foods using the FO process, an ideal DS must exhibit certain criteria, which include (i) membrane and material compatibility, (ii) propensity for membrane fouling, (iii) product compatibility,

DOI: 10.1201/9781003657873-4

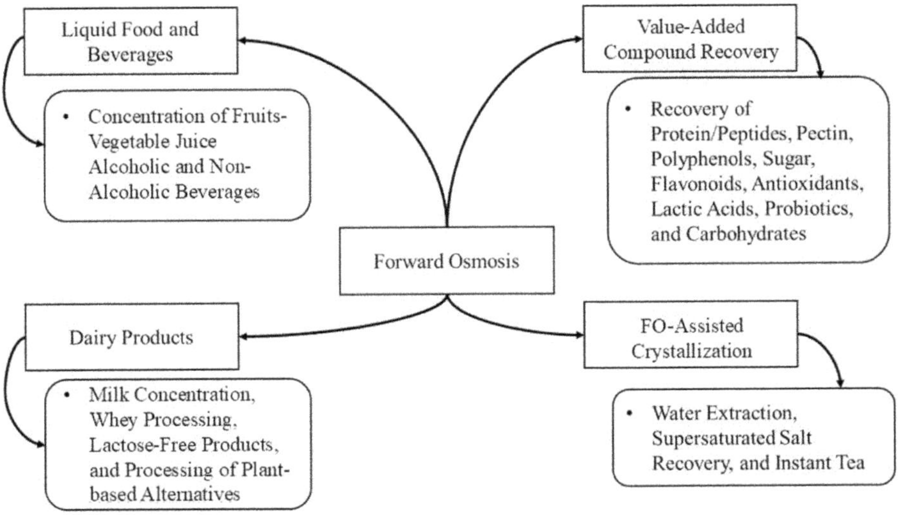


**FIGURE 4.1** Overview of forward osmosis applications in food processing

(iv) product compatibility and food safety concerns, and (v) cost for DS replenishment (4). In food processing industries, FO stands out for its superior energy efficiency, addressing crucial concerns about operational and environmental impact. Reportedly utilizing osmotic gradient instead of thermal or high-pressure methods, FO cuts energy usage by 50–70% compared to traditional evaporators (5). A pilot plant illustrated efficiency, demonstrating FO-based tomato paste concentration that required only 0.3 kWh $m^{-3}$ of energy, in contrast to thermal evaporators' 10 kWh $m^{-3}$ (6). FO systems often employ food-grade DSs, such as sugar syrups or salt brines, which can be restored through low-energy techniques like dilution or gentle heating. In some instances, the draw solute becomes an integral part of the end-product. The craft brewing sector exemplifies this, using FO to concentrate wort with brewing sugars as the DS, thereby optimizing production and decreasing water consumption by 30% (7).

FO can also play a vital role in wastewater management. It can concentrate food processing effluents rich in organic matter, sugars, and oil for resource recovery. For instance, in olive oil production, FO processes wastewater containing polyphenols and organic acids, reclaiming 80% of water for reuse and concentrating antioxidants for cosmetic or pharmaceutical extraction (8,9).

### 4.1.1 Concentration of Fruit Juice

The FO process for fruit-juice concentration is applied for the preservation of the delicate flavour, aroma, appearance, and mouthfeel of freshly squeezed juices in the concentrate (25). The studies indicate that juice concentrated by the FO process retains significantly higher levels of phenolic compounds and essential aromas compared to thermally concentrated juices. For instance, raspberry (26) and watermelon

**TABLE 4.1**
**Summary of forward osmosis in liquid food and beverage processing industries**

| Feed Solution | Draw Solution | FO Membrane (area) | Initial Flux, L $m^{-2}$ $h^{-1}$ | Critical Findings | Ref. |
|---|---|---|---|---|---|
| *Kokum (Garcinia indica)* | 6 M Sodium chloride | CTA-FO (0.114 $m^2$) | 17.5 | The anthocyanin extract was concentrated approximately 54 times (from 49.63 mg/L to 2.69 g/L). | (10) |
| Hibiscus flower | 6 M Sodium chloride | CTA-FO (0.114 $m^2$) | 20.11 | TAC in hibiscus flower increases from 461.39 mg/kg to 3967.90 mg/kg and TSS increases from 3.1 °Brix to 24.4 °Brix. | (11) |
| Grape juice | 1 M Sodium chloride | CTA-FO (0.002 $m^2$) | 19.1 ± 0.03 | | (12) |
| Blueberry juice | 1 M Citric acid | TFC-FO (0.002 $m^2$) | 18.14 ± 2.81 | TAC in blueberry juice increased by about 26% and TPC by about 21%. | (13) |
| Pomegranate juice | 6 M Sodium chloride | TFC-FO (0.013 $m^2$) | 8.81 | With fourfold concentration, the anthocyanin content increases from 414.4 mg/kg to 1731.9 mg/kg with RSF of 1.03 g/$m^2$ h | (14) |
| Red beetroot juice | 6 M Sodium chloride | CTA-FO (0.013 $m^2$) | 12.2 | Beetroot juice concentrated to about 12-fold with about three times slower betalain degradation when compared to thermal evaporation. | (15) |
| Pear juice | 6.4 M Sodium lactate | CTA-FO (0.003 $m^2$) | 10.0 | Compared to vacuum evaporation, the FO process can concentrate pear juice from 9.6° Brix to 62° Brix accompanied with a superior colour/flavour profile and enhanced antioxidant capacity. | (16) |
| Grape juice | 4 M Sodium diacetate | TFC-FO (0.002 $m^2$) | 27.78 | Using the FO process, grape juice could be concentrated to 54.0 °Brix | (17) |
| Apple juice | 4 M Sodium chloride | TFC-FO (0.002 $m^2$) | 20.7 ± 0.62 | In a single-stage FO process, apple juice was concentrated to about 5.45-fold. | (18) |

*(Continued)*

**TABLE 4.1** *(Continued)*
**Summary of forward osmosis in liquid food and beverage processing industries**

| Feed Solution | Draw Solution | FO Membrane (area) | Initial Flux, L $m^{-2}$ $h^{-1}$ | Critical Findings | Ref. |
|---|---|---|---|---|---|
| Liquid egg white | 4 M Sodium chloride | CTA-FO (0.002 $m^2$) | 15.6 ± 0.17 | In a single FO process, the liquid egg white was concentrated to about 2.1-fold. | (19) |
| De-alcoholization of beer | 1.5 M Magnesium chloride hexahydrate | CTA-FO (0.002 $m^2$) | 12.5 ± 0.8 | FO reduced the alcohol content (from 5% to 0.5%). | (20) |
| De-alcoholization of kiwi wine | 4 M Magnesium chloride | CTA-FO (0.002 $m^2$) | 11.10 | FO reduced the alcohol content of kiwi wine to 0.45% v/v, including phenol content (91.59 μg/mL), and VOCs (2735.36 μg/mL). | (21) |
| Freshly brewed tea extract | Multicomponent inorganic salt solution** | CTA-HFFO (2.3 $m^2$) | 16.91 | Using a single FO process, freshly brewed tea extract was concentrated to about 2.67-fold while maintaining essential tea components such as caffeine and catechins. | (22) |
| Orange peel waste extract | 1 M Sodium chloride | CTA-FO (0.004 $m^2$) | 6.77 | The FO process increases phenolic compound concentration by a factor of 2.03. | (23) |
| Coconut water | Potassium lactate (60 ± 1 °Brix) | CTA-SWFO (0.52 $m^2$) | 7.3 ± 0.4 | Under refrigerated conditions, coconut water concentrated (23 °Brix) using FO and HPP with acidification was stable for 18 weeks. | (24) |

CTA: cellulose triacetate; TFC: thin-film composite; TAC: total anthocyanin content; TPC: total phenolic content; FO: forward osmosis; VOC: volatile organic compound; HFFO: hollow-fibre forward osmosis; SWFO: spiral-wound forward osmosis; HPP: high-pressure processing
** (mixture of 0.25 M NaCl, 0.5 M $Na_2SO_4$, 1 M $MgCl_2$, and 0.25 M KCl)

(27) juice concentrated through FO exhibit superior retention of essential volatile aroma compounds for maintaining the characteristics of fresh flavour without significant nutritional loss.

The quality of watermelon juice concentrated by the FO process using 70% glycerol as a DS was evaluated (28). The freshly produced watermelon juice was concentrated to 65 °Brix, without significant degradation of key nutritional components such as lycopene and citrulline or its sensory properties.

Pineapple juice concentrated via the FO process exhibits reduced pulp degradation, maintaining texture and enhancing consumer acceptance. A similar study demonstrated the application of the FO process for the concentration of pineapple juice (60 °Brix) without significant loss of ascorbic acid. The author revealed the sucrose (40% w/w)–sodium chloride (12% w/w) combination can minimize the RSF (0.58%) with higher flux (1.18 L $m^{-2}$ $h^{-1}$) during the FO process.

Using a 6.0 M NaCl solution as the DS, the FO process was employed to concentrate grape juice to 54 °Brix, resulting in a 6.8-fold increase in anthocyanin concentration—from 104.8 mg $L^{-1}$ to 715 mg $L^{-1}$ (29). However, using the FO process alone, it was not possible to attain the desired final concentration due to the inherent limitations of the process, such as CP, membrane fouling, and other transport properties of the solutions. A study demonstrated that by combining FO with evaporation, grape juice can be potentially concentrated up to 65.7 °Brix. However, some degradation of anthocyanins was observed, attributed to thermal effects during the evaporation stage (12). Additionally, the author revealed that the combined FO–evaporation process would require about 23% lower energy cost than the evaporation process alone.

Grapefruit juice was successfully concentrated using the FO process with a commercial TFC FO membrane and a 2.0 M glucose solution as the draw solute, without any degradation of essential nutrient compounds such as vitamins and minerals (30). The author revealed that even though water flux was low (13.1 L $m^{-2}$ $h^{-1}$), the RSF of glucose molecules would positively impact the concentrated juice.

Apple juice was concentrated by a factor of 4.25 using a TFC FO membrane, with potassium sorbate—a commonly used preservative—employed as the DS (31). The final concentrate contained very low levels of potassium sorbate (0.45 g $L^{-1}$), ensuring the product's safety while maintaining quality.

The feasibility of concentrating blueberry juice using the FO process was demonstrated with food additives such as citric acid, sodium benzoate, and potassium sorbate serving as DSs (32). The author revealed that food additives have a protective effect on bioactive substances of juice, and thus, the RSF caused by these DS has a positive impact on the final product.

A single-run FO process using a CTA membrane and 6.4 M sodium lactate as the draw solute achieved a 6.5-fold concentration of pear juice (16). Compared to vacuum evaporation, the FO concentrates offer juice concentrates with higher antioxidant capacity, richer aroma, and flavour.

Pomegranate juice was concentrated using the FO process with 6 M NaCl as the DS (14). The transmembrane flux during the FO process decreases from 8.81 L $m^{-2}$ $h^{-1}$ to 0.55 L $m^{-2}$ $h^{-1}$, with 1.03 g $m^{-2}$ $h^{-1}$ RSF. The pomegranate juice was concentrated by fourfold (i.e. 15.2 to 61.3 °Brix), and anthocyanin content also increased from 414.4 mg $kg^{-1}$ to 1731.9 mg $kg^{-1}$.

The aforementioned studies suggested the application of polymeric membranes for fruit-juice concentration using the FO process. Despite all these studies, the water flux reduction due to membrane fouling by organic components and fruity flesh is still an essential issue in FO, which needs to be addressed, though the impact of fouling in FO is smaller than in RO because of its lower operating pressure. Several reports have shown how fouling can be reduced by ultrasonic cleaning and hydraulic flushing (239, 313). The development of highly fouling-resistant membranes is still a major challenge. Recent studies reported the potential of Na-ZSM-5 membrane for FO operations as an inorganic membrane material (33,34). A recent study demonstrated that the concentration of fruit juices such as apple, orange, and pineapple was achieved using a tubular Na-ZSM-5 FO membrane (35). The study revealed that, based on a molecular sieving effect of the zeolite membrane, the fruit juices were successfully concentrated while maintaining their quality without contamination by the DS. In terms of fouling, the Na-ZSM-5 membrane exhibited stable dehydration performance for fruit-juice concentration over 120 h without fouling due to pectin, possibly because of the combination of the molecular sieving property and anionic framework of zeolite, which inhibits the formation of a cake layer on the membrane surface. Additionally, the membrane exhibited high heat resistance, with no deterioration up to 78.85°C.

### 4.1.2 Concentration of Vegetable Juices

The FO process is used for the concentration of vegetable juices for extending shelf life while maintaining nutritional, sensory, and organoleptic properties. The FO process operates under mild conditions and can effectively retain vitamins, antioxidants, and phytochemicals (such as carotenoids, chlorophyll, and polyphenols). Similar to fruit juices, vegetable juices, such as tomato, carrot, and spinach, contain valuable bioactive compounds that are susceptible to degradation under high-temperature conditions. The gentle operating conditions of the FO process, which operates under low-pressure conditions and ambient temperatures, help to retain the delicate flavours, colours, vitamins, and antioxidants found in vegetable juices.

The FO process has proven effective in concentrating tomato juice while preserving both its flavour and nutritional content, particularly lycopene, making it well-suited for producing high-quality tomato juice concentrate (36). Similarly, black carrot juice was concentrated from 10.4 to 43.4 °Brix using 1.1 M magnesium chloride ($MgCl_2$) as the DS. However, a noticeable colour change was observed in the final FO concentrate, likely due to interactions between magnesium ions ($Mg^{2+}$) and anthocyanin pigments (37).

### 4.1.3 Concentration of Coffee and Tea Extracts

The application of the FO process for the concentration of non-alcoholic beverages (such as tea and coffee) has gained significant interest due to its ability to preserve the delicate flavour, aroma, and bioactive compounds. Coffee and tea extracts are rich sources of thermosensitive bioactive compounds, including caffeine, polyphenols, essential oils, and antioxidants. For instance, polyphenols in green tea, such as catechins, are known for their antioxidant properties and health benefits. However, these compounds are prone to oxidation and degradation when exposed to elevated temperatures.

The mild operating conditions of the FO process help minimize the risk of thermal degradation, thereby preserving the integrity of valuable bioactive compounds (38). Utilising this advantage, freshly brewed black tea extract was successfully concentrated fivefold using a multicomponent inorganic DS in combination with a commercial HFFO membrane (22). The study revealed that compared to vacuum evaporation, the FO process was more efficient in retaining the essential tea components (such as epicatechin, epigallocatechin, epicatechin gallate, epigallocatechin gallate, and caffeine).

### 4.1.4 Concentration of Alcoholic Beverages

Numerous research groups studied the feasibility of separating alcohol from water, using the FO process. De-alcoholization is usually accomplished using biological or membrane separation processes. The biological approach (fermentation) is controlled in such a way as to limit ethanol formation, whereas the separation (membrane separation) uses a membrane to selectively remove the alcohol. In the FO process, the water/alcohol separation relies on the preferential affinity and diffusivity through the FO membrane of one of the compounds.

The FO process has been shown to effectively reduce the alcohol content in beer from 5.0 vol.% to 0.5 vol.% using a commercial CTA membrane (20 $cm^2$) and a 1.5 M $MgCl_2 \cdot 6H_2O$ solution as the DS. However, key beer properties—such as turbidity, salinity, and colour—were negatively impacted, leading to a deterioration in the product's organoleptic qualities. Notably, an increase in turbidity was observed in the de-alcoholized beer, attributed to heightened ionic strength resulting from RSF of $MgCl_2$ at a rate of 2.5 g $m^2$ $h^{-1}$(39).

The effect of FO on the de-alcoholization of kiwi wine was investigated using a CTA membrane with an area of 20 $cm^2$ and a 4 M $MgCl_2$ solution as the draw solute, successfully reducing the alcohol content from 5.5% v/v to 0.45% v/v (21). The study revealed that de-alcoholization technology improved the phenolic compound content (94.59 μg $mL^{-1}$), volatile organic compound content (2735.36 μg $mL^{-1}$), and antioxidant activity of kiwi wine.

### 4.1.5 Concentration of Herbal and Functional Beverages

Polyphenols are natural compounds in fruits, vegetables, coffee, wine, and tea. Dietary polyphenols offer various health-promoting activities, including anti-inflammatory, antioxidant, anti-carcinogenic, antidiabetic, and neuroprotective potentials (40). With the advancement of the food industry, there has been an increasing demand for concentrated polyphenol-rich foods (such as fruit/vegetable juice) by the dewatering process to reduce the costs associated with packaging, freezing, storage, and transportation. Reportedly, the stability and bioavailability of different polyphenols can be influenced by heat treatment (above 50°C). Non-thermal membrane technology, such as the FO process, has a high rejection rate for functional components and is highly suitable for liquid concentration. A recent study reported the impact of the FO process on nine essential dietary polyphenols, including 3,4-dihydroxybenzoic acid (found in olives and barley), p-coumaric acid (blueberries, garlic, coffee, and honey), gallic acid (pomegranate, apples, blueberries, and strawberries), caffeic acid (coffee, tea, red wine, kale, and spinach), ferulic acid (carrot, tomato, spinach, orange, and pomegranate), sinapic acid (mustard seed, grape, pomegranate, apple, and red wine), chlorogenic acid (green coffee, black tea, apple, pear, and tomato), phloretin 2′-β-D-glucoside (apple cider), and epigallocatechin gallate (green tea and black tea) (41). The author revealed that the FO system could retain the polyphenols with a greater molecular size in the liquid food concentration process. Compared to the CTA membrane, the TFC membrane with a higher negative charge can selectively exhibit higher polyphenol rejection (>99%).

### 4.1.6 Sugar and Syrup Concentration

The feasibility of concentrating clarified sugarcane juice (osmotic coefficient ~ 1.01) using the FO process was demonstrated by employing sea bitter as the DS, which exhibited an osmotic coefficient ranging from 1.41 to 3.24 (42). In the given study, the author reported that sucrose concentration was found to increase from 10.5% to 40.6%, with an average flux of 13.0 L $m^{-2}$ $h^{-1}$ (membrane area, 0.0057 $m^2$). The study highlights the significance of high osmotic pressure and its influence on permeate flux. However, the influence of RSF on the final concentrate quality was not discussed. Using a 6 M NaCl solution as the osmotic agent, sugarcane juice was efficiently dewatered from 17.6 to 31.7 °Brix with a transmembrane flux of 0.79 L $m^{-2}$ $h^{-1}$, employing a flat-sheet CTA membrane with an area of 0.0144 (43). The study revealed that about 179.3 mg $L^{-1}$ of osmotic agent (NaCl) was transferred to FS (sugarcane concentrate). This may result in an alteration in nutritional quality and consumer acceptability of the FO concentrated product. Furthermore, the importance of selecting an appropriate membrane and DS has been emphasized as a key factor in minimizing RSF during the FO process (44). However, high organic content (including sucrose, fructose, glucose, protein, and polysaccharides) in liquid food may cause severe membrane fouling and CP, especially at higher concentrations of sugar solution.

## 4.2 APPLICATION OF FO PROCESS IN DAIRY PROCESSING INDUSTRIES

In dairy industries, concentration of milk and other dairy streams is a common unit operation used to prepare dairy products with extended shelf life and reduced cost associated with transportation and storage costs (45). Thermal evaporation is widely used to concentrate milk. The conventional thermal methods are energy-intensive and have a negative impact on colour, flavour, and nutrient degradation (46). Non-thermal concentration such as RO has been adopted as an alternative to thermal evaporation for the concentration of liquid milk. However, due to high operating pressure, RO is severely prone to membrane fouling, thereby limiting the final product concentration (47). Several practical applications of FO for the concentration of milk and dairy by-products are summarized in *Table 4.2*.

### 4.2.1 Concentration of Milk

Concentration of milk is a standard operation in the dairy industry. The concentration of milk is an essential step to increase solid levels in most dairy products, such as yoghurt, cheese, and milk powder. Concentration of milk is achieved by thermal evaporation, vacuum evaporation, cryo-concentration, and RO. Evaporation is energy intensive and has a negative impact on the colour, taste, flavour, and nutritional value of milk (48,49). High investment cost and high energy consumption limit the application of cryo-concentration in large-scale production. To address these challenges, RO has been adopted by some dairy processors as an alternative to the thermal evaporation process. While the RO process avoids phase change and reduces energy

**TABLE 4.2**
**Application of the FO process for the concentration of dairy and dairy by-products**

| Feed Solution | Membrane (Area, $m^2$) | Draw Solution | Initial Flux ($L\ m^{-2}\ h$) | Concentration Factor | Ref. |
|---|---|---|---|---|---|
| Skimmed goat milk | Flat sheet (0.002 $m^2$) | 4 M Sodium chloride | 4.10 ± 0.54 | 2.0 | (53) |
| Skim milk | CTA-FO Flat sheet (0.002 $m^2$) | 2 M Sodium chloride | 23.20 | 2.41–2.51 | (54) |
| | TFC-FO Flat sheet (0.002 $m^2$) | | 13.65 | | |
| HTST pasteurized whole milk | Spiral-wound CTA-FO (0.5 $m^2$) | 60 °Brix potassium lactate | 3.5 | 2.35 ± 0.08 | (58) |
| Whey protein solution | CTA-FO (0.006 $m^2$) | 4.5 M Sodium chloride | | 1.22 ± 0.04 | (59) |
| Whey solution | TFC-HFFO (0.011 $m^2$) | 0.5 M Sodium chloride | 11.7 | 1.5 | (60) |
| Whey solution | CTA-FO (0.014 $m^2$) | 3 M Sodium chloride | 25 | ~2.14 | (61) |
| Skim milk | HFFO (2.3 $m^2$) | Delactosed permeate (36.4 % w/w dry matter) | ~ (24,25) | ≤ 1.90 | (52) |
| Acid whey | SW CTA-FO (0.5 $m^2$) | Potassium lactate (60° Brix) | 3.6 ± 0.7 | ~3.0 | (62) |

HTST: high temperature short time; CTA: cellulose triacetate; TFC: thin-film composite; HFFO: hollow-fibre forward osmosis; RSF: reverse solute flux

consumption, it is highly susceptible to membrane fouling, thereby limiting the final product concentrations (50). Within the dairy industry, FO is gently used to concentrate milk and whey proteins without denaturation, with up to a 15–20% increase in cheese yield and production of protein-rich nutritional supplements (48,51).

The mild operating condition of FO not only preserves the structural integrity and nutritional properties of milk proteins but also aligns with the dairy industry's sustainability and cost-effectiveness goals. Recent studies suggest that FO can potentially concentrate skim milk from 9% to 18% dry matter using delactosed permeate as DS (52). Furthermore, FO (5–10 kWh) requires significantly lower energy compared to RO (9–20 kWh) when removing one tonne of water at a pilot scale (51).

In another study, the feasibility and effects of the FO process for the concentration of SGM were assessed and compared with thermal evaporation (53). The study also evaluated the process performance, membrane fouling, and membrane cleaning methods of the FO process to concentrate SGM. Through a single-stage FO process, the total solid content of raw SGW can be increased from 9.07% to 18.78%. Membrane fouling can be mitigated by improving cross-flow rate, and osmotic backwash proved to be the most effective cleaning approach to restore membrane performance with a 100% flux recovery rate. Furthermore, in terms of economic and energy evaluation, FO with a pilot scale could be promising in the milk concentration of low cost (\$2.69 $m^{-3}$) with a low energy consumption of 4.80 kWh $m^{-3}$.

The FO–MD hybrid process has emerged as a promising approach for concentrating skim milk, enabling DS recovery while simultaneously generating high-quality clean water (54). At the pilot scale, this system was reported to consume nearly 50% less energy compared to conventional evaporators. Moreover, the strategic utilization of low-grade waste heat available in dairy plants was highlighted to further reduce energy consumption costs associated with the process.

In cheese production, the concentration of milk is a crucial process to improve cheese yield and reduce overall production cost (55). Thermal evaporation and pressure-driven membrane processes (such as UF and RO) are energy-intensive and risk protein denaturation due to high temperature and pressure. The mild operating condition of FO not only preserves the structural integrity and nutritional properties of milk proteins but also prevents heat damage to vital nutrients and enzymes essential for the high quality of the cheese produced. Reportedly, the pre-concentration reduces the volume of milk required for cheese processing, thereby enhancing the logistical efficiency and minimizing the cost (related to transportation and storage). During cheese production, the integration of FO in dairy facilities can optimize both their operational efficiency and product quality, positioning FO as a leading innovation in the modernization of cheese manufacturing.

### 4.2.2 Dairy Processing By-Products

"Whey", the liquid fraction drained from curd during cheese production, is a rich source of nutritional and functional ingredients for food and pharmaceutical industries (56). Cheese whey is composed of around 93% water, 0.8% protein, 0.3% fat, 4.8% lactose, and 0.5% ash (57). Whey protein (a rich source of essential amino acids, functional peptides, and antioxidants) is claimed to confer benefits against a wide range of metabolic disorders such as cancer, cardiovascular complications, hypertension, diabetes, obesity, and phenylketonuria.

While being the most abundant by-product in the dairy industry, whey is considered problematic organic wastewater. Due to low solid content, low pH, and high lactose and mineral content, concentration of whey is an essential process to enhance its utility and value. If raw whey is disposed of untreated, it can significantly contribute to wastewater pollution due to its high biological oxygen demand (BOD). Concentrating whey facilitates efficient recovery of specific protein and lactose concentrations for the processing of energy drinks, protein bars, and infant formulas.

### 4.2.3 Lactic Acid Recovery from Dairy Wastewater

Lactic acid is produced in large quantities in dairy industries as a result of the fermentation of lactose by lactic acid bacteria (LAB). Lactic acid is widely used in food formulations as a preservative and flavour enhancer and in the production of biodegradable plastics (polylactic acid). However, it is often lost in wastewater streams during dairy processing. In particular, the production of dairy products like cheese, yoghurt, and buttermilk generates significant quantities of lactic acid through the natural fermentation of lactose by LAB. Additionally, in some cases, the dairy products may undergo unintended spoilage or souring due to excessive bacterial activity, leading to high levels of lactic acid. These spoilt products may be discarded, resulting in a loss of lactic acid and other nutrients. If these spoilt products are not repurposed or processed, the lactic acid they contain is effectively wasted. Thus, instead of discarding acid whey or effluent, dairy industries can adopt processes to recover lactic acid, thereby reducing waste and creating value-added by-products from what was traditionally considered waste. Thus, an efficient recovery technology can help reduce the wastage by reclaiming lactic acid from these streams. FO provides an efficient solution for concentrating lactic acid from dairy wastewater, allowing for its recovery and reuse. By concentrating lactic acid, FO helps reduce the environmental impact of dairy wastewater disposal while also providing a valuable by-product that can be repurposed in various industries. The FO process ensures that lactic acid is concentrated without degradation, preserving its functional properties for use in food products or as a raw material for producing biodegradable plastics. This contributes to the circular economy, as dairy waste is converted into valuable resources that can be reused in other production processes.

The application of FO in the recovery of organic compounds from food by-products offers several environmental and economic advantages. First, it reduces the volume of waste that needs to be treated or disposed of, which decreases the environmental footprint of food processing industries. Second, it allows the recovery of high-value compounds like proteins, peptides, bio-phenols, pectin, and lactic acid, which can be repurposed into new products, providing additional revenue streams for food processors. Furthermore, the FO process is energy-efficient and operates under mild conditions, which not only preserves the functional properties of the recovered compounds but also reduces the risk of membrane fouling. This ensures a longer lifespan for FO membranes and decreases operational costs, making it a highly cost-effective solution for food processing waste management.

### 4.2.4 Whey Protein Concentration and Fractionation

In dairy industries, whey processing is one of the most important sectors. Formerly treated as waste, whey has gained high value for human consumption due to its high nutritional value and protein concentration. Whey consists of all nine essential amino acids, while it is low in carbohydrate and fat.

The feasibility of concentrating whey solution was demonstrated using an integrated FO–RO membrane process, with NaCl employed as the DS (63). The study revealed that, compared to the conventional UF–RO membrane process, FO is a sustainable method of concentrating whey solution (64). For the given study, different types of membranes were investigated, including CTA-FO membrane and PA-based RO membrane, of which the CTA membrane exhibited better performance due to the thinner support structure and lower ICP (64). Concentration of whey solution was also achieved using a CTA-FO membrane with ammonia as the DS, which helped in reducing RSF (65). Furthermore, a HF TFC-FO membrane was used to concentrate whey protein from 6% to 18% by mass, using NaCl as the DS. During the 22-hour filtration process, the water flux declined from 12 to 7.5 L $m^{-2}$ $h^{-1}$ (60). The author hypothesized that this flux reduction was primarily due to the CP effect in the FS due to mass transfer limitations in the boundary layer near the membrane surface, with the increase in the osmotic pressure of the retained protein reducing the flux. During the FO process, whey, as FS, induces a severe initial flux decline due to instantaneous protein fouling of the membrane. However, intermediate cleaning with water is sufficient to recover water flux and to prevent initial fouling (66). Though irreversible fouling is suppressed due to salting in/out effects, flux decline is a result of reversible fouling only (67).

The FO process has shown strong potential for concentrating cheese whey, as demonstrated in a pilot-scale filtration system operated in batch mode (68). The study revealed that the process was capable of concentrating cheese whey with a concentration factor of approximately 2.7 by increasing the total solids content of whey from 6.5% to 18%, though the study provides important information regarding flux performance at pilot scale. However, this method is only applicable when the system is allowed to approach an osmotic equilibrium and where no fouling is encountered.

### 4.2.5 FO in Lactose-Free Product Processing

Lactose intolerance is a widespread condition where individuals lack the enzyme lactase, which is necessary for the digestion of lactose, the sugar found in milk and dairy products. Consequently, the demand for lactose-free dairy products is on the rise. Due to its lower energy consumption compared to traditional membrane processes (such as RO) and evaporation, FO offers an attractive alternative for concentrating and processing lactose-free products while maintaining their quality and nutritional value. For lactose-free product processing, FO can be applied to concentrate lactose-free milk, whey, and other dairy alternatives without causing damage to heat-sensitive nutrients or altering the products' taste and texture.

Lactose-free milk can be concentrated using the FO process without the need for heat or pressure. Additionally, by concentrating lactose-free milk through FO, manufacturers can create lactose-free milk powders, which are often used as ingredients in other lactose-free products. Furthermore, FO can be used to concentrate lactose-free whey without denaturing the essential milk proteins. This is crucial for producing high-quality WPI or WPC conventionally used in sports nutrition and functional foods.

### 4.2.6 Processing of Plant-Based Dairy Alternatives

With rising awareness of health, sustainability, and ethical concerns, there has been a significant surge in the demand for plant-based dairy alternatives. Plant-based milk essentially lacks certain components normally associated with mammalian milk, such as cholesterol, antigens, saturated fatty acids, and lactose, while at the same time being a good source of minerals, non-allergenic proteins, and essential fatty acids. In recent times, plant sources have been accepted as functional foods and nutraceuticals, as they are a rich source of health-beneficial bioactive components such as vitamins, minerals, dietary fibre, and antioxidants. The plant-based dairy market is experiencing robust growth, with projections indicating continued expansion due to increasing adoption across all age groups and geographies. FO offers distinct advantages, including lower energy consumption, preservation of nutritional value, and retention of flavour, all of which are critical in maintaining the quality of plant-based alternatives. FO offers several distinct advantages over traditional thermal and pressure-driven membrane processes. The FO process requires lower energy input, preserves the nutritional integrity of heat-sensitive components, and retains the natural flavour and texture of plant-derived ingredients. These characteristics are especially valuable in maintaining the sensory and nutritional quality of plant-based dairy alternatives, which are often sensitive to heat and mechanical stress.

This section explores the potential of FO application for the concentration of popular plant-based dairy alternatives, highlighting its potential to meet both industry demands and consumer expectations in this rapidly evolving market.

#### 4.2.6.1 Concentration of Almond Milk

Almond (*Prunus dulcis*) is a concentrated source of nutrients, notably rich in α-tocopherol, a potent antioxidant that plays a key role in combating oxidative stress. However, traditional concentration methods such as evaporation can degrade heat-sensitive compounds, leading to loss of flavour and nutritional quality. FO offers a gentle alternative by enabling the concentration of almond milk at ambient or low temperatures, thereby preserving vital nutrients such as vitamin E, polyphenols, and antioxidants, while maintaining its clean and delicate taste profile.

#### 4.2.6.2 Concentration of Coconut Milk

Coconut milk is widely used in beverages, cooking, and dairy replacement products. Coconut milk is sensitive to heat, and excessive heat exposure during concentration processes can cause loss of flavour, texture, and nutritional value. FO concentrates coconut milk without heating, thereby helping in preserving the fatty acid composition and essential flavour compounds and thereby resulting in a concentrated coconut milk with a naturally creamy texture and extended shelf life suitable for use in both commercial and culinary formulations. Coconut milk was concentrated up to two times, corresponding to ~50% recovery using AQPs inside TM tubular FO membranes (by Aquaporin A/S) (69). The author revealed that as the degree of concentration increases, the possibility of fouling also increases, thereby resulting in the requirement of frequent cleaning.

#### 4.2.6.3 Concentration of Soy Milk

Soy milk, one of the most popular plant-based dairy alternatives, is valued for its high protein content and rich supply of vitamin B, isoflavones, and minerals. However, processing soy milk often involves concentration steps that can lead to the loss of heat-sensitive compounds such as proteins and flavonoids. In this context, FO provides a viable solution by operating under mild conditions, thereby retaining essential amino acids, vitamins, and flavonoids, and helping to preserve the nutritional quality and functionality of soy milk for yoghurt, tofu, and protein-rich beverages.

#### 4.2.6.4 Concentration of Other Plant-Based Dairy Alternatives

In addition to almond, coconut, and soy milk, other plant-based products such as chickpea, walnut, rice milk, and oat milk are gaining popularity. These products often require concentration for commercial use in beverages, soups, and sauces.

Oat milk is a plant-based milk alternative made from whole oat grains and water. Oat milk is naturally free of lactose and rich in a variety of nutrients and phytochemicals (70). Chickpea milk is rich in plant-based protein, fibre, and micronutrients such as iron, magnesium, and folate. Walnut milk is gaining popularity due to its high omega-3 fatty acid content and antioxidants (such as polyphenols and ellagic acid) and its potential in supporting cardiovascular and cognitive health.

Minimal operating conditions of the FO process can potentially help in preserving the sensitive protein structure and ensure minimal oxidative degradation of bioactive compounds while minimizing the development of beany/grassy off-flavours often triggered by heat-based processing. Utilizing the FO process to partially dewater these milks not only improves shelf life but also provides a natural creamy texture to the milk without the need for thickeners/stabilizers.

While FO is well-suited for these applications, the presence of starch and fats in plant-based milk can affect the long-term performance of the FO membrane. Additionally, the appropriate selection of the draw solute is also crucial for determining the overall efficiency of the given process. Post-processing removal/recovery of draw solute must be efficient and non-disruptive to the final product. The flavour stability during concentration must be closely monitored, particularly for the lipid-rich products (like walnut milk), which are sensitive to oxidation even under low temperatures.

Overall, as the demand for high-protein, allergen-friendly, and clean-label plant-based beverages continues to rise, compared to conventional concentration techniques, FO stands out as a promising tool for sustainable processing, capable of delivering nutrient-dense, minimally processed products that meet modern consumer preferences. The continued innovation in membrane materials, hybrid FO systems, and DS recovery will further support the adoption of FO in processing emerging plant-based dairy alternatives.

## 4.3 ORGANIC VALUE-ADDED COMPOUNDS

Organic value-added compounds such as antioxidants, colourants, dietary fibres, proteins, peptides, and flavour can be recovered either directly from food products or their by-products. FO offers an innovative and sustainable method to recover these compounds while maintaining their functional properties. In this section, the viability of the FO process to recover organic compounds from food by-products like slaughterhouse wastewater (SWW), olive mill wastewater (OMWW), and dairy effluents is discussed, focusing on proteins, peptides, bio-phenols, pectin, and lactic acid.

### 4.3.1 Protein and Peptide Recovery from Slaughterhouse Wastewater

SWW is known to contain a complex composition, consisting of oil, fats, proteins, fibres, blood, nutrients, and pathogenic microorganisms (71). Conventional treatment (such as screening, dissolved air flotation, sedimentation, activated sludge, and anaerobic lagoons) of SWW often results in the waste of valuable proteins and peptides. Enriching and separating proteins still remain technically and economically challenging because most proteins are heat-sensitive and chemically unstable. However, FO provides an energy-efficient alternative that concentrates these proteins and peptides while ensuring that their functional and structural properties are preserved. One of the key advantages of FO is its ability to operate under mild conditions. This ensures that proteins do not undergo denaturation, which is often a risk in high-temperature processes. By preserving the bioactive properties of these proteins and peptides, they can be repurposed into protein-rich by-products suitable for animal feed, bioactive peptide extraction, and other industrial applications. This not only reduces waste but also transforms SWW into a sustainable source of valuable proteins and peptides, contributing to a circular economy (Table 4.3).

The FO process in SWW recovery not only enhances sustainability by reducing the amount of waste that enters water treatment systems but also provides an economically viable method to capture these valuable compounds. With a growing demand for alternative protein sources and bioactive peptides in food, cosmetics, and pharmaceuticals, the application of FO in SWW treatment shows considerable promise.

### 4.3.2 Recovery of Protein from Tuna Wastewater

In tuna processing industries, the cooking juice, a by-product of tuna canning, contains valuable proteins. FO technology has been employed to concentrate and recover these proteins from tuna cooking juice. Studies indicate that FO can increase protein concentration by up to 9%, with a permeate flux of 2.54 L $m^{-2}$ $h^{-1}$ (72). The use of FO minimizes protein degradation compared to heat-based methods, thus preserving the bioactivity of the recovered protein, which can be used in food or nutraceutical

**TABLE 4.3**
**Summary of studies related to the concentration of organic value-added compounds from food and food products using the forward osmosis process**

| Organic Value-Added Compounds | Source | Membrane Type | Draw Solution | Initial Flux, $Lm^{-2}h^{-1}$ | Concentration Factor | Ref. |
|---|---|---|---|---|---|---|
| Proteins | Tuna cooking juice | CTA-FO | 0.5–2.0 M NaCl | 5 | 1.6 | (74) |
| Pectin | Apple peel pomace | CTA-FO | 3.0 M NaCl | 7.8 | 2.1 | (75) |
| Polyphenols | Carob pulp | CTA-FO | K-lactate 60°Brix | 13 | 16.2 | (76) |
| Biophenol | Olive mill wastewater | CTA FO | 3.0 M $MgCl_2$ | 6.5 | > 2.0 | (77) |
| Polyphenols and melanoidins | Molasses distillery wastewater | TFC-FO | 4.0 M $MgCl_2$ | 4 | > 2.0 | (78) |
| | | CTA-FO and TFC-FO | 3.0 M $MgCl_2$ | 4–6 | ~4.0 | (79) |
| Xylose | | TFC-FO | 1.0–3.0 M $MgCl_2$/$LaCl_3$ | 25 | 6 | (80) |
| Sugar/ | Rice straw | TFC-FO | $MgSO_4$ (saturated) | - | 3.6 | (81) |
| Sugar | Pretreated rice straw | TFC-FO | 3.6 M TEA | 4 | 8.1 | (82) |
| Succinic acid | Fermentation broth | TFC-FO | 5.0 M NaCl | 16 | 4.5 | (83) |
| Lactic acid | Fermentation broth | TFC-FO | 5.0 M NaCl | 13 | 3.8 | (83) |
| Ethanol | Fermentation broth | TFC-FO | 5.0 M NaCl | 22 | 5.5 | (83) |

TFC: thin-film composite; CTA: cellulose triacetate

applications. Despite the challenge of membrane fouling due to the organic and inorganic content of the wastewater, FO offers a more sustainable solution with lower energy requirements compared to traditional methods such as evaporation or RO.

### 4.3.3 Pectin Recovery from Plant-Based Waste Streams

Pectin, a natural biodegradable polymer, is found in the cell walls of many plants and is widely used in the pharmaceutical, food, and cosmetic industries due to its gelling, thickening, and stabilizing properties (73). Pectin is especially important as an ingredient in products such as jams, jellies, and fruit preserves, as well as in drug delivery systems. FO presents a promising method for the recovery of pectin from waste streams, particularly from fruit and vegetable processing by-products. Unlike pressure-driven processes, FO operates under lower pressures, minimizing the risk of fouling and degradation of pectin's functional properties. The gentle nature of the FO process helps retain the sensory and nutritional properties of pectin, ensuring its high quality for use in food and pharmaceutical products. One of the most significant advantages of FO for pectin recovery is its low energy consumption and its capacity to concentrate pectin solutions without altering their molecular structure. This ensures that the recovered pectin maintains its efficacy as a gelling agent, thickener, or stabilizer, making FO a highly effective method for pectin recovery from plant-based waste streams.

FO represents a transformative technology for the recovery of organic value-added compounds from food by-products, offering a sustainable and efficient alternative to traditional methods. Whether it is recovering proteins and peptides from SWW, bio-phenols from OMWW, pectin from plant-based waste, or lactic acid from dairy wastewater, FO enables the preservation of these compounds' functional properties while contributing to environmental sustainability and economic viability.

By reducing waste and reclaiming valuable resources, FO supports the principles of the circular economy, where waste is minimized and by-products are repurposed into useful materials. As demand for sustainable food processing solutions continues to grow, FO is poised to play a critical role in transforming waste streams into sources of high-value compounds, furthering both environmental and economic goals in the food industry.

### 4.3.4 Recovery of Polyphenols from Food Processing Wastewater

#### 4.3.4.1 Phenolic Compound Recovery from Olive Mill Wastewater

The olive oil industry generates significant volumes of OMWW, a by-product known for its high organic load and toxicity due to the presence of phenolic compounds, organic acids, and other contaminants. OMWW poses a serious environmental challenge, but it also contains valuable bio-phenols, which are known for their antioxidant properties. FO technology has proven to be an effective method for managing OMWW. Studies show that utilizing a 0.02 $m^2$ CTA-FO membrane can achieve up to a 71% reduction in wastewater volume, with a remarkable 98% rejection rate of contaminants, including bio-phenols and ions (77). Taking into account the circular economy, the feasibility of the FO process for concentrating OMWW using fermentation brine from the tabletop as DS was demonstrated (84). The study revealed that with a volume reduction of 53%, the concentration of phenolic compounds was 77.43%. This makes FO a highly effective tool not just for wastewater volume reduction but also for the recovery of bio-phenols, which can be used in nutraceuticals, cosmetics, and food products due to their antioxidant activities. By using FO, the energy and fouling potential are minimized compared to other methods like RO or thermal evaporation. FO's minimal energy requirements and its ability to retain the bioactive properties of phenolic compounds make it a highly sustainable and efficient method for OMWW management. This process not only mitigates environmental hazards but also transforms OMWW from a waste product into a source of valuable antioxidant compounds.

#### 4.3.4.2 Recovery of Polyphenols from Wine Industries Wastewater

During wine processing, wastewaters are generated during both processing and fermentation. Wine industries' wastewater consists of diverse polyphenolic compounds, including anthocyanins, flavonoids, and tannins, which possess notable antioxidant and antimicrobial characteristics (73,85). These valuable polyphenols frequently remain unrecovered due to limitations in conventional wastewater treatment approaches. The FO process presents a suitable solution, proficiently concentrating these bioactive compounds while operating at ambient temperatures to ensure minimal degradation of their functional properties. With proper utilization of the FO process, wineries can effectively reclaim these polyphenols, enabling their reuse in dietary supplements, functional foods, natural preservatives, and cosmetics. Thus, implementation of FO processes within the wine industries not only supports sustainable wastewater management but also offers significant economic advantages through the valorization of previously under-utilized polyphenols.

#### 4.3.4.3 Recovery of Polyphenols from Brewery Wastewater

Brewery wastewater is another significant source of polyphenolic compounds, particularly flavonoids and antioxidants derived from barley and hops (86). These polyphenolic compounds possess notable antioxidant, antimicrobial, and health-promoting attributes, providing ample opportunities for commercial application (87). However, conventional wastewater treatment processes typically fail to recover these beneficial substances, leading to their loss and environmental concerns. FO offers an innovative and efficient solution to concentrate and recover these polyphenols from brewery wastewater. The gentle operating conditions of FO ensure minimal degradation of the antioxidant and bioactive properties of the recovered polyphenols. Utilizing FO technology in brewery wastewater management not only facilitates effective recovery and potential commercialization of polyphenols but also significantly reduces wastewater volumes and environmental pollution, supporting industry sustainability goals and resource optimization.

#### 4.3.4.4 Recovery of Polyphenols from Coffee and Tea Processing Wastewater

Wastewater from coffee and tea processing industries is characterized by high concentrations of polyphenols, caffeine, and other antioxidants. Conventional treatment methods typically fail to adequately recover these valuable compounds, leading to lost economic opportunities and environmental waste. FO technology provides a suitable method for the efficient concentration and extraction of polyphenols and caffeine, ensuring minimal exposure to heat and mechanical stress, thus preserving their antioxidant properties and structural integrity. Studies indicate that FO effectively concentrates these compounds, maintaining their functionality and commercial viability. The recovered polyphenols and caffeine can be reintegrated into various industries, including food additives, nutraceutical formulations, and pharmaceutical products, thereby enhancing economic sustainability and environmental stewardship within the coffee and tea industries (22).

#### 4.3.4.5 Recovery of Polyphenols from Citrus Processing Wastewater

The citrus processing industry generates substantial volumes of wastewater containing valuable polyphenols such as hesperidin and naringin, recognized for their strong antioxidant properties. Traditional extraction methods often involve high temperatures and harsh conditions that degrade these sensitive compounds, reducing their commercial value. FO technology provides a compelling alternative, operating under ambient conditions to concentrate citrus wastewater gently and effectively. This approach significantly preserves the biological activity and quality of recovered polyphenols, facilitating their incorporation into dietary supplements, functional foods, pharmaceuticals, and cosmetic products. By utilizing FO technology, citrus processors can transform wastewater into a sustainable source of high-quality antioxidants, enhancing both economic profitability and environmental sustainability [refer].

#### 4.3.4.6 Recovery of Polyphenols from Pulses and Soya Bean Wastewater

Wastewaters from pulse and soybean processing industries contain valuable polyphenolic compounds, such as isoflavones, which offer significant nutritional and health benefits. Traditional wastewater treatment methods typically do not recover these compounds efficiently, resulting in loss of beneficial nutrients and increased environmental impact. FO technology presents an innovative and sustainable method to concentrate and recover these bioactive polyphenols without compromising their antioxidant properties and nutritional value. Research has shown that FO effectively preserves the health-promoting activities of these compounds, allowing their reintegration into the food supply chain as functional ingredients or dietary supplements. Employing FO technology in pulses and soybean wastewater management supports a circular economy approach, optimizing resource utilization, enhancing nutritional value, and significantly reducing environmental burdens.

In summary, FO technology emerges as a transformative approach for the recovery of valuable polyphenols from diverse food processing wastewater streams. Offering significant advantages in terms of energy efficiency, sustainability, and preservation of bioactive properties, FO provides a robust solution that aligns with contemporary demands for sustainable resource management, environmental protection, and economic profitability across various sectors within the food processing industry.

### 4.3.5 Recovery of Flavonoids, Polyphenols, and Antioxidants from Fruit and Vegetable Waste

Fruit and vegetable processing industries generate substantial amounts of waste, consisting of certain valuable bioactive components such as flavonoids, polyphenols, and antioxidants. Polyphenols consist of molecules with one or more phenol groups, including phenolic acids, flavonoids, stilbenes, and lignans. Among them, phenolic acids (such as hydroxybenzoic and hydroxycinnamic) and flavonoids (oligomeric tannins) are the main phenolic components in the human diet. For instance, the polyphenols extracted from spinach (luteolin and quercetin), tea (caffeine and catechins), and orange (ferulic acid and vanillic acid) residues may be reused in the food and pharmaceutical industries. Conventionally, polyphenols are extracted from agri-food wastes using various techniques, including solvent extraction, ultrasound-assisted extraction, microwave-assisted extraction, and supercritical fluid extraction. Traditionally, extracted polyphenols are concentrated using a pressure-driven membrane process such as NF or RO. However, FO has emerged as a gentle and efficient method for the concentration and purification of bioactive substances from food waste (88,89).

The feasibility of concentrating OMWW using a single-step FO process was demonstrated with 3.7 M $MgCl_2$ as the DS. This approach achieved a 71% volume reduction, over 98% rejection of essential components (such as bio-phenols and ions), and less than 1% w/v of draw solute back-diffused into the FS (90). The FO process was used to concentrate phenolic compounds from olive oil washing wastewater, employing table olive fermentation brine as the DS. This effluent, generated during table olive processing, serves as a cost-effective and sustainable DS, eliminating the need for purchasing or regenerating synthetic alternatives (91).

An integrated UF–FO membrane process was employed for the purification and concentration of phenolic compounds from orange peel extract. The UF step served as a pretreatment, effectively removing up to 50% of pectin prior to FO concentration (92). Furthermore, the UF permeate was used as FS for the concentration of the FO process against 1 M NaCl as a DS. The author revealed that, using the FO process, the polyphenols were concentrated until 358±10 mg $L^{-1}$, achieving a concentration factor of 2.03 with SRSF ranging from 0.82 to 0.89 g $L^{-1}$.

Wastewater generated by corn starch processing plants exhibits high chemical oxygen demand and elevated concentrations of nitrogen and phosphorus. The feasibility of recovering both phosphorus and water from the effluent of secondary sedimentation tanks in corn starch processing wastewater was explored using a combined FO–hydroxyapatite (HAP) process. Operational parameters were optimized to enhance recovery efficiency (93). In the FO unit, the phosphorus concentration increased by 5.9-fold, with a water recovery rate of 80%. In the HAP-induced crystallization unit, phosphorus removal efficiency reached 98%. Reportedly, by employing FO technology, phosphorus was effectively concentrated, and the diluted DS (potassium and nitrogen fertilizer) was suitable for direct application in irrigation systems.

The FO process was applied for dewatering sugarcane molasses distillery wastewater while simultaneously concentrating colour-imparting constituents such as melanoidins, which are key contributors to both colour and antioxidant activity. This was achieved using AQP-based biomimetic membranes and $MgCl_2.6H_2O$ as the DS (94).

### 4.3.6 Collagen and Gelatine Recovery from Meat and Fish By-Products

The recovery of collagen and gelatine from meat and fish by-products gained significant attention due to their wide range of applications in food, pharmaceutical, and cosmetic industries. Collagen, a fibrous protein found in connective tissue of animals, is a primary source of gelatine. Traditionally, collagen and gelatine are extracted through processes that involve thermal or chemical treatments, which may cause degradation of sensitive bioactive components. FO, on the other hand, is a membrane-based separation process that operates under mild conditions, promising an alternative for the recovery of collagen and gelatine from meat and fish by-products, preserving the functional properties of these proteins while minimizing the environmental impact. Collagen and gelatine are highly sensitive to heat, and traditional extraction methods that involve high temperatures can lead to denaturation and loss of functional properties. The mild operating conditions can potentially help in maintaining the molecular integrity of collagen and gelatine, thereby ensuring that their bioactive properties are retained.

Additionally, conventional collagen extraction often involves the use of harsh chemicals (such as acid and alkali) to break down the convective tissues and release the proteins. These chemicals are not only harmful to the environment, but they also require extensive wastewater treatment, whereas FO, on the other hand, typically uses inorganic salt solutions or food-grade osmotic agents as DSs, thereby reducing the need for chemical inputs and making the process more environmentally friendly.

### 4.3.7 Recovery of Probiotics from Dairy Wastewater

Dairy processing industries use large volumes of water to produce various dairy products and generate wastewater. Dairy effluent consists of a large amount of nutrients (including protein, lipid, and carbohydrate), and recovery of these nutrients will not only provide economic benefits but also improve the efficiency of wastewater treatment. During the production of yoghurt, kefir, cheese, and other probiotic-rich dairy products, a substantial amount of fermented dairy wastewater is generated. This wastewater typically consists of a high concentration of probiotics, beneficial microorganisms widely recognized for their health-promoting properties, along with lactose, proteins, and organic acids. Conventional wastewater treatment processes often neglect the recovery of these valuable probiotics, resulting in their loss and leading to increased environmental load. FO has emerged as a promising technology for the efficient recovery and concentration of probiotics from fermented dairy wastewater, offering benefits in terms of preserving microbial viability, energy efficiency, and environmental sustainability.

A recent study discussed the implementation of the FO process to recover lactose from whey, a wastewater stream generated in dairy processing industries (95). Notably, using the FO process, lactose is concentrated from 3.95% to 49.5%, and the FO process exhibited remarkable efficiency, achieving a specific energy consumption of less than 8 kWh per tonne of water removal. The study revealed that this process, in turn, benefits the entire economics of the dairy industry by reducing the quantity of wastewater generated and to be treated in the effluent treatment plant (ETP), affecting the water and energy consumption of the dairy industry, and indirectly providing various environmental benefits.

An integrated UF–FO membrane process has been proposed for the recovery of proteins and carbohydrates from dairy wastewater (96). Reportedly, the UF membrane shows almost complete protein recovery (~98%), while the FO process resulted in 65% carbohydrate retention. The recovery of protein and carbohydrate from wastewater could be potentially used for various processes, including animal fodder.

### 4.3.8 Recovery of Natural Food Colours from Plant Processing Waste

Natural food pigments (including anthocyanin, betalains, carotenoids, and chlorophylls) extracted from plant sources have significant commercial value in the food and beverage industry due to increasing consumer preference for natural, healthy, and sustainable products (97). Food processing industries generate a substantial waste stream rich in natural pigments, presenting unique opportunities for recovery and valorization.

The FO process employs a mild operational condition that helps preserve the chemical structure and functional integrity of natural pigments, such as anthocyanin from *Hibiscus rosa-sinensis* flower (11) and betalains from beetroot (*Beta vulgaris*) (15). The osmotic gradient-driven membrane process allows concentration of these pigments without significant thermal degradation or chemical alternatives, maintaining their vivid colours and bioactive properties. Reportedly, the recovered natural colourants exhibited excellent stability, purity, and bioactivity, suitable for immediate incorporation into natural food formulations, supplements, and nutraceutical products.

The adoption of FO technology for recovering natural food colours from plant processing waste aligns with global sustainability goals, promoting resource efficiency and waste minimization while delivering high-quality, market-ready natural colourants. This method not only enhances economic value through the creation of new revenue streams but also supports industry sustainability and environmental compliance.

In summary, FO technology represents an innovative, sustainable, and economically attractive method for recovering natural food colours from plant processing waste, highlighting its potential to contribute significantly to the sustainable development of the food processing industry.

## 4.4 FO-BASED CRYSTALLIZATION IN FOOD APPLICATIONS

### 4.4.1 Principles of FO-Assisted Crystallization

The FO process is an osmotically driven membrane process that has gained significant attention due to its energy efficiency, gentle operation, and potential for sustainable water recovery and resource extraction. In FO-assisted crystallization, the unique properties of FO membranes are leveraged to concentrate an FS, leading to supersaturation and crystallization of dissolved solutes. The two primary steps in FO-assisted crystallization include water extraction from the FS and the subsequent crystallization of solutes once supersaturation is achieved.

The FO process relies on a semipermeable membrane that allows water to pass from a low-concentration FS to a higher-concentration DS, driven by the osmotic pressure gradient between the two solutions. Unlike RO, FO does not require high-pressure pumps, making it a low-energy process suitable for a wide range of applications. In the FO-assisted crystallization process, the membrane provides distinct heterogeneous nucleation sites, which diminish the risks of membrane fouling even though the crystal growth occurs on the membrane surface. It is a combination of mass transfer operation coupled with the initiation of nucleation. The formation of nuclei of critical size depends on several factors, such as surface roughness, contact angle, membrane porosity, and zeta potential. It is important to understand that the solution bearing the target species for crystallization is characterized by random collision and motion of the solutes in different directions. They have a 50% probability of being dissolved and a 50% probability of assimilation leading to nucleation (98).

Crystallization occurs when a solute in the FS reaches a concentration beyond its solubility limit (i.e. supersaturation). This leads to the formation of nuclei, which then grow into larger crystals. The formation of crystals can occur in two ways, that is, by direct crystallization (the solute crystallizes directly in the FS after it has been concentrated by FO) and by indirect crystallization (the concentrated FS is then transferred to a separate crystallization unit, where the solutes are further processed to achieve crystal formation).

The application of FO-assisted crystallization for the recovery of rifaximin crystals from pharmaceutical effluents was reported (99). The study demonstrated the efficient recovery of antibiotic (rifaximin) crystals by combining membrane separation processes, including UF, vacuum distillation, FO, and MD (Figure 4.2).

FO-based crystallization is gaining interest in the food industry, particularly in salt and mineral recovery, sugar recovery, and the extraction of organic acids, as it offers several benefits, including lower energy consumption, minimal heat exposure, and the ability to handle heat-sensitive products without degradation.

### 4.4.2 FO Crystallization for Salt and Mineral Recovery

One of the most promising applications of FO-based crystallization is the recovery of salts and minerals from wastewater, brine, and food processing streams. Salts, including NaCl, $CaSO_4$, and $MgSO_4$, are often present in food processing wastewater, industrial brines, and other effluents. The recovery of these salts through crystallization processes can contribute to zero-liquid discharge (ZLD) systems and resource recovery, as well as minimize environmental pollution.

In this process, FO membranes extract water from the FS (which could be wastewater or brine) by drawing it through a high-concentration DS. As water is removed, the concentration of salts in the FS increases, and upon reaching supersaturation, the

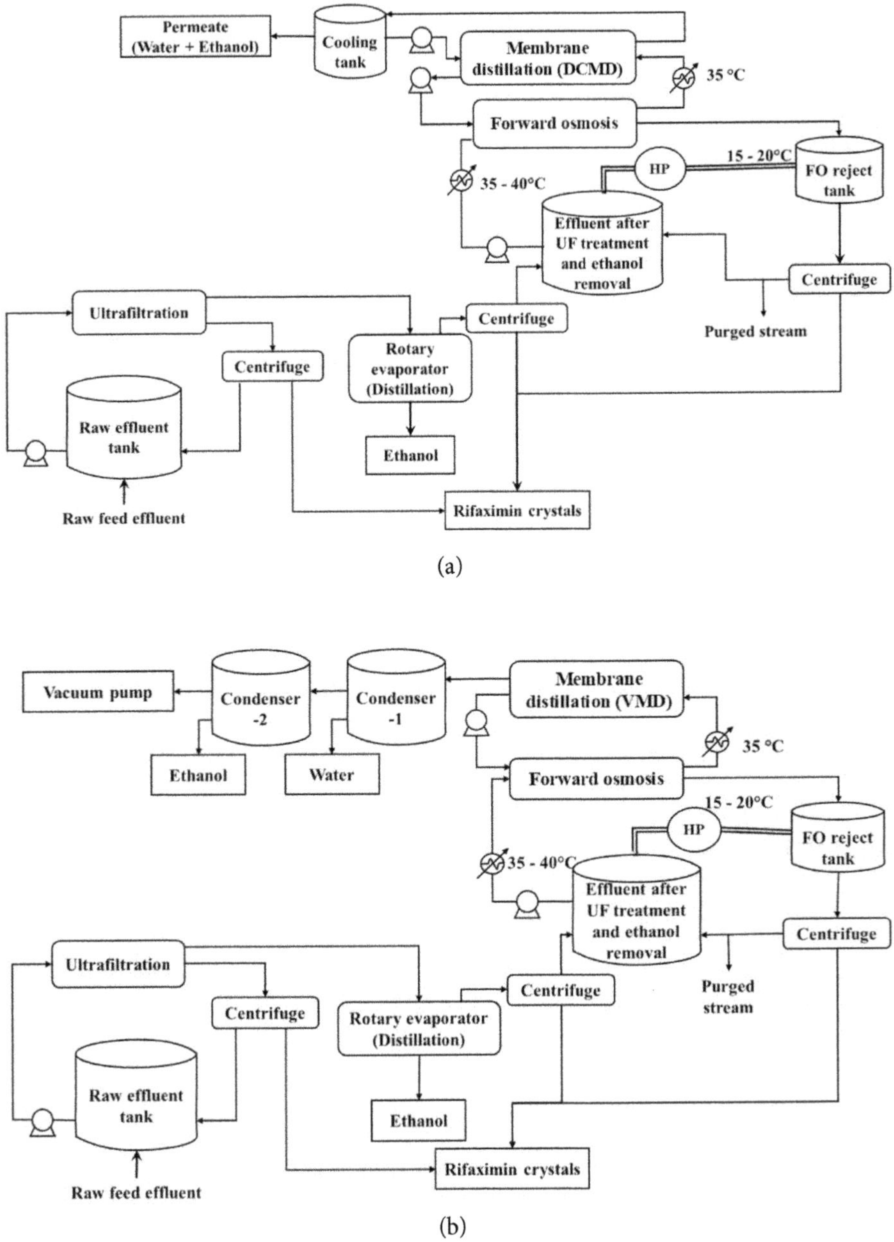


**FIGURE 4.2** Recovery of rifaximin crystals from pharmaceutical effluents using (a) FO–direct contact membrane distillation (DCMD) and (b) FO–vacuum membrane distillation (VMD) process

salts begin to crystallize. These crystals can then be separated and purified, making the process suitable for recycling salts or extracting valuable minerals.

For instance, NaCl (table salt) can be recovered from seawater brine using FO membranes, with the final product being crystalline NaCl that can be used in food processing or other industrial applications. Similarly, magnesium sulphate (Epsom salt) and gypsum can be recovered from brines and used in applications ranging from food production to pharmaceuticals.

In the production of fermented foods like pickles and soy sauce, FO can be used to recover and recycle salts used in the fermentation process, reducing waste and improving the sustainability of the process. FO can also be used to concentrate and recover mineral-rich water, which can then be used as a mineral supplement in food or beverage formulation.

### 4.4.3 FO Crystallization for Sugar Recovery

Sugar recovery is another area where FO-assisted crystallization has shown promise, particularly in juice concentration and the production of sugar crystals from sucrose. Traditional methods of sugar extraction and purification often involve high-temperature and energy-intensive processes, which can degrade sensitive compounds in the sugar or result in undesirable flavours.

In this process, the FO membrane concentrates sugar solutions such as fruit juices or sugar syrups, removing water and preserving the delicate flavours and nutrients. As concentration increases, the sugar solution becomes supersaturated, leading to the formation of sugar crystals. These can be separated and purified, resulting in high-purity crystalline sugar that can be used in food manufacturing. FO can help recover high-purity sugar crystals from juice concentrates or syrups, suitable for use in various food products (100).

### 4.4.4 FO Crystallization for Organic Acids

Organic acids such as citric acid, lactic acid, and acetic acid are key ingredients in a wide range of food and beverage products. The recovery of organic acids from fermentation broth and food waste streams is an important application of FO-based crystallization. The gentle process helps avoid the degradation of bioactive compounds while enabling the efficient extraction of valuable organic acids.

The FO process can be efficiently utilized to concentrate fermentation broths consisting of organic acids, making it possible to recover these acids in their crystalline form. By removing water from the FS using FO, the concentration of organic acid increases, and once the supersaturation is reached, crystallization occurs. The crystallized organic acids can then be separated and purified further based on application. For instance, demonstrated recovery of bio-based succinic acid (with 90.52% purity) using the FO-assisted crystallization technique (101). Lactic acid is an essential component in dairy fermentation and is widely used as a preservative and flavour enhancer. FO-based crystallization can recover lactic acid from fermentation broths, facilitating sustainable production processes.

### 4.4.5 FO Crystallization for Cold-Water-Soluble Instant Tea

On a commercial scale, instant tea is manufactured by concentrating tea extract using vacuum-drying, freeze-drying, or spray-drying techniques. During the manufacturing of cold-water-soluble instant tea, volatile chemicals with desirable aroma notes can escape, thus affecting the aroma profile of instant tea and instant tea-based products, thereby researching on developing novel technologies for preparing tea concentrates that are not detrimental to heat-sensitive biological compounds. The study revealed that instant tea is made using the integrating FO-based crystallization technique (Figure 4.3). In this study, the freshly brewed tea extract was concentrated to 93.67% using a mixed-component inorganic DS. The study suggests that the recovered tea crystals' cold and hot water solubilities were observed to be 88.11% and 94.75%, respectively, thereby suggesting that the final product can be either directly used as a ready-to-drink (RTD) tea product or further purified to isolate essential tea components.

FO-based crystallization is a sustainable and energy-efficient process that holds significant potential in the food industry. It allows for efficient recovery of salts, sugars, and organic acids while preserving the nutritional and sensory qualities of the FS. While challenges like fouling and supersaturation control remain, ongoing research and development are addressing these issues, making FO-based crystallization a promising solution for food processing applications. With continued advancements in membrane technology, DS recovery, and process optimization, FO-based crystallization will likely become an integral part of sustainable food production in the future.

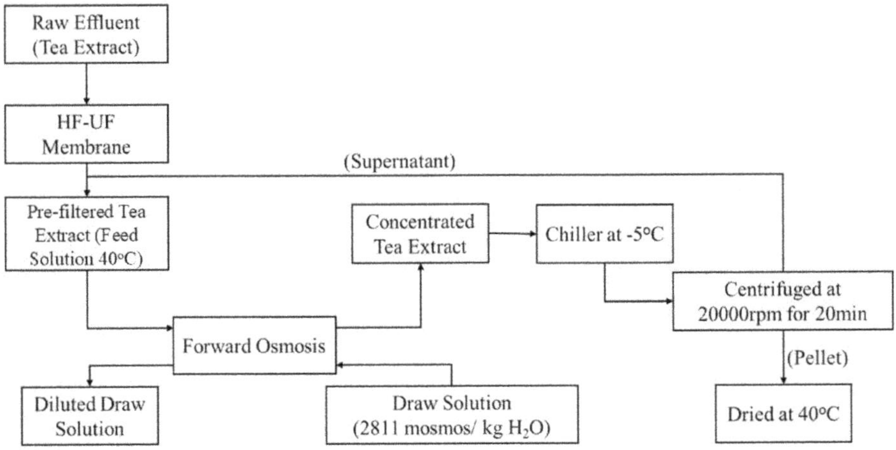


**FIGURE 4.3** Process flow diagram of the FO-based crystallization process for the preparation of instant tea

## 4.5 FO APPLICATION ADOPTED IN THE INDUSTRY

Although still an emerging technology, FO has already transitioned from the laboratory to several industrial applications. These early deployments—though limited in scale—demonstrate FO's practical viability and highlight its potential for broader adoption:

### 4.5.1 Desalination and Brine Concentration

FO has been piloted both as a stand-alone process and within hybrid desalination systems. For instance, Modern Water has operated full-scale plants where FO extracts water from seawater or brine, followed by regeneration of the DS. One installation in Oman successfully produced drinking water using an FO–RO hybrid, showing FO's ability to treat high-salinity feeds with reduced fouling (102). Similarly, Oasys Water has deployed FO for industrial wastewater brine concentration (e.g. oil and gas produced water), achieving high water recovery from extremely saline streams that would overwhelm conventional RO. In these cases, FO reduces liquid waste volume and recovers water with lower energy demand than thermal evaporation.

### 4.5.2 Food and Beverage Industry Trials

Pilot and semi-commercial trials have begun in the food sector. In brewing, FO has been tested for wort concentration using sugar-based DSs compatible with beer production. This approach avoids heat, preserves flavour compounds, and has shown up to 30% water savings while reducing energy consumption compared to traditional boiling. In the dairy industry, FO has been piloted for concentrating milk and whey. These trials suggest FO can retain protein quality better than UF or evaporation by operating at lower temperatures, though large-scale adoption is still pending.

### 4.5.3 Niche and Humanitarian Uses

FO has also been commercialized in specialized markets, such as emergency water purification. Personal hydration devices—often marketed for military or disaster relief—use FO membranes to draw clean water into a sugar or electrolyte solution, producing safe drinking fluid without external power. These products demonstrate FO's simplicity and reliability under field conditions.

### 4.5.4 Hybrid Systems in Development

FO is increasingly being tested as part of integrated treatment trains. Examples include using FO as a pretreatment for high-fouling feeds such as fruit juice or biomass hydrolysates, followed by RO or evaporation. These FO–RO hybrids aim to improve throughput and reduce fouling, though full commercialization remains in progress.

These early implementations confirm that FO can operate continuously and economically at an industrial scale. They also provide valuable lessons in module design, membrane durability, and cleaning strategies. Importantly, FO has already proven effective in niches that leverage its unique strengths—treating high-salinity solutions and preserving heat-sensitive products.

While challenges remain—notably membrane cost, draw solute recovery, and system integration—the trajectory of FO commercialization suggests that food and dairy processing may be among the first industries to achieve widespread adoption. Lessons learnt from these pioneers will guide future developments and support broader acceptance of FO in mainstream applications, as explored further in Chapters 6 and 7.

## REFERENCES


1. Park HW, Baek J, Kim WJ. Forward osmosis and direct contact membrane distillation: Emerging membrane technologies in food and beverage processing. Innov Food Sci Emerg Technol. 2024 May;93:103626.
2. Blandin G, Ferrari F, Lesage G, Le-Clech P, Héran M, Martinez-Lladó X. Forward osmosis as concentration process: Review of opportunities and challenges. Membranes (Basel). 2020;10(10):1–40.
3. Pei J, Gao S, Sarp S, Wang H, Chen X, Yu J, et al. Emerging forward osmosis and membrane distillation for liquid food concentration: A review. Compr Rev Food Sci Food Saf [Internet]. 2021 Mar 1;20(2):1910–36. Available from: https://doi.org/10.1111/1541-4337.12691
4. Beldie AA, Dumpler J, Moraru CI. Draw solution selection and process parameters impact the performance of forward osmosis for nonthermal concentration of liquids. J Food Eng. 2025 Nov;400:112655.
5. Rahbari-Sisakht M, Ismail AF. A comprehensive review of pressure and osmosis driven membrane processes: Processes, characteristics and materials. Desalination. 2025 Apr;598:118427.
6. Zhan M, Kim Y, Hong S. Comprehensive review of osmotic dilution/concentration using FO membranes for practical applications. Desalination. 2021 Nov;515:115190.
7. Derwish G, O. Sharif A, Al-Aibi S, A. Al-Najar J. Comparison the kinetics of The uptake of sugars from the draw solution. Eng Technol J. 2016 Mar 1;34(3):441–51.
8. Gebreyohannes AY, Curcio E, Poerio T, Mazzei R, Di Profio G, Drioli E, et al. Treatment of olive mill wastewater by forward osmosis. Sep Purif Technol. 2015 Jun;147:292–302.
9. Cifuentes-Cabezas M, Pavani A, Vincent-Vela MC, Mendoza-Roca JA, Álvarez-Blanco S. Concentration of phenolic compounds from olive washing wastewater by forward osmosis using table olive fermentation brine as draw solution. Environ Technol Innov. 2023 May;30:103054.
10. Nayak CA, Rastogi NK. Forward osmosis for the concentration of anthocyanin from Garcinia indica Choisy. Sep Purif Technol. 2010;71:144–51.
11. Bal A, Shilpa HN, Debnath S, Rastogi NK. Extraction of anthocyanin from Hibiscus rosa-sinensis and concentration by forward osmosis membrane process. Innov Food Sci Emerg Technol. 2024 Aug;96:103782.

12. Tavares HM, Tessaro IC, Cardozo NSM. Concentration of grape juice: Combined forward osmosis/evaporation versus conventional evaporation. Innov Food Sci Emerg Technol. 2022 Jan 1;75:102905.
13. Chu H, Zhang Z, Zhong H, Yang K, Sun P, Liao X, et al. Athermal concentration of blueberry juice by forward osmosis: Food additives as draw solution. Membranes (Basel). 2022 Aug 21;12(8):808.
14. Trishitman D, Negi PS, Rastogi NK. Concentration of pomegranate juice by forward osmosis or thermal evaporation and its shelf-life kinetic studies. Food Chem [Internet]. 2023 Jan 15 [cited 2022 Sep 7];399:133972. Available from: https://linkinghub.elsevier.com/retrieve/pii/S0308814622019343
15. Trishitman D, Negi PS, Rastogi NK. Concentration of beetroot juice colorant (betalains) by forward osmosis and its comparison with thermal processing. LWT. 2021 Jun;145:111522.
16. Li Z, Xiao S, Xiong Q, Wu C, Huang J, Zhou R, et al. Assessment of highly concentrated pear juice production through single-run forward osmosis using sodium lactate as the draw solute. J Food Eng. 2022 Nov 1;333:111122.
17. Zhang K, An X, Bai Y, Shen C, Jiang Y, Hu Y. Exploration of food preservatives as draw solutes in the forward osmosis process for juice concentration. J Memb Sci. 2021 Oct 1;635:119495.
18. Wang H, Zhang Y, Ren S, Pei J, Li Z. Athermal concentration of apple juice by forward osmosis: Process performance and membrane fouling propensity. Chem Eng Res Des. 2022 Jan;177:569–77.
19. Pei J, Pei S, Wang W, Li S, Youravong W, Li Z. Athermal forward osmosis process for the concentration of liquid egg white: Process performance and improved physicochemical property of protein. Food Chem [Internet]. 2020;312:126032. Available from: https://www.sciencedirect.com/science/article/pii/S0308814619321776
20. Catarino M, Mendes A, Madeira L, Ferreira A. Beer dealcoholization by reverse osmosis. Desalination. 2006;200(1–3):397–9.
21. Huang J, Ren Y, Wang X, Li H, Wang Y, Zhang J, et al. Dealcoholization of kiwi wine by forward osmosis: Evaluation of membrane fouling propensity and product quality. Chem Eng Res Des. 2022 Feb 1;178:189–98.
22. Bardhan A, Senthilmurugan S, Mohanty K. Optimisation of multi-component inorganic salt composition as draw solute for preparation of concentrated tea extract using forward osmosis process. Food and Bioproducts Processing [Internet]. 2023;138: 126–38. Available from: https://doi.org/10.1016/j.fbp.2023.01.007
23. Alonso-Vázquez P, Isola A, Sánchez-Arévalo CM, Cuartas-Uribe B, Vincent-Vela MC, Álvarez-Blanco S. Concentration of phenolic compounds from an orange peel waste extract using a combination of ultrafiltration and forward osmosis. Sep Purif Technol. 2025 Jul;360:131228.
24. Punzalan MEH, Marcelo PA, Padilla-Zakour OI. Nonthermal preservation of coconut water by forward osmosis concentration and high pressure processing. LWT. 2025 May;223:117693.
25. Bhattacharjee C, Saxena VK, Dutta S. Fruit juice processing using membrane technology: A review. Innov Food Sci Emerg Technol. 2017 Oct;43:136–53.
26. Wrolstad RE, McDaniel MR, Durst RW, Micheals N, Lampi KA, Beaudry EG. Composition and sensory characterization of red raspberry juice concentrated by direct-osmosis or evaporation. J Food Sci. 1993 May 26;58(3):633–7.
27. Milczarek RR, Sedej I. Aroma profiling of forward-osmosis watermelon juice concentrate and comparison to fresh fruit and thermal concentrate. LWT. 2021 Nov;151:112147.
28. Milczarek RR, Olsen CW, Sedej I. Quality of watermelon juice concentrated by forward osmosis and conventional processes. Processes. 2020 Nov 28;8(12):1568.

29. Nayak CA, Valluri SS, Rastogi NK. Effect of high or low molecular weight of components of feed on transmembrane flux during forward osmosis. J Food Eng. 2011 Sep;106(1):48–52.
30. Kim DI, Gwak G, Zhan M, Hong S. Sustainable dewatering of grapefruit juice through forward osmosis: Improving membrane performance, fouling control, and product quality. J Memb Sci. 2019 May;578:53–60.
31. An X, Hu Y, Wang N, Zhou Z, Liu Z. Continuous juice concentration by integrating forward osmosis with membrane distillation using potassium sorbate preservative as a draw solute. J Memb Sci [Internet]. 2019;573(December 2018):192–9. Available from: https://doi.org/10.1016/j.memsci.2018.12.010
32. Chu H, Zhang Z, Zhong H, Yang K, Sun P, Liao X, et al. Athermal concentration of blueberry juice by forward osmosis: Food additives as draw solution. Membranes (Basel). 2022 Aug 21;12(8):808.
33. Sakai M, Seshimo M, Matsukata M. Hydrophilic ZSM-5 membrane for forward osmosis operation. J Water Process Eng. 2019 Dec;32:100864.
34. Sakai M, Nomura Y, Matsukata M. Strategy of development on zeolite membrane for forward osmosis operation: Effects of structural parameter of support layer and Si/Al ratio of ZSM-5 layer on water flux. J Chem Eng Japan. 2023 Dec 31;56(1).
35. Negishi E, Sakai M, Matsukata M. Fouling-resistant Na-ZSM-5 zeolite membrane for fruit juice concentration in forward osmosis process. Appl Food Res. 2024 Dec;4(2):100487.
36. Petrotos KB, Quantick P, Petropakis H. A study of the direct osmotic concentration of tomato juice in tubular membrane—Module configuration. I. The effect of certain basic process parameters on the process performance. J Memb Sci. 1998;150(1):99–110.
37. Roda-Serrat MC, Schytt-Nielsen JF, Braekevelt S, Dzhanzefova T, Joernsgaard B, Norddahl B, et al. Processing of black carrot juice by nanofiltration and forward osmosis. Chem Eng Trans. 2021;87:547–52.
38. Bardhan A, Subbiah S, Mohanty K. Modelling and experimental validation of osmotic driven energy efficient process for tea solution concentration. Environ Technol Innov [Internet]. 2020;20:101065. Available from: https://doi.org/10.1016/j.eti.2020.101065
39. Ambrosi A, Motke MB, Souza-Silva ÉA, Zini CA, McCutcheon JR, Cardozo NSM, et al. Beer dealcoholization by forward osmosis diafiltration. Innov Food Sci Emerg Technol. 2020 Jul;63:102371.
40. Rana A, Samtiya M, Dhewa T, Mishra V, Aluko RE. Health benefits of polyphenols: A concise review. J Food Biochem. 2022 Oct 13;46(10).
41. Pei J, Wang W, Wang Y, Wang H, Bucs SS, Vrouwenvelder JS, et al. Fate of polyphenols in forward osmosis. J Memb Sci. 2021 Mar;621:118993.
42. Mondal D, Nataraj SK, Rami Reddy AV, Ghara KK, Maiti P, Upadhyay SC, et al. Four-fold concentration of sucrose in sugarcane juice through energy efficient forward osmosis using sea bittern as draw solution. RSC Adv [Internet]. 2015;5(23):17872–8. Available from: http://dx.doi.org/10.1039/C5RA00617A
43. Shalini HN, Nayak CA. Forward osmosis membrane concentration of raw sugarcane juice. In: Regupathi I, Shetty KV, Thanabalan M, editors. Recent advances in chemical engineering. Singapore: Springer; 2016. p. 81–8.
44. Rackemann D, Lam R, Shi C, Marasinghege C, Millar G, Alexander D. Forward osmosis for the concentration of sugarcane juice. Sugar Industry. 2023 May 5;309–14.
45. Christiansen MV, Skibsted LH, Ahrné L. Control of viscosity by addition of calcium chloride and glucono-δ-lactone to heat treated skim milk concentrates produced by reverse osmosis filtration. Int Dairy J. 2021 Mar;114:104916.

46. Lin Y, Kelly AL, O'Mahony JA, Guinee TP. Effect of heat treatment, evaporation and spray drying during skim milk powder manufacture on the compositional and processing characteristics of reconstituted skim milk and concentrate. Int Dairy J. 2018 Mar;78:53–64.
47. Moejes SN, van Wonderen GJ, Bitter JH, van Boxtel AJB. Assessment of air gap membrane distillation for milk concentration. J Memb Sci. 2020 Jan;594:117403.
48. Beldie AA, Moraru CI. Forward osmosis concentration of milk: Product quality and processing considerations. J Dairy Sci. 2021 Jul;104(7):7522–33.
49. Lin Y, Kelly AL, O'Mahony JA, Guinee TP. Effect of heat treatment, evaporation and spray drying during skim milk powder manufacture on the compositional and processing characteristics of reconstituted skim milk and concentrate. Int Dairy J. 2018 Mar;78:53–64.
50. Moejes SN, van Wonderen GJ, Bitter JH, van Boxtel AJB. Assessment of air gap membrane distillation for milk concentration. J Memb Sci. 2020 Jan;594:117403.
51. Chen GQ, Artemi A, Lee J, Gras SL, Kentish SE. A pilot scale study on the concentration of milk and whey by forward osmosis. Sep Purif Technol [Internet]. 2019;215(October 2018):652–9. Available from: https://doi.org/10.1016/j.seppur.2019.01.050
52. Blais HN, Schroën K, Tobin J. Concentration of skim milk by forward osmosis using delactosed permeate as an innovative draw solution. Int Dairy J. 2023 Feb;137:105510.
53. Cao X, Zhang P, Xian Y, Zhang Y, Muratkhan M, Youravong W, et al. Nonthermal concentration of skimmed goat milk by forward osmosis: The insights into process performance and product quality. Chem Eng Res Des. 2024 Apr;204:450–60.
54. Zhou M, Cao X, Yang L, Wang W, Wang H, Li Z. A green approach for non-thermal concentration of skim milk by forward osmosis combined with membrane distillation for draw solution regeneration. Chem Eng Res Des. 2024 Oct;210:469–80.
55. Taivosalo A, Kriščiunaite T, Stulova I, Part N, Rosend J, Sõrmus A, et al. Ripening of hard cheese produced from milk concentrated by reverse osmosis. Foods. 2019 May 15;8(5):165.
56. Smithers GW. Whey and whey proteins—From 'gutter-to-gold'. Int Dairy J. 2008 Jul;18(7):695–704.
57. Nazir A, Khan K, Maan A, Zia R, Giorno L, Schroën K. Membrane separation technology for the recovery of nutraceuticals from food industrial streams. Trends Food Sci Technol. 2019 Apr;86:426–38.
58. Beldie AA, Moraru CI. Forward osmosis concentration of milk: Product quality and processing considerations. J Dairy Sci. 2021 Jul;104(7):7522–33.
59. Oymaci P, Offeringa PE, Borneman Z, Nijmeijer K. Effect of osmotic pressure on whey protein concentration in forward osmosis. Membranes (Basel). 2021 Jul 29; 11(8):573.
60. Wang YN, Wang R, Li W, Tang CY. Whey recovery using forward osmosis—Evaluating the factors limiting the flux performance. J Memb Sci. 2017 Jul;533:179–89.
61. Aydiner C, Topcu S, Tortop C, Kuvvet F, Ekinci D, Dizge N, et al. A novel implementation of water recovery from whey: "Forward–reverse osmosis" integrated membrane system. Desalination Water Treat. 2013 Jan;51(4–6):786–99.
62. Menchik P, Moraru CI. Nonthermal concentration of liquid foods by a combination of reverse osmosis and forward osmosis. Acid whey: A case study. J Food Eng. 2019 Jul;253:40–8.
63. Aydiner C, Topcu S, Tortop C, Kuvvet F, Ekinci D, Dizge N, et al. A novel implementation of water recovery from whey: "Forward–reverse osmosis" integrated membrane system. Desalination Water Treat. 2013 Jan;51(4–6):786–99.

64. Aydiner C, Sen U, Topcu S, Ekinci D, Altinay AD, Koseoglu-Imer DY, et al. Techno-economic viability of innovative membrane systems in water and mass recovery from dairy wastewater. J Memb Sci. 2014 May;458:66–75.
65. Seker M, Buyuksari E, Topcu S, Babaoglu DS, Celebi D, Keskinler B, et al. Effect of pre-treatment and membrane orientation on fluxes for concentration of whey with high foulants by using NH3/CO2 in forward osmosis. Bioresour Technol. 2017 Nov;243:237–46.
66. Wang YN, Wang R, Li W, Tang CY. Whey recovery using forward osmosis—Evaluating the factors limiting the flux performance. J Memb Sci. 2017 Jul;533:179–89.
67. Oymaci P, Offeringa PE, Borneman Z, Nijmeijer K. Effect of osmotic pressure on whey protein concentration in forward osmosis. Membranes (Basel). 2021 Jul 29;11(8):573.
68. Artemi A, Chen GQ, Kentish SE, Lee J. Pilot scale concentration of cheese whey by forward osmosis: A short-cut method for evaluating the effective pressure driving force. Sep Purif Technol. 2020 Nov 1;250:117263.
69. Pedersen PA, Bjørkskov FB, Alvisse S, Hélix-Nielsen C. From channel proteins to industrial biomimetic membrane technology. Faraday Discuss. 2018;209:287–301.
70. Yu Y, Li X, Zhang J, Li X, Wang J, Sun B. Oat milk analogue versus traditional milk: Comprehensive evaluation of scientific evidence for processing techniques and health effects. Food Chem X. 2023 Oct;19:100859.
71. Bustillo-Lecompte CF, Mehrvar M. Slaughterhouse wastewater characteristics, treatment, and management in the meat processing industry: A review on trends and advances. J Environ Manage. 2015 Sep;161:287–302.
72. Rastogi NK. Applications of forward osmosis process in food processing and future implications. In: Current trends and future developments on (Bio-) membranes [Internet]. Elsevier Inc.; 2020. p. 113–38. Available from: http://dx.doi.org/10.1016/B978-0-12-816777-9.00005-8
73. Sharma S, Wani KM, Mujahid SM, Jayan LS, Rajan SS. Review on pectin: Sources, properties, health benefits and its applications in food industry. J Future Foods. 2026 Mar;6(2):205–19.
74. Khongnakorn W, Youravong W. Concentration and recovery of protein from tuna cooking juice by forward osmosis. J Eng Sci Technol. 2016;11(7):962–73.
75. Andrzejewski A, Krajewska M, Zheng L, Nghiem LD, Oleskowicz-Popiel P, Prochaska K, et al. Pectin recovery from apple pomace by forward osmosis—Assisted technology. J Memb Sci. 2024 Jul;706:122956.
76. Wang J, Martínez-Hernández A, de Lamo-Castellví S, Romero MP, Kaade W, Ferrando M, et al. Low-energy membrane-based processes to concentrate and encapsulate polyphenols from carob pulp. J Food Eng. 2020 Sep;281:109996.
77. Gebreyohannes AY, Curcio E, Poerio T, Mazzei R, Di Profio G, Drioli E, et al. Treatment of olive mill wastewater by forward osmosis. Sep Purif Technol. 2015 Jun;147:292–302.
78. Singh N, Petrinic I, Hélix-Nielsen C, Basu S, Balakrishnan M. Concentrating molasses distillery wastewater using biomimetic forward osmosis (FO) membranes. Water Res. 2018 Mar;130:271–80.
79. Singh N, Petrinic I, Hélix-Nielsen C, Basu S, Balakrishnan M. Influence of forward osmosis (FO) membrane properties on dewatering of molasses distillery wastewater. J Water Process Eng. 2019 Dec;32:100921.
80. Kim DI, Choi J, Hong S. Evaluation on suitability of osmotic dewatering through forward osmosis (FO) for xylose concentration. Sep Purif Technol. 2018 Jan;191:225–32.
81. Zhang Y, Nakagawa K, Shibuya M, Sasaki K, Takahashi T, Shintani T, et al. Improved permselectivity of forward osmosis membranes for efficient concentration of pretreated rice straw and bioethanol production. J Memb Sci. 2018 Nov;566:15–24.
82. Shibuya M, Yasukawa M, Sasaki K, Tanaka Y, Takahashi T, Kondo A, et al. Up-concentration of sugars in pretreated-rice straw by an osmotic pressure-driven method. Biochem Eng J. 2017 May;121:13–6.

83. Garcia-Aguirre J, Alvarado-Morales M, Fotidis IA, Angelidaki I. Up-concentration of succinic acid, lactic acid, and ethanol fermentations broths by forward osmosis. Biochem Eng J. 2020 Mar;155:107482.
84. Cifuentes-Cabezas M, Pavani A, Vincent-Vela MC, Mendoza-Roca JA, Álvarez-Blanco S. Concentration of phenolic compounds from olive washing wastewater by forward osmosis using table olive fermentation brine as draw solution. Environ Technol Innov. 2023 May;30:103054.
85. Latessa SH, Hanley L, Tao W. Characteristics and practical treatment technologies of winery wastewater: A review for wastewater management at small wineries. J Environ Manage. 2023 Sep;342:118343.
86. Barbosa-Pereira L, Bilbao A, Vilches P, Angulo I, LLuis J, Fité B, et al. Brewery waste as a potential source of phenolic compounds: Optimisation of the extraction process and evaluation of antioxidant and antimicrobial activities. Food Chem. 2014 Feb; 145:191–7.
87. Tatullo M, Simone GM, Tarullo F, Irlandese G, De Vito D, Marrelli M, et al. Antioxidant and antitumor activity of a bioactive polyphenolic fraction isolated from the brewing process. Sci Rep. 2016 Oct 27;6(1):36042.
88. Alonso-Vázquez P, Valle C, Sánchez-Arévalo C, Cuartas-Uribe BE, Vincent-Vela MC, Bes-Piá A, et al. Separation of phenolic compounds from canned mandarin production wastewater by ultrafiltration and nanofiltration. J Water Process Eng. 2024 Mar;59:105041.
89. Montenegro-Landívar MF, Tapia-Quirós P, Vecino X, Reig M, Granados M, Farran A, et al. Recovery of natural polyphenols from spinach and orange by-products by pressure-driven membrane processes. Membranes (Basel). 2022 Jun 28;12(7):669.
90. Gebreyohannes AY, Curcio E, Poerio T, Mazzei R, Di Profio G, Drioli E, et al. Treatment of olive mill wastewater by forward osmosis. Sep Purif Technol. 2015 Jun;147:292–302.
91. Cifuentes-Cabezas M, Pavani A, Vincent-Vela MC, Mendoza-Roca JA, Álvarez-Blanco S. Concentration of phenolic compounds from olive washing wastewater by forward osmosis using table olive fermentation brine as draw solution. Environ Technol Innov. 2023 May;30:103054.
92. Alonso-Vázquez P, Isola A, Sánchez-Arévalo CM, Cuartas-Uribe B, Vincent-Vela MC, Álvarez-Blanco S. Concentration of phenolic compounds from an orange peel waste extract using a combination of ultrafiltration and forward osmosis. Sep Purif Technol. 2025 Jul;360:131228.
93. Wan L, Yun S, Zhang L, Xing BS, Xiao B, Zhai J, et al. Recovery phosphorus and water resources from corn deep-processing wastewater using a combined process of forward osmosis and hydroxyapatite. Environ Technol Innov. 2025 May;104277.
94. Singh N, Petrinic I, Hélix-Nielsen C, Basu S, Balakrishnan M. Concentrating molasses distillery wastewater using biomimetic forward osmosis (FO) membranes. Water Res. 2018 Mar;130:271–80.
95. Agarwal V, Kharade S, Achari M, Pawar H, Sarode D. Forward osmosis: An energy-efficient approach for the treatment and recovery of resources from the dairy industry effluent. J Sci Ind Res (India). 2024 Sep;83(9):943–54.
96. Lakra R, Choudhury S, Basu S. Recovery of protein and carbohydrate from dairy wastewater using ultrafiltration and forward osmosis processes. Mater Today Proc. 2021;47:1400–3.
97. Magalhães D, Gonçalves R, Rodrigues CV, Rocha HR, Pintado M, Coelho MC. Natural pigments recovery from food by-products: Health benefits towards the food industry. Foods. 2024 Jul 19;13(14):2276.
98. Das P, Singh KKK, Dutta S. Insight into emerging applications of forward osmosis systems. J Ind Eng Chem [Internet]. 2019 Dec;72:1–17. Available from: https://doi.org/10.1016/j.jiec.2018.12.021

99. Deka P, Verma VK, Chandrasekaran A, Neog AB, Bardhan A, Raidongia K, et al. Performance of novel sericin doped reduced graphene oxide membrane for FO based membrane crystallization application. J Memb Sci [Internet]. 2022 Oct 15 [cited 2022 Aug 24];660:120884. Available from: https://linkinghub.elsevier.com/retrieve/pii/S0376738822006299
100. Akhtar A, Singh M, Subbiah S, Mohanty K. Sugarcane juice concentration using a novel aquaporin hollow fiber forward osmosis membrane. Food Bioproducts Processing. 2021 Mar;126:195–206.
101. Law JY, Mohammad AW, Tee ZK, Zaman NK, Jahim JM, Santanaraj J, et al. Recovery of succinic acid from fermentation broth by forward osmosis-assisted crystallization process. J Memb Sci. 2019 Aug;583:139–51.
102. Francis L, Ogunbiyi O, Saththasivam J, Lawler J, Liu Z. A comprehensive review of forward osmosis and niche applications. Environ Sci (Camb). 2020;6(8):1986–2015.

# 5 Economic Analysis and Market Trend

As the food industry and other sectors are exploring sustainable and energy-efficient alternatives for processing, the FO process continues to gain attention due to its potential for lower energy consumption, high water recovery rates, and minimal heat exposure. This section presents an in-depth economic analysis of FO technology, compares it with competing technologies, examines investment scenarios and return on investment (ROI), and explores global market trends, regulatory considerations, and safety standards as they pertain to the adoption and application of FO.

## 5.1 DETAILED COST–BENEFIT ANALYSIS OF FORWARD OSMOSIS TECHNOLOGY

A comprehensive CBA is essential for evaluating the economic feasibility and long-term sustainability of implementing FO technology in food processing and other industries. The analysis takes into account the various costs and potential benefits associated with FO systems, enabling businesses to assess whether the technology offers value in specific applications (*Figure 5.1*):

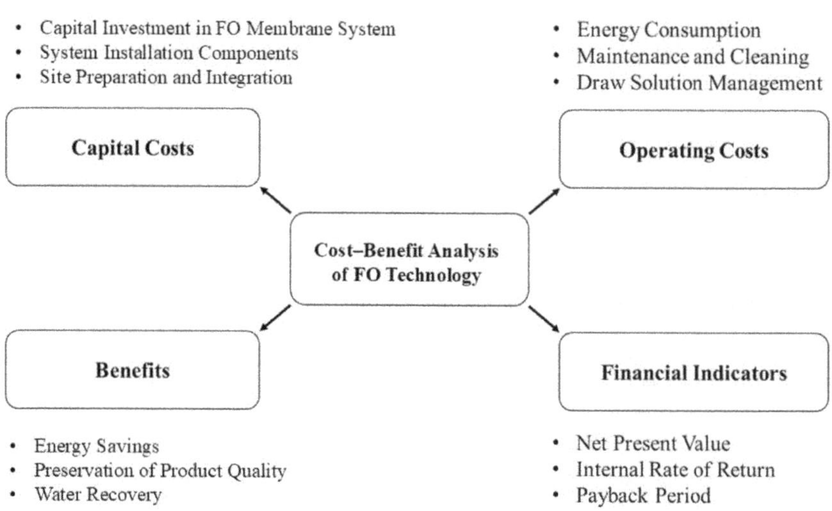


**FIGURE 5.1** Key aspects of cost–benefit analysis for FO technology

DOI: 10.1201/9781003657873-5

### 5.1.1 Capital Costs

The initial investment in FO membrane systems can be significant, particularly when high-performance, anti-fouling membranes are employed. These advanced membranes typically have a higher upfront cost compared to traditional membranes (such as RO or UF). The capital cost also includes the installation of the complete FO system, which typically involves pretreatment systems, DS management components, pumps, and other essential equipment.

Additionally, depending on the application, site preparation and system integration might contribute further to the capital costs, particularly in large-scale industrial settings. While the capital investment for FO systems may seem high initially, their long-term savings potential often makes them an attractive choice.

### 5.1.2 Operating Costs

One of the most significant advantages of FO technology is its low energy consumption compared to traditional pressure-driven systems like RO. FO operates at lower pressures, reducing energy needs. However, some energy is still required to maintain the osmotic pressure gradient and manage the DS. As a result, the primary operating costs for FO systems stem from:

- Energy Use for Water Extraction: FO requires minimal power compared to conventional desalination processes like RO and evaporation.
- DS Management: The cost of maintaining, regenerating, and replacing the DS is another key factor. Since the DS needs periodic replacement or regeneration, particularly in food processing applications where safety is a concern, ensuring that it remains food-safe adds a layer of cost.
- Maintenance and Cleaning: Although membrane fouling is less of an issue in FO systems compared to RO, regular maintenance and periodic cleaning are still necessary to preserve membrane performance. These activities incur additional costs, including labour, cleaning chemicals, and downtime for system maintenance. A comprehensive overview of all cost factors relevant to the specified application of FO and technoeconomic analysis is provided in Table 5.1.

### 5.1.3 Benefits of Forward Osmosis Technology

Compared to conventional methods (such as RO and evaporation), the primary benefit of FO technology is its energy savings. FO requires much lower energy input, which makes it particularly appealing for industries with high energy demands.

Preservation of Heat-Sensitive Components: In food processing industries, the preservation of heat-sensitive compounds like vitamins, proteins, and flavours is critical. FO enables concentration of liquids under mild operating conditions, which helps in retaining the product's nutritional quality and sensorial properties, reducing the need for additives or preservatives.

**TABLE 5.1**
**Effect of individual process elements on cost distribution**

| S. No. | Process Element | Effect on Overall Cost |
|---|---|---|
| 1 | Membrane cost | It is one of the key elements whose cost affects the overall process. The membranes have a definite shelf life, so this expenditure is periodic. |
| 2 | Capital cost | This is a one-time investment and entails considerable expenditure, which includes different process components, assembling, and fabrication costs. |
| 3 | Operating cost | This is a periodic cost involved in operation, which includes the cost of electricity, chemicals, and membrane cleaning cycles. |
| 4 | Maintenance cost | This is also a key aspect for any industrial-scale operation. |
| 5 | Membrane properties | Membrane properties like thickness, porosity, and selectivity, which affect the overall yield, invariably affect the process economics. |
| 6 | Manpower | Operating a membrane-based system requires a skilled workforce, which contributes to the overall cost. However, the modular design warrants a relatively smaller number of operators/engineers. |

Water Recovery and Reuse: FO systems are also capable of recovering significant amounts of water from waste streams, making them ideal for industries that require water recycling, such as food processing, wastewater treatment, and desalination. This makes FO a highly cost-effective and sustainable solution for water-intensive industries, further supporting its economic value.

### 5.1.4 Financial Indicators

Several key financial indicators can be used to evaluate the economic viability of FO technology:

1. Net present value (NPV) estimates the total costs of implementing an FO system against the savings and revenue over its operational life. A positive NPV indicates the technology offers a financial benefit over time.
2. Internal rate of return (IRR) is a metric used to evaluate the profitability of the investment. A higher IRR suggests the FO technology delivers a competitive ROI relative to other potential investments.
3. Payback period refers to how quickly the initial capital investment in an FO system can be recovered through savings and increased productivity. Thereby, shorter payback periods make FO more attractive to businesses that need a rapid ROI.

While initial capital costs are high, the long-term financial benefits of FO systems, especially in terms of energy savings, water recovery, and product quality preservation, offer a compelling ROI. Industries with high energy costs or those requiring sustainable water management are likely to experience faster payback periods and better financial outcomes.

In conclusion, FO technology offers significant economic benefits in terms of energy savings, water recovery, and preservation of product quality. However, the initial capital investment and operational costs must be carefully assessed based on specific applications.

While initial capital costs are high, the long-term financial benefits of FO systems are significant, especially in terms of energy savings, water recovery capabilities, and the ability to preserve product quality. However, the initial investment and operating costs must be carefully evaluated, considering factors such as DS management and system maintenance. The long-term cost savings and potential for increased product quality and sustainability make FO technology an attractive option for industries aiming to reduce operational costs and environmental impact. Thus, a thorough understanding of costs, savings, and financial indicators is crucial for business considering the adoption of FO technology.

## 5.2 TECHNO-ECONOMIC ANALYSIS OF FORWARD OSMOSIS TECHNOLOGY

TEA is a systematic evaluation method that combines technical performance and economic feasibility of a technology, process, or product. In practice, TEA usually includes

1 Technical Modelling: process design, material balances, energy requirements, and efficiency metrics.
2 Economic Modelling: capital expenditure (CAPEX), operating expenditure (OPEX), maintenance costs, revenues, and product value.
3 Comparisons: benchmarking with existing or conventional technologies (e.g. FO vs. evaporation vs. RO).
4 Scenario/Sensitivity Analysis: testing how changes in key parameters (e.g. energy price, membrane cost, and lifetime) affect outcomes.

*Figure 5.2* provides a stepwise framework for the TEA of FO applied to liquid food concentrates. The analysis begins with scope definition, identifying the application and baseline technologies (evaporation and RO). It then examines technical performance, focusing on membrane flux, solute rejection, recovery efficiency, and fouling/cleaning behaviour. The next step, economic assessment, quantifies CAPEX (membranes, pumps, and modules) and OPEX (energy, labour, and maintenance). Reported production costs for FO using AQP-based membranes range from $0.42 to $0.60 per m$^3$, significantly lower than many hybrid FO systems and competitive with TFC FO membranes (1). For food applications, costs may be reduced further because food-compatible draw solutes such as sugars and salts can be directly integrated into the product, thereby reducing or eliminating regeneration costs.

FO performance is then compared with conventional methods: It offers 30–50% energy savings relative to evaporation, superior nutrient and flavour retention, and greater concentration potential than RO (2). The framework also emphasizes risks and uncertainties, such as high membrane costs, draw solute regeneration requirements,

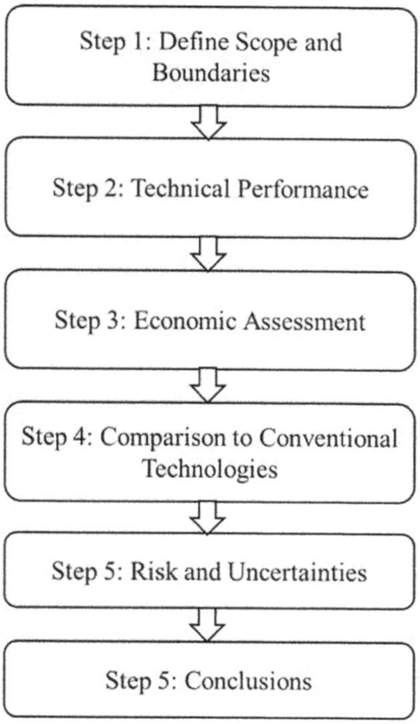


**FIGURE 5.2** Techno-economic analysis of the forward osmosis process for application in the liquid food and dairy processing industries

and challenges in scaling up to industrial capacity. Finally, the conclusion highlights FO's viability for high-value food and beverage products, where quality preservation and energy efficiency are paramount, while noting that further research and development in membranes and DSs are needed to strengthen economic competitiveness.

In summary, TEA is not only a cost and performance evaluation but also a decision-support tool. Applied to FO technology, it demonstrates both the promise of reduced energy use and improved product quality and identifies the critical areas where innovation is required to make FO a commercially attractive alternative to evaporation and RO in the food and beverage industry.

## 5.3 COMPARATIVE ANALYSIS WITH COMPETING TECHNOLOGIES

To evaluate the potential and benefits of FO technology in food processing industries, it is essential to compare it to established competing technologies such as RO, UF, and evaporation. Each of these traditional methods has unique advantages and limitations in food processing contexts, particularly in terms of cost, energy efficiency, and impact on product quality. This section focuses specifically on how FO addresses critical issues in food processing industries, emphasizing applications involving concentration, preservation, and gentle processing of sensitive products.

### 5.3.1 Reverse Osmosis

RO systems have long been utilized within the food processing industries for water purification and concentration tasks. Although highly effective for desalination and removal of small molecules such as salts and certain contaminants, RO typically requires substantial energy input to achieve a high-pressure gradient across membranes. This energy-intensive operation not only raises operational cost but also risks damaging the sensitive nutritional components such as protein, vitamins, and volatile flavour compounds. Furthermore, the RO system often produces significant volumes of waste brine, posing environmental challenges within food processing operations.

FO provides significant distinct advantages due to its lower operational pressure, significantly reduced energy requirements, and minimal thermal exposure, whereas these characteristics help maintain product integrity, especially in processes where preservation of nutritional and sensory properties is crucial.

### 5.3.2 Ultrafiltration

UF systems are commonly employed for separating larger particles, such as proteins, macromolecules, and colloids within food processing applications such as dairy production, juice clarification, and protein concentration. Compared to both FO and RO, UF systems are relatively inexpensive, as they operate at lower operating pressure, thereby reducing the overall energy cost.

However, UF technology has limitations in applications that require the concentration of smaller solutes, such as sugar, salts, and certain bioactive compounds. Additionally, UF membranes tend to foul rapidly when exposed to feed streams with high solid content or complex matrices, thereby necessitating frequent cleaning and maintenance, which results in higher operating costs and downtime. In contrast, FO is an osmotically driven membrane process with minimal fouling potential, resulting in extended membrane life, enabling effective concentration of small solutes, making it especially valuable in producing high-quality concentrated food products with minimal maintenance.

### 5.3.3 Evaporation

Evaporation is traditionally used in food processing industries for concentrating liquid food and dairy products. Although evaporation can effectively reduce the water content, it is a highly energy-intensive process and requires considerable heat input. The elevated thermal exposure often results in significant degradation of thermoslabile components (such as aroma, vitamins, and proteins) and thus negatively impacts the product quality and consumer appeal.

Overall, evaporation processes typically result in significant losses of volatile flavours and aroma compounds, compromising the sensory profile of finished products. FO operating at ambient or low temperatures mitigates these disadvantages by maintaining the nutritional quality and delicate sensory attributes of food products, providing a superior method for concentration tasks in applications demanding high-quality outcomes.

**TABLE 5.2**
**Summary of comparative analysis of forward osmosis with competitive liquid food concentration technique (1,3)**

| | Forward Osmosis | Reverse Osmosis | Ultrafiltration | Evaporation |
|---|---|---|---|---|
| | | **Economic Considerations in Technology Adoption** | | |
| Capital cost (CAPEX) | Moderate—specialized membranes and draw solute systems | High—robust membrane systems and pressure vessels | Low to moderate—relatively simple systems | Very high—high-grade stainless-steel vessels and stream generators |
| Operating cost (OPEX) | Low—minimal energy input and reduced fouling costs | Moderate to high—high energy for pressure generation and fouling management | Low—energy efficient but requires regular cleaning | Very high—heat energy-intensive, scaling issues |
| Scalability | Medium—modular, limited industrial scale-up so far | High—widely deployed infrastructure and expertise | High—well established in pretreatment and clarification | High—used in most large-scale juice plants |
| ROI outlook | Promising in the long term due to energy savings and high-value product retention | Stable but sensitive to energy and maintenance costs | Fast ROI in pre-concentration or clarification settings | ROI depends on by-product valorization or energy |
| | | **Energy Cost and Environmental Impact** | | |
| Energy use (kWh $m^{-3}$) | ~1–2 (without water recovery) | ~3–6 | ~0.2–1.5 | ~40–70 (via thermal evaporation) |
| Environmental impact | Low—no phase change and low emissions | Moderate—electric load intensive | Low—minimal emission | High—fossil fuel dependence unless waste heat is used |

### 5.3.4 Advantages of Forward Osmosis

FO offers numerous advantages over traditional separation technologies in liquid food processing industries:

1. Lower Energy Consumption: FO is an osmotic-driven membrane process, thereby eliminating the need for high-pressure pumps or high-temperature heating.
2. Gentle Operating Condition: A mild operating condition is ideal for preserving heat-sensitive nutrients, vitamins, proteins, and aromatic compounds, ensuring superior nutritional and sensory product profiles.
3. Reduced Environmental Impact: Compared to RO and evaporation processes, FO significantly reduces waste generation, supporting sustainability and compliance with environmental regulations.
4. FO facilitates a higher rate of water recovery from processing streams, enhancing water efficiency and reducing water treatment costs, a critical factor in resource-sensitive food processing operations.

In summary, FO distinctly addresses key challenges in food processing industries, providing efficient, cost-effective, and sustainable solutions. Its unique operational advantages including low-energy consumption, preservation of product quality, and environmental sustainability make FO a compelling choice for liquid food and beverage processing industries.

## 5.4 INVESTMENT SCENARIOS AND RETURN ON INVESTMENT

### 5.4.1 Initial Investment and Operating Cost

FO technology represents an advanced approach to concentration and separation processes in food processing industries. The capital and operational expenditure for implementing FO can vary greatly depending on factors such as the scale of operation, the type of FS, and the complexity of the system. For example, in industries like the liquid food and beverage processing industries, FO offers a promising ROI over both medium- and long-term horizons. While the initial investment may appear significant, the energy savings and water recovery offered by FO can provide a substantial reduction in operational costs, ultimately leading to a quick return on capital.

#### 5.4.1.1 Capital Expenditure (CAPEX)

The initial CAPEX for implementing FO technology in food processing can be broken down into several components:

1. Membrane Systems: FO membranes are typically more expensive than those used in RO or UF systems and account for roughly 30% of the total capital investment (4). However, their longer lifespan and reduced fouling potential (due to their low-pressure operation) can make them a more cost-effective option.

2 Pretreatment and System Integration: Depending on the type of food products being processed, additional systems such as pretreatment for removing large particles or DS units may be required, adding to the capital costs.
3 Installation Costs: Installation of an FO system for large-scale industrial applications can involve significant costs, especially when new infrastructure is required to integrate FO into existing operations. For example, the juice concentration industry may need to adapt its existing systems to accommodate the low-pressure operation of FO membranes.

#### 5.4.1.2 Operating Costs

In the FO process, the primary operating costs include the following:

1 Energy Consumption: The most notable advantage of FO is its low energy consumption. Compared to the RO process, FO operates at low pressure, which typically requires high-pressure pumps. This low energy input makes it an ideal solution for sustainable food production, particularly in industries like dairy, where large volumes of liquid need to be concentrated without compromising product quality.
2 DS Management: The DS in the FO process must be periodically regenerated or replaced, depending on the type of solution used. DS recovery technology has improved in recent years, making this process more economical and thereby reducing long-term operational cost.
3 Maintenance and Cleaning: Membrane fouling is a common challenge in most filtration systems. However, because FO operates at lower pressures, fouling in FO systems tends to be less severe than in other membrane technologies. Additionally, similar to any membrane process, periodic cleaning and maintenance are necessary, which incur additional labour and chemical costs.

### 5.4.2 Return-on-Investment Calculation

ROI for implementing FO technology in food processing industries can be assessed by considering both direct and indirect benefits:

#### 5.4.2.1 Direct Benefits

The key direct benefit of FO technology is its reduced energy consumption compared to traditional processes like RO and evaporation. By operating at low pressures, FO can dramatically cut down on energy costs, making it highly attractive for energy-intensive sectors such as the liquid food and dairy processing industries.

Another significant direct benefit is water recovery. FO can recover large volumes of water from waste streams, which can be reused in other processes. This water recycling capability is particularly valuable in food production, where water usage is substantial, and water management is becoming an increasingly important issue.

FO's lower operating pressure and energy consumption lead to lower operational costs in the long run. Unlike evaporation, which requires high heat and results in significant energy expenditure, FO reduces the energy costs associated with the concentration process.

#### 5.4.2.2 Indirect Benefits

One of the most crucial indirect benefits in the food industry is the preservation of product quality. FO allows food manufacturers to concentrate liquids such as fruit juices, dairy, and plant-based milks without exposing them to high temperatures. This results in better retention of heat-sensitive components such as vitamins, proteins, and flavours. Products such as plant-based milk retain their original taste and nutritional value better than those processed by traditional thermal methods.

In many regions, food safety and sustainability are under increasing scrutiny. FO technology's ability to operate with minimal waste generation and high-water recovery helps companies comply with environmental regulations, such as ZLD standards.

By reducing energy consumption and promoting water reuse, FO technology helps businesses lower their carbon footprint, making it an appealing solution for companies seeking to meet their corporate sustainability goals.

### 5.4.3 Investment Scenarios

For small-scale operations like niche food production (such as concentrated juices and organic food processing), FO can provide a quick ROI by significantly reducing energy costs and improving product quality. Small-scale operations benefit greatly from reduced energy costs, minimal heat exposure, and enhanced product quality. These advantages enable smaller companies to rapidly recover their initial investment. Additionally, the ability to concentrate beverages without compromising their nutritional content and flavour helps enhance brand value, enabling a premium price for the final product.

In large-scale applications, such as desalination or wastewater treatment, longer payback periods may occur, but the environmental benefits and cost savings are substantial in the long run. While the initial investment is higher due to larger FO systems and additional infrastructure, the long-term benefits—including energy savings, water reuse, and product quality preservation—make FO technology a worthwhile investment. In these large-scale settings, FO can significantly improve operational efficiency and sustainability, especially for multi-product lines or diverse beverage portfolios.

## 5.5 GLOBAL MARKET TRENDS, REGULATORY CONSIDERATIONS, AND SAFETY STANDARDS

### 5.5.1 Global Market Trends

In the food and beverage processing industries, FO is primarily used as a non-thermal concentration technology and for innovative water reuse strategies. FO leverages osmotic pressure differences to draw water out of an FS across a semi-permeable membrane. In food and beverage processing industries, the FO market is poised for significant expansion in the coming decade.

The global market trend for FO technology is experiencing rapid growth, driven by industries such as food and beverage, water treatment, pharmaceuticals, and environmental management. The global FO market across all applications was estimated at around \$0.62 billion in 2024 and is projected to grow to roughly \$2.1–2.2 billion by 2034, indicating a robust compound annual growth rate (CAGR) in the 13% range. Based on region, the global food market for food and beverage processing industries can be segmented as North America, Europe, Asia-Pacific, South America, and the Middle East and Africa (Table 5.3).

- North America commands a substantial share of the global FO market in food and beverage processing, driven by consumer preferences for natural, minimally processed, and nutrient-rich products. FO's capability to preserve both sensory attributes and nutritional quality aligns well with the region's clean-label and health-centric trends. Additionally, stringent environmental regulations and aggressive sustainability targets are pushing manufacturers to adopt low-energy, low-waste technologies, making FO an increasingly attractive solution for both product concentration and wastewater treatment.
- Europe stands out as a mature and innovation-driven FO market, supported by its robust dairy industry, premium beverage segments, and strong regulatory emphasis on sustainable practices. Countries like Germany, France, Italy, and the Netherlands are actively adopting FO for applications such as whey protein concentration, de-alcoholization of beer and wine, and organic beverage production. EU environmental policies promoting energy efficiency, water reuse, and emissions reduction have catalysed the integration of FO in commercial-scale operations. Furthermore, European FO membrane manufacturers and system integrators are collaborating closely with food processors to deploy hybrid and scalable FO solutions.
- Asia-Pacific is the fastest-growing regional market for FO, propelled by rapid industrial growth, expanding food manufacturing capabilities, and rising demand for high-quality, health-conscious products. Nations such as China, India, Japan, South Korea, and Southeast Asian countries are investing in advanced processing technologies for dairy, juices, plant-based beverages, and RTD formats. The region's growing middle class and preference for natural, functional products further bolster FO adoption for nutritional preservation and flavour integrity. Importantly, water scarcity across many parts of Asia makes FO's energy-efficient and water-reuse features particularly valuable, with governments actively encouraging adoption through sustainability programmes.
- South America represents an emerging market for FO, primarily supported by growing juice, dairy, and alcoholic beverage industries. Countries like Brazil, Argentina, and Chile—with their strong agricultural outputs—are increasingly exploring FO for juice concentration, wine clarification, and dairy processing. As manufacturers respond to both export market demands and local consumer health trends, FO's ability to preserve natural taste, colour, and nutrition without thermal treatment becomes a compelling advantage. While still in the early stages, the region's fruit abundance and sustainability focus are likely to accelerate FO uptake in the coming years.
- The Middle East & Africa (MEA) region holds untapped potential for FO in food and beverage processing, largely due to acute water scarcity and a

**TABLE 5.3**
**Global forward osmosis market for the food and beverage processing industry (by region, 2019–2035, USD thousand)**

| | 2019-h | 2020-h | 2021-h | 2022-h | 2024-b | 2035-p | CAGR (2025–2035) |
|---|---|---|---|---|---|---|---|
| **North America** | | | | | | | |
| • Fruit and vegetable juice concentration | 1326.077 | 1252.410 | 1353.587 | 1475.788 | 1781.064 | 5221.951 | 10.188% |
| • Dairy product processing | 1046.596 | 988.054 | 1070.789 | 1162.666 | 1400.643 | 4029.377 | 9.968% |
| • Brewery industry application | 417.203 | 391.669 | 423.282 | 454.547 | 542.011 | 1446.313 | 9.187% |
| • Coffee and tea extract concentration | 635.225 | 600.192 | 649.795 | 705.639 | 851.476 | 2451.120 | 9.955% |
| • Liquid food clarification and purification | 875.678 | 823.943 | 889.579 | 962.686 | 1150.208 | 3170.047 | 9.550% |
| • Others | 185.274 | 173.417 | 188.990 | 200.973 | 237.316 | 597.133 | 8.606% |
| • Total | 4486.054 | 4229.685 | 4576.022 | 4962.298 | 5962.719 | 16915.941 | 9.833% |
| **Europe** | | | | | | | |
| • Fruit and vegetable juice concentration | 1247.708 | 1177.395 | 1271.483 | 1386.371 | 1674.924 | 4900.163 | 10.149% |
| • Dairy product processing | 950.615 | 896.723 | 971.076 | 1054.535 | 1271.846 | 3652.964 | 9.934% |
| • Brewery industry application | 374.102 | 350.840 | 378.831 | 406.663 | 485.258 | 1288.350 | 9.117% |
| • Coffee and tea extract concentration | 589.979 | 556.983 | 602.530 | 654.378 | 790.501 | 2271.232 | 9.918% |
| • Liquid food clarification and purification | 846.674 | 796.030 | 858.798 | 929.515 | 1111.957 | 3061.019 | 9.521% |
| • Others | 199.044 | 186.322 | 202.700 | 216.112 | 256.046 | 653.672 | 8.737% |
| • Total | 4208.122 | 3964.293 | 4285.417 | 4647.574 | 5590.533 | 15827.399 | 9.795% |
| **Asia-Pacific** | | | | | | | |
| • Fruit and vegetable juice concentration | 2362.241 | 2231.916 | 2408.644 | 2625.775 | 3169.730 | 9346.423 | 10.231% |
| • Dairy product processing | 1793.847 | 1694.285 | 1833.506 | 1990.761 | 2399.112 | 6946.058 | 10.018% |
| • Brewery industry application | 716.430 | 672.786 | 725.981 | 779.336 | 929.367 | 2490.379 | 9.214% |
| • Coffee and tea extract concentration | 1110.665 | 1049.874 | 1134.951 | 1232.417 | 1487.618 | 4308.656 | 10.003% |
| • Liquid food clarification and purification | 1579.313 | 1486.699 | 1602.805 | 1734.481 | 2073.203 | 5751.876 | 9.603% |
| • Others | 353.861 | 331.539 | 360.571 | 383.982 | 454.170 | 1161.177 | 8.753% |
| • Total | 7916.358 | 7467.099 | 8066.457 | 8746.751 | 10513.201 | 30004.569 | 9.880% |

| | | | | | | | |
|---|---|---|---|---|---|---|---|
| **South America** | | | | | | | |
| • Fruit and vegetable juice concentration | 555.697 | 521.266 | 563.496 | 606.931 | 723.540 | 1982.041 | 9.533% |
| • Dairy product processing | 412.256 | 386.593 | 416.885 | 449.608 | 535.145 | 1436.163 | 9.296% |
| • Brewery industry application | 157.893 | 147.193 | 156.628 | 168.701 | 198.642 | 489.195 | 8.405% |
| • Coffee and tea extract concentration | 247.860 | 232.633 | 251.333 | 270.314 | 322.298 | 862.561 | 9.245% |
| • Liquid food clarification and purification | 356.253 | 332.961 | 356.966 | 384.481 | 453.872 | 1166.921 | 8.881% |
| • Others | 76.598 | 71.228 | 75.946 | 81.604 | 95.246 | 224.266 | 7.974% |
| • Total | 1806.557 | 1691.874 | 1821.253 | 1961.638 | 2328.743 | 6161.147 | 9.159% |
| **Middle East and Africa** | | | | | | | |
| • Fruit and vegetable juice concentration | 359.267 | 335.393 | 361.940 | 387.186 | 544.616 | 1188.899 | 9.053% |
| • Dairy product processing | 259.310 | 242.013 | 260.047 | 279.091 | 327.927 | 839.331 | 8.831% |
| • Brewery industry application | 98.686 | 91.549 | 97.041 | 104.009 | 120.869 | 282.315 | 7.923% |
| • Coffee and tea extract concentration | 156.233 | 145.940 | 157.114 | 168.148 | 197.971 | 505.136 | 8.779% |
| • Liquid food clarification and purification | 231.114 | 214.979 | 229.664 | 246.155 | 286.864 | 703.530 | 8.419% |
| • Others | 50.961 | 47.175 | 49.445 | 53.614 | 61.813 | 140.193 | 7.618% |
| • Total | 1150.570 | 1077.049 | 1155.251 | 1238.203 | 1451.006 | 3660.404 | 8.692% |

*Source:* *International Council of Beverages Associations, Indian Beverage Association, The Food and Beverage Association of America, International Food Information Council, Government Records, Journals, Expert Interviews, and MRFR Analysis*

*h: historic; b: base; p: predicted/forecasted; CAGR: compound annual growth rate*

growing emphasis on energy-efficient technologies. Countries such as the UAE, Saudi Arabia, and South Africa are modernizing their food industries to meet both domestic needs and export standards. FO's low energy usage and high water recovery rates make it an ideal fit for these markets. Applications in juice concentration, dairy processing, and de-alcoholization are gaining momentum, supported by pilot projects and collaborative industry initiatives. While adoption is still nascent, the strategic alignment with regional sustainability goals indicates promising growth for FO technologies.

Between 2023 and 2035, the global FO market for food and beverage processing is projected to grow significantly across all regions, led by Asia-Pacific with the highest CAGR of 9.88%, followed by North America (9.83%) and Europe (9.80%), while South America and the Middle East and Africa also show steady growth. This growth is primarily driven by increasing demand for sustainable and energy-efficient processing technologies in fruit and vegetable juice concentration, dairy product processing, and coffee and tea extract concentration—segments that consistently exhibit the highest CAGR across regions. Asia-Pacific's rapid industrial expansion and evolving consumer preferences for health-oriented products make it the dominant market, whereas North America and Europe benefit from mature industries and stringent environmental regulations. Though smaller in size, South America and MEA are emerging as promising markets due to their agricultural strengths and growing focus on water reuse and low-waste technologies.

As the demand for sustainable solutions increases, FO technology's ability to deliver low-energy, high-efficiency results makes it a leading choice for industries seeking to minimize operational costs and reduce their environmental footprint.

1 Technological Advancements: Significant advancements in membrane materials, DS management, and process integration are enhancing the competitive edge of FO systems in the food sector. In food and beverage processing industries, FO technology revolves around specialized membranes, system design, and DS innovations that tailor the process to sensitive food streams. Incorporation of advanced membrane materials such as CTA, polybenzimidazole (PBI), and carbon nanotubes–polyamide (PA) composites, offering enhanced selectivity, durability, and flux performance. This market encompasses various membrane types, which are pivotal in optimizing the FO process. These innovations are particularly suited to managing sensitive food streams where product integrity is essentially important. Spiral-wound and HF configurations are two primary module configurations of FO membranes used in food and beverage processing applications, of which spiral-wound membrane modules are generally preferred over HF for food concentration due to their better resistance to fouling and ease of cleaning, factors crucial for maintaining hygiene and processing efficiency in food-grade applications. Additionally, optimized system designs and DS formulations contribute to improved process efficiency and scalability.
2 Sustainability Focus: Growing concerns about resource scarcity, water conservation, and carbon footprint reduction are driving the adoption of

energy-efficient and sustainable technologies like FO. Food manufacturers are progressively turning to FO technology for effective water recovery and achieving ZLD targets, thus ensuring compliance with stringent environmental regulations.

3 Market Players: In food concentration segments, key commercial players in FO markets include Aquaporin and Aromatec Pte. Ltd. (Singapore). These companies are at the forefront of delivering specialized FO systems tailored for food and beverage applications, leveraging proprietary membrane technologies and integration expertise to meet the evolving demands of the industry. Aquaporin leverages its proprietary biomimetic membrane technology based on AQP proteins to enhance water transport efficiency while maintaining excellent selectivity. This technology is particularly advantageous in food and beverage applications, where product purity and energy efficiency are critical. Aromatec specializes in FO system integration and tailored solutions for food concentration focusing on preserving flavour, aroma, and nutritional value during processing. This company collaborates closely with regional partners and research institutes to advance the deployment of FO technology in the Asia-Pacific markets. Both companies play a vital role in scaling commercial FO adoption, driving innovation, and setting benchmarks for performance and sustainability in food sectors.

### 5.5.2 Regulatory Considerations

As with any technology used in food production, FO systems must meet strict food safety and environmental regulations:

1 Food Safety Regulations: FO systems used in food processing must comply with food safety standards set by regulatory agencies such as the FDA (US), EFSA (Europe), and other national food safety bodies. This includes ensuring that the membranes and DSs are food-safe and do not contaminate the product.

2 Environmental Impact: FO is highly suited for applications where waste minimization and water recycling are essential. In many countries, food processing companies must adhere to regulations governing wastewater discharge and ZLD systems. FO's ability to recover water from waste streams positions it as a technology of choice for companies seeking to meet environmental and regulatory standards.

### 5.5.3 Safety Standards

Safety is paramount in food processing, especially when applying FO technology. Companies must adhere to stringent safety protocols to ensure that the systems used in food applications meet both food safety and water quality standards:

1 Food-Grade Membranes: Membranes used in FO must be made from food-grade materials to ensure that no harmful chemicals leach into food

products. This is particularly important in food processing, where consumer health is a top priority.

2 DS Management: The DS must be carefully chosen to ensure it is non-toxic, safe, and effective. Additionally, companies must ensure that any residual DS is adequately recovered or removed from the final product to meet food safety standards.

FO technology continues to revolutionize the food processing industry by offering a sustainable, cost-effective, and energy-efficient solution for concentrating and separating food products. While the capital investment required to implement FO systems can be substantial, the long-term benefits—including energy savings, water recovery, and improved product quality—make it a valuable investment for food processing companies. Chapter 6 explores emerging trends and prospects in FO, highlighting innovations and market developments that are expected to further enhance its adoption and impact in liquid food processing, ensuring that FO continues to play a central role in transforming food production processes and helping companies achieve both economic and environmental goals.

## REFERENCES

1. Dsilva Winfred Rufuss D, Hosseinipour E, Arulvel S, Davies PA. Techno-economic feasibility of advanced aquaporin-based hollow fiber forward osmosis membrane in industrial wastewater treatment. J Water Process Eng. 2024 Nov;67:106155.
2. Mahto A, Aruchamy K, Meena R, Kamali M, Nataraj SK, Aminabhavi TM. Forward osmosis for industrial effluents treatment—sustainability considerations. Sep Purif Technol. 2021 Jan;254:117568.
3. Im SJ, Jeong S, Jeong S, Jang A. Techno-economic evaluation of an element-scale forward osmosis-reverse osmosis hybrid process for seawater desalination. Desalination. 2020 Feb;476:114240.
4. Osipi SR, Secchi AR, Borges CP. Cost analysis of forward osmosis and reverse osmosis in a case study. In: Current trends and future developments on (Bio-) membranes. Elsevier; 2020. p. 305–24.

# 6 Emerging Trends and Future Prospects in Forward Osmosis for Liquid Food Processing

## 6.1 EMERGING TRENDS

The application of FO in the liquid food processing industry is evolving rapidly, driven by innovations in membrane materials, advancements in DSs, and the integration of FO with other technologies. With the growing demand for sustainable, energy-efficient, and cost-effective food processing methods increasing, FO technology is poised to play a significant role in transforming the way liquid food and beverages are concentrated and processed. This section explores some of the key emerging trends that are shaping the future of the FO in food processing industries.

### 6.1.1 Innovations in Membrane Material and Design

In the FO process, the appropriate selection of the membrane directly impacts the efficiency, selectivity, and sustainability of the overall process. The recent development in membrane material and designs has the potential to revolutionize the capabilities of the FO systems. A few key innovations in this area include the following.

#### 6.1.1.1 High-Performance Membranes

The recent advances in hydrophilic membrane materials and nanocomposite membranes have significantly enhanced the efficiency of FO systems. GO and oxidized multi-walled carbon nanotube interlayers were introduced onto electrospun PVDF nanofibres, providing efficient water transfer channels for water molecules and contributing to a reduced thickness of the PA layer (1). Therefore, the mass transfer resistance decreased, exhibiting superior FO performance. A novel in situ growth strategy was proposed for incorporating a ZIF as an interlayer within TFC membranes, aiming to enhance solvent permeance (2). A modified PAN-based membrane was developed by incorporating CNTs and the UiO-66-$NH_2$ MOF (3). Furthermore, an ultrathin polyester-amide HF membrane was synthesized by incorporating a PVA hydrogel layer into the PA layer over a PSf HF membrane support (4). The author demonstrates the potential of the developed polyester-amide composite HFFO for concentrating natural anthocyanin dye. The ultrathin polyester-amide layer was formed by incorporating a PVA hydrogel layer, followed by interfacial polymerization (IP) of piperazine and TMC. The incorporation of the PVA hydrogel layer

DOI: 10.1201/9781003657873-6

reduced the AL thickness from 96 nm to 12 nm and exhibited 99.9% rejection of anthocyanin natural dye, with an enhanced osmotic water flux of 43 $L\,m^{-2}\,h^{-1}$ using a 4.27 M NaCl DS.

A PA-based biocomposite FO membrane was fabricated using an eggshell membrane as the SL (5). The PA layer was deposited by the IP technique, and the composite membrane exhibited osmotically driven water flux. Furthermore, the biocomposite FO membrane was developed by surface coating with GO, resulting in the stable formation of a polyamide (PA) layer. The author revealed that the fabricated eggshell composite membrane with a low structural parameter (~138 μm) reached 46.19 $L\,m^{-2}\,h^{-1}$ in AL-FS mode using 2.0 M NaCl as a DS.

#### 6.1.1.2 Anti-Fouling Surfaces

Fouling, especially in high-solid content feeds like fruit juice and dairy products (such as milk), is a major issue for FO membranes. Fouling, caused by the accumulation of organic or inorganic substances on the membrane surface, reduces water flux and operational efficiency and increases the need for frequent cleaning and maintenance. Recent advancements in fouling technology have led to significant improvements in membrane longevity and performance. The anti-fouling strategies for modification of FO membranes are summarized as follows (6):

##### *6.1.1.2.1 Enhancing Hydrophilicity of FO Membranes*

A widely known methodology for improving the anti-fouling strategy of FO membranes involves the enhancement of hydrophilicity. This can be achieved by incorporating materials containing hydrophilic functional groups (including -OH, -CHO, -COOH, and $-NH_2$) such as PDA, zwitterions, and others (7). Alternatively, materials can be functionalized by adding hydrophilic functional groups. If the material displays high hydrophilicity, it will form a hydrated layer on the FO membrane surface through hydrogen bonding with water molecules, thereby exhibiting enhanced resistance to adhesion of foulants (8). Hydrophilic materials are typically loaded on the membrane surface or in the membrane matrix. For instance, incorporating hydrophilic UiO-66-$(COOH)_2$ into the PA AL of an FO membrane decreased the water flux loss from 27% (pristine membrane) to 18% in the presence of BSA foulant (9). The surface of the FO membrane modified with GO exhibited enhanced hydrophilicity, resulting in reduced SA adhesion (10).

A high-performance non-woven FO membrane was fabricated using a blend of CTA and PS through the phase inversion technique (11). The study revealed that higher PS content slightly reduced the membrane's hydrophilicity, resulting in elevated water flux and exhibiting a low fouling tendency during brine concentration.

##### *6.1.1.2.2 Improving the Ridge–Valley Surface Morphology of FO Membrane*

The surface morphology of the conventional FO membrane, comprising ridges and valleys, provides an environment conducive to the accumulation of pollutants on the membrane surface. A reduction in surface roughness serves to mitigate the adhesion of hydrophobic pollutants on FO membranes. For the formation of ridge–valley morphology, during the IP process, an MPD aqueous solution will penetrate and be stored in the substrate pores as it flows through them. Subsequently, the TMC/n-hexane solution is added, whereby the organic solution flows above the substrate pores and rapidly reacts with the MPD solution within the substrate, resulting in the formation of an irregular and wrinkled PA AL.

One method for constructing a smoother FO membrane surface is to add an interlayer between the SL and the PA AL. The interlayer prevents the MPD solution from penetrating into the basement pores, thus ensuring the formation of a flat and defect-free AL. The second method entails the preparation of a new, smooth layer on the primary AL. Reportedly, the potential candidate for this layer includes the TA-metal chloride layer (12) and the ultra-smooth hydrogel layer (13–15). The most common materials used to prepare ultra-smooth hydrogel layers are PVA and SA.

###### *6.1.1.2.3 Reducing the S-Parameter of the Support Layer*

The ICP phenomenon has been shown to reduce the concentration gradient inside the SL, which in turn reduces the effective osmotic pressure through the membrane. In the FO membrane, the S-parameter is used to estimate the intensity of ICP. It has been demonstrated that by increasing the porosity and mean pore diameter of the SL and reducing its thickness, the S-parameter of the SL will be reduced, thus alleviating the ICP phenomenon (16). The mitigation of ICP contributes to improving the effective osmotic pressure difference, maintaining a stable water flux, and reducing adhesion of foulants to the membrane surface (17).

###### *6.1.1.2.4 Enhancing the Electronegativity of the FO Membrane*

The enhancement of electronegativity in the FO membrane can effectively mitigate membrane fouling through the principle of electrostatic repulsion. An increase in the zeta potential on the membrane surface elevates the energy barrier between the contaminants and the membrane. The FO membranes were rendered with a less negative zeta potential value through the use of an overall electrically neutral zwitterion (95,391). Consequently, the accumulation of the fouling cake layer was prevented by electrostatic repulsion of organic and biological contaminants on the membrane surface.

###### *6.1.1.2.5 Using Specific Anti-Fouling Agents in Conjugation with PA Active Layer*

The mitigation of both inorganic and biological foulants can be achieved by incorporating specific anti-fouling agents in conjugation with the PA AL. The application of a scale inhibitor (such as organic phosphonic acid) can hinder the growth of gypsum crystals and effectively resist the inorganic fouling on FO membrane surfaces (18). A TFC membrane modified with DTPMP was developed, exhibiting only a 15% reduction in water flux during gypsum fouling tests (19). The incorporation of antibacterial agents, including Ag, and nitrogen-doped carbon dots (N-C-dots) has been demonstrated to effectively prevent bacterial adhesion and colonization on the FO membrane surface (20). Silver-based materials possess substantial and potent antibacterial properties. MOFs are formed by the co-ordination polymerization of silver cores and organic linkers, offering effective control over the release of silver ions, thereby ensuring sustained antimicrobial activity of FO membranes. A silver MOF (Ag-MOF) membrane was developed by depositing Ag-MOFs onto the surface of an FO membrane through covalent binding (21). Furthermore, in order to improve the anti-fouling ability of the FO membrane, it is expected to improve the hydrophilicity of the FO membrane, accompanied by the incorporation of antibacterial materials.

N-C-dots/PEI nanocomposite-grafted FO membrane has been shown to mitigate 19% of the biological contamination on the surface (20). The N-C-dots/PEI modification strategy has proven to be highly effective, establishing a new field for the development of efficient mitigation of biological fouling under visible light. This finding provides a novel concept for the selection of FO membrane modification materials to inhibit biological fouling in the future and is expected to enhance the modification effects of other antibacterial agents.

##### *6.1.1.2.6 Using Modified Material as Sacrificial Structures*

The amide bond (CO-NH-) on the PA AL is susceptible to external influences, rendering it a sensitive site, which is susceptible to attack by foulants (such as active chlorine). The grafted materials such as organic phosphonic acids and sulfonamide substances exhibit a greater propensity to react with foulants on the membrane surface than with the amide bonds. This approach can effectively prevent the destruction of the PA AL structure and maintain the operational stability of the FO membrane.

For instance, sulfonamide substances, such as sulfathiazole, 4-aminobenzenesulfonamide, and 2-aminosulfamide hydrochloride, could act as chlorine sacrifices by grafting onto the AL through reaction with its remaining carboxyl group. Among them, the sulfathiazole-modified membrane exhibited the greatest resistance to chlorine (22). The utilization of modification materials as sacrificial protective layers effectively mitigates the detrimental impacts of foulants on the PA AL, thereby enhancing the versatility of the FO membranes.

#### 6.1.1.3 Membrane Architectures

The design and structure of FO membranes, including HF and flat-sheet configurations, are being optimized to improve the flow dynamics and mass transfer efficiency, which directly impact water recovery and concentration performance.

### 6.1.2 Advancements in Draw Solutions

After the FO membrane, DS is another critical component of the FO process, as it drives the osmotic pressure that allows for water transfer through the membrane. The advancements in the development of novel DSs are central to improving the efficiency and sustainability of FO systems in food processing. This includes the following:

#### 6.1.2.1 Food-Safe and Sustainable Draw Solutions

There is an increasing interest in the use of food-grade, biodegradable, and non-toxic DSs that can be used safely in food applications. The natural substances such as sucrose, salt, and organic acids are becoming popular alternatives to synthetic solutions, thereby ensuring that the final product remains safe for consumption.

#### 6.1.2.2 Draw Solution Regeneration

New methods for regenerating DSs are improving the economic efficiency of the FO systems by reducing the need for frequent replacements. Techniques, such as adsorption–desorption and solvent recovery, are being developed to make the process more cost-effective.

### 6.1.3 Integration with Other Technologies

Though the FO process reportedly shows prospective potential in various niche applications, it is generally accepted by the scientific community that a stand-alone FO process appears to be less attractive and competitive. To advance the FO process and to realize its prospective application in various industries, innovative designs of hybridization and integration involving the FO process and other technologies have been actively proposed and sought by the research community. This section discusses the hybrid/integrated FO process based on three main purposes, namely process enhancement, DS regeneration, and fouling mitigation (23).

#### 6.1.3.1 Process Enhancement

FO technology demonstrates significant potential in food processing; however, its full benefits are often realized when integrated with complementary membrane processes. Such integrations enhance efficiency, improve product quality, and mitigate operational challenges such as fouling and DS recovery.

- FO–MD Integration: Coupling FO with MD enables simultaneous juice concentration and DS regeneration. For instance, in apple juice concentration, the FO–MD system increased soluble solids from 10.6 to 45.1 °Brix, producing high-quality juice with minimal nutrient loss. The process also safely concentrated potassium sorbate, supporting microbial stability. Despite these advantages, the additional thermal energy required for MD introduces cost and energy burdens that must be carefully managed (24).
- FO–UF Integration: Integration of FO with UF has been explored for sugarcane juice concentration. The UF pretreatment step reduced fouling, allowing FO to operate more effectively for subsequent concentration. This approach highlights the role of hybrid processes in improving membrane longevity and operational reliability (25).
- FO–RO Integration: In dairy applications, combining FO with RO has shown particular promise. Greek yoghurt acid whey (GAW) was first concentrated by RO from 6.6 to 19.6 °Brix and subsequently by FO to 40.2 °Brix (26). This approach overcame the limitations of RO—such as CP and fouling—while achieving concentration levels comparable to thermal evaporation but with lower thermal stress on the product.

Overall, integrating FO with MD, UF, or RO provides distinct process enhancements, including higher concentration factors, DS regeneration, and reduced fouling. However, these advantages are offset by increased energy consumption and capital investment requirements. Future research should focus on optimizing energy-efficient DS recovery systems and scaling integrated FO processes for industrial applications.

#### 6.1.3.2 Draw Solution Regeneration

The FO process generates two streams, concentrated FS and diluted DS, where the final products are determined by the purpose of having the concentration–dilution process. On the other hand, the aim of the FO process is to extract and recover the clean water in the FS. As the product of a simple FO system is not pure, but rather the water is trapped in the diluted DS, to obtain a pure water product, FO needs to be part of an integrated process with a second step aimed at both reconcentrating the DS and producing a pure water product. Except for a few DS (such as seawater, food additives, and fertilizer solution), the majority of DS need to be regenerated. Furthermore, depending on the type of DS, the FO process can be integrated with the corresponding DS regeneration process.

Integration of solar energy with the FO process enhances overall sustainability and can significantly minimize the overall energy consumption of the system. For instance, an integrated FO–MD system powered by solar energy reduced the specific energy consumption to 1.1 kWh $m^{-3}$ when compared to 7.06 kWh $m^{-3}$ for an unmodified MD system. Based on the literature, solar integration with FO systems can be broadly categorized into four main approaches:

- Photovoltaic (PV) system directly converts solar energy to power using solar cells. This electricity can be used to power several components of integrated systems such as pump (that circulates FS and DS), auxiliary equipment of the integrated FO process, and the DS recovery process. While the operational and maintenance needs of a PV system are relatively small, this approach requires high initial cost, significant land area, and substantial energy input.
- Solar thermal systems harness solar energy to produce heat, which can be used in several ways such as preheating FS to improve water flux, DS regeneration by providing thermal energy needed for processes (such as MD), and thermally responsive DS. Since solar thermal system uses heat directly without conversion to electricity, this system can be more effective than PV systems. In certain applications, solar thermal energy can be converted into mechanical energy to power high-pressure pumps in an RO system, which can be coupled with FO. Solar ponds act as a dual-purpose system, functioning as both solar energy collectors and thermal energy storage (TES) units. A solar pond is capable of regenerating DS below 70°C.
- Photovoltaic thermal (PVT) collectors generate electricity for powering FO system components and produce energy for preheating the FS or DS regeneration. Reportedly, since this system can use the thermal component to remove excess heat, the overall efficiency can be improved through the interplay between the demand for electricity and thermal energy. Furthermore,

PVT systems can minimize the need for separate PV panels and thermal collectors, which has the potential to reduce the total costs and land area requirements.

- Solar-driven interfacial evaporation utilizes solar energy to concentrate DS in FO systems. The photothermal material in this process first absorbs sunlight and converts it into heat at the localized water–material interface, causing water in the diluted DS to evaporate. Consequently, this process leaves concentrated solutes and restores the osmotic pressure gradient. Fresh water can also be recovered through condensation. The performance of solar-driven interfacial evaporation depends on advanced photothermal materials such as polypyrrole (PPy)-modified sponges and carbon-based nanofluids such as potassium-functionalized CNFs, which improve the light absorption and heat conversion.

A key consideration of this approach is the interplay between water permeation in FO and evaporated water, which is controlled by the initial DS concentration and light intensity. Higher DS concentration results in a higher initial flux but faster dilution when evaporation is not well-managed. On the other hand, an excessive light intensity results in excessive light evaporation, whereas inadequate light intensity results in reduced water flux due to slow evaporation and dilution. Thus, dynamic management of both concentration and light intensity is required to optimize water flux stability.

Despite these advantages, there are certain challenges in scaling solar-driven FO systems, which need to be addressed for efficient DS regeneration. One significant challenge for scaling solar-driven FO systems is the thermal stability of draw solutes, making them unsuitable for solar thermal regeneration. While lower critical solution temperature (LCST) materials offer an energy-efficient alternative due to their low-cost thermal energy requirements and lower energy consumption for phase separation compared to water vaporization, a major issue with LCST material is incomplete phase separation of the draw solute from the water, even after heating above the LCST, which may lead to a small amount of the draw solute remaining in the water-rich phase, which requires purification cost.

Moreover, while the presence of hydrophilic groups in the LCST material can lead to a higher water flux, it also causes the problem of draw solute retention in the water-rich phase. This is undesirable, since it can compromise the purity of the treated water. Lastly, despite the intent to utilize the low-cost solar thermal energy for regeneration, the production cost of certain LCST materials, such as functionalized MnPs, remains high.

#### 6.1.3.3 Fouling Mitigation

Generally, FO demonstrates much lower fouling propensity and higher fouling reversibility than RO, which can be attributed to the lack of applied hydraulic pressure. These properties have enabled FO to be used for the handling of various low-quality water sources, including landfill leachate, municipal wastewater, coal mine wastewater, and anaerobic digestate. Despite this, fouling is still prevalent in the FO process as the phenomenon of flux decline and is normally associated with the deposition of suspended impurities on the membrane surface that blocks the passage of water and leads to CP (27). Fouling arrests the capability of the membrane (flux and retention of impurities) and incurs additional costs associated with membrane cleaning for restoration of membrane performance and membrane replacement expenses. Hence, to alleviate the issue of membrane fouling, pretreatment processes can be integrated prior to the FO process to minimize the fouling issues in the other membrane processes.

Mitigation and prevention of FO membrane fouling can also be achieved indirectly through the design of a hybrid FO process. For instance, when the biological treatment processes of osmotic membrane bioreactor (OMBR) have been modified into biofilm or entrapped cell forms, the fouling propensity of the FO process has also been minimized. Reportedly, when the bacteria grew in a fixed place (biofilm and polymeric matrix), the amount suspended in the bioreactor was also reduced. Furthermore, the continuous draw out of wastewater from the bioreactor helped to remove a portion of the dispersed cells. The change in the bioreactor form reduced the presence of suspended cells, subsequently minimizing the deposition of foulants on the FO membrane.

Overall, even though the potential of FO has been proven through the integrated process, there are some areas of study that need to be further investigated, prior to implementation at a commercial level:

- Most of the performance data for the integrated FO process are limited to lab-scale testing, and there is still a lack of comprehensive data from pilot-scale or real applications of the integrated FO process with DS recycle to convince the industry to adopt this concept in their process. Furthermore, the flux decline appeared to be significant, which indicated the need for frequent cleaning. Though FO is claimed to be easily cleaned, the impact of membrane fouling and cleaning on the overall performance and cost needs to be explored.
- Though the reported performance of integrated FO processes is quite encouraging, the associated cost has been left out with no data on the improvement of overall expenditure. For instance, pretreatment processes installed prior to FO can arrest fouling issues, but the benefits in terms of cost benchmarking need to be investigated.
- The feasibility of long-term operation of integrated or hybrid FO processes depends on overcoming challenges related to membrane performance and DS management. Earlier studies highlighted the limitations of FO membranes, which—until around 2020—generally suffered from low flux and fouling susceptibility. However, recent advancements have led to the development of high-flux, fouling-resistant FO membranes with improved solute rejection and durability, thereby enhancing their suitability for continuous operation. In parallel, viable draw solute regeneration strategies, such as the integration of MD and heat pump-assisted recovery, have demonstrated technical feasibility at the pilot scale. These advancements indicate that FO is steadily progressing towards industrial application, with several recent projects validating its potential in food and beverage processing.

- The success of the integrated FO process is also reliant on the quality of the FO membrane. Though remarkable advances have been achieved in terms of membrane fabrication, it is still not sufficient to support industrial-scale application of the FO process. Challenges remain in the aspect of control over membrane properties (such as thickness, porosity, pore structure, water permeability, selectivity, and anti-fouling) that will influence the practicality of the upscaled integrated FO process in terms of flux and membrane fouling. Incorporation of nanoparticles seems to be able to improve membrane properties, yet it is challenged with inherent difficulties in terms of leakage, cost, and mass production of FO membranes incorporated with nanoparticles.
- Another major hindrance for the widespread application of the integrated FO process is the availability of draw solute, which can generate high osmotic pressure with minimal RSF and can be efficiently regenerated with minimal energy consumption. Though a significant amount of work has been done on the exploration of draw solutes for the FO process, the ideal draw solute for upscaling applications needs to be investigated. Existing draw solute is economically not feasible, where either the material is costly or the cost associated with regeneration is too.

Considering all the challenges mentioned earlier, almost all are related to the economic aspects of the FO process. The economic feasibility of the integrated FO process remains a topic of dispute, as the comprehensive cost analyses for real applications are limited. Different views of economic feasibility have been reported in the literature. For example, the FO process utilizing fertilizer as a DS, operating at 20 L $h^{-1}$, consumed 0.396 kWh $m^{-3}$ of energy, which was lower than the conventional activated sludge plant (0.5647 kWh $m^{-3}$) and the conventional MBR (1.0465 kWh $m^{-3}$) process (28–31). However, the finding was limited to lab-scale testing using saline brackish water (5000 mg $L^{-1}$ NaCl) and without the need for DS regeneration, where the relative size of an actual plant receiving a real sample might result in a different performance in the long run. Furthermore, the CP issue was apparent in the testing, as reflected by the FO flux decline. This indicates the possibility of higher expenditure for long-time operation in real applications.

Despite positive simulation results supporting the economic feasibility of the FO process for the integrated FO process, there was no evidence from large-scale demonstration that the integrated FO–RO process can consume less energy than the conventional liquid food RO concentration process. Alternatively, the cost of the integrated FO process could be potentially reduced if the regeneration of draw solute is not required, or the energy required could be derived from a low-cost or renewable source such as solar energy or waste heat (32,33).

### 6.1.4 Integration of Artificial Intelligence in Forward Osmosis

The term "AI" was first coined in 1956 during a Dartmouth University conference (34). The main development in the AI field took place in the past two decades, prompted by substantially improved computing power and the need to analyse increasingly large and complex volumes of data. Today, AI is considered one of the key technologies within Industry 4.0, as its application is growing in many fields including intelligent search, pattern recognition, big data analytics, and smart environments. The primary advantage of AI techniques over other methods is in their ability to model and solve complex non-linear relationships between input and output variables. AI enables the creation of real-time process monitoring systems, which can model and estimate multiple inputs in water treatment. The AI-enhanced models allow you to save time and overcome complicated laboratory-based process measurements and can be built without assumptions about factor interaction and data distribution. They are considered "black box" models since they rely on learning mechanisms based on data rather than on applying numeric or mathematical equations. Other beneficial features of AI models are self-diagnosis and autonomous learning, which make them capable of automatic control and feedback. Such models are capable of conducting independent analysis, making predictions based on the input dataset, and enabling real-time action for equipment used and/or researchers for timely decision-making. This makes AI models applicable beyond laboratory settings and pilot plants. Finally, AI models are flexible and adaptable for working with new datasets, which is a useful feature in the constantly changing technology environments for liquid food and dairy processing industries.

AI applications in water treatment, and especially membrane-based processes, have been extensively researched during the past decade. Studies applied a broad range of techniques to estimate process set-up and performance parameters, including membrane choice and fouling, water flux estimates, solute recovery, polarization processes, and removal of matter from solutions. AI, encompassing ML, deep learning, and related computational techniques, offers powerful tools to model, predict, and optimize FO processes. By learning from data, AI models can uncover hidden patterns and relationships, enabling more accurate predictions and efficient process control. This section delved into the application of FO in FO processes, highlighting its role in modelling, optimization, fouling prediction, and real-time control, with a focus on food processing applications.

- Support vector machine (SVM) is a widely used ML technique applicable to both classification and regression tasks, grounded in the principles of statistical learning theory (Figure 6.1). In binary classification, SVMs excel at segregating data points into discrete categories by identifying the optimal hyperplane within a feature space. The optimization involves maximizing the margin, the spatial interval between the hyperplane and the nearest points representing the respective classes, known as support vectors. By employing diverse kernel functions like linear, polynomial, radial basis function (RBF), and sigmoid, SVMs accommodate complex decision boundaries, addressing both linearly and non-linearly separable datasets.
- Decision trees (DTs), a widely employed and potent ML technique, demonstrated efficacy in both classification and regression tasks. They operate within the realm of supervised learning, catering to binary and multi-class classifications, as well as regression analyses. At each node, decisions are

made based on attributes or features, leading to branches representing different outcomes. Through iterative processes, DTs are constructed by selecting features that partition the data into homogeneous groups, guided by criteria such as entropy or Gini impurity. Employing a top-down, greedy strategy for feature selection, the algorithm aims to maximize information gain or minimize impurities, resulting in informative splits at each node. This adaptable approach offers a transparent method for classification and regression tasks. However, DT has potential drawbacks, including susceptibility to overfitting and sensitivity to changes in the dataset. Figure 6.2 represents the DT development process and workflow.

- Random forest (RF) is applicable for both classification and regression. It entails constructing numerous DTs during the training phase. Additionally, RF provides metrics for assessing the significance of features, revealing the relative importance assigned to different attributes throughout the prediction process. Understanding the importance of features in model performance can aid in the selection of relevant features. Optimization of RF's performance involves adjusting various hyperparameters, including the number of trees in the forest, the maximum depth of each tree, and the number of features considered at each split. This technique is powerful and widely acclaimed in the field of ML due to its ability to handle complex data and deliver accurate predictions.

#### 6.1.4.1 AI Techniques in FO Process Modelling and Optimization

The FO process has a high degree of complexity and variability due to the many potential inputs and outputs involved. Therefore, AI techniques have been considered as a possible solution to model the FO process through simulation, predict its

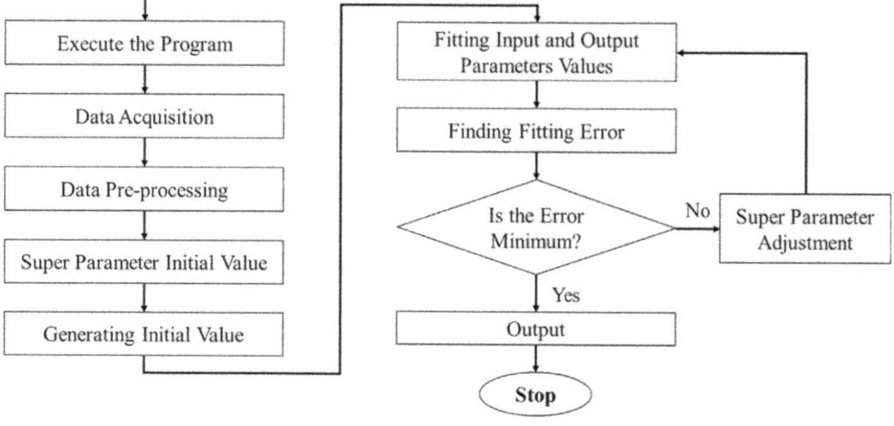


**FIGURE 6.1** Workflow of the support vector machine

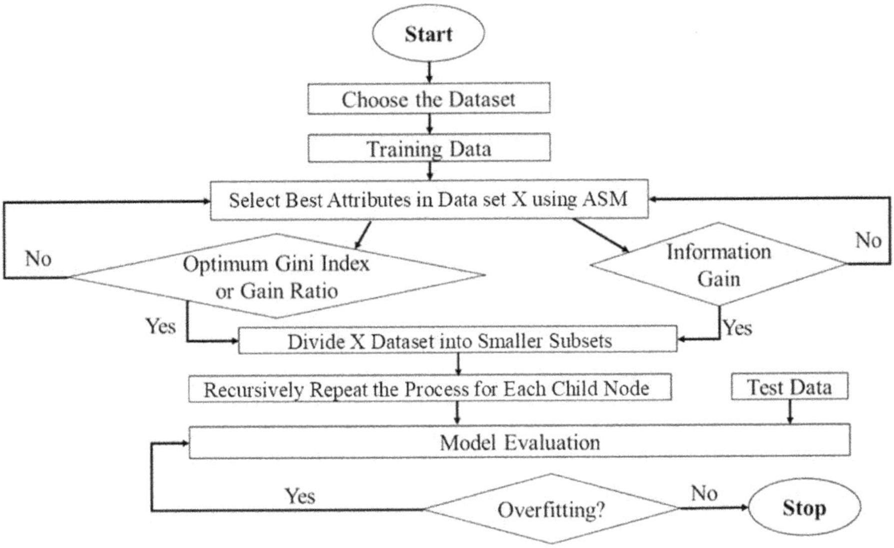


**FIGURE 6.2** Working flow of the decision tree

performance, and identify the optimal combinations of inputs and conditions for the process performance.

Application of pure AI techniques for FO process modelling refers to investigations of stand-alone single techniques with a set of specific parameters. In a typical process, an AI algorithm is created and trained by processing a supplied dataset. The model then provides outputs, which can be compared with the real results of an experiment. The model's accuracy then can be assessed numerically by using the $R^2$ and error values to the degree of fit.

Using the Taguchi normal approach, an FO process based on four inputs (FS, DS, velocities, and temperature) with RSF as a response parameter was developed (35). The model successfully identified the optimum combination of inputs to minimize RSF. A deep learning protocol was proposed to optimize energy efficiency in an OMBR system while maintaining high-quality effluent standards. The study concluded that such an AI-optimized system could achieve up to 34% energy savings when implemented at full scale (36).

The aforementioned studies did not measure the predictive power of their models against experimental data, nor did they compare the performance of AI techniques against traditional models. Later studies, however, demonstrated that AI techniques both demonstrated excellent predictive results in the FO system and compared favourably to conventional methods like empirical and mathematical modelling. An ANN-based protocol was shown to enhance water flux, reduce RSF, and minimize ICP, achieving a high prediction accuracy with an $R^2$ value of 0.9859 (37). An optimally trained ANN model utilizing nine input parameters demonstrated a high predictive capability for permeate flux in an FO process, achieving an $R^2$ value of 0.973 (38).

A deep learning model developed to predict flux in an FO membrane module achieved significantly lower prediction error (3.87%) compared to a traditional theoretical model (10.6%) when using concentration, velocity, and pressure of the solutions as input parameters (39). An ANN model developed for predicting antibiotic transfer through an FO membrane, using eight input parameters, demonstrated high prediction accuracy with $R^2$ values of 0.998, 0.993, and 0.990 for partial, full, and principal ANN configurations, respectively (40). These predictive values were much stronger than the value demonstrated by the traditional dimension reduction approach ($R^2 = 0.6980$).

High predictive power of the AI-based model was also demonstrated in the OMBR studies involving FO systems. A recurrent fuzzy neural network (RFNN) protocol was employed as part of an intelligent detection system to predict membrane permeability in an OMBR (41). The model demonstrated a high predictive power of 0.9432 based on $R^2$. The author argued that such a level of prediction renders the RFNN suitable for detecting FO membrane fouling in OMBR. An OMBR–clarifier system was investigated for simultaneous reduction of salt accumulation and membrane fouling, utilizing an RNN-based model for predictive analysis and process optimization (42). The model demonstrated a high level of long-term prediction in water flux, conductivity, and FO membrane lifespan.

Though examined AI models demonstrated stronger predictive power over traditional methods, in order to understand whether any of the AI models are superior to other AI models and under what specific conditions, several studies attempted to use different AI techniques to distinguish the predictors of FO system performance. Four AI techniques—artificial neural networks, extreme gradient boosting, categorical boosting, and RF—along with a linear regression (LR) approach were used to predict ICP in the FO process based on secondary data extracted from published studies (43). The author reported that all the AI models substantially outperformed the LR approach ($R^2 = 0.48$) in terms of prediction power, thereby concluding that all AI models demonstrated good ICP prediction and generalizability. Similarly, for comparative analysis, micropollutant removal in the FO process was simulated using seven distinct AI models, including LR, ANN, K-nearest neighbour (KNN), DT, ensemble of trees (ET), SVM, and adaptive neuro-fuzzy inference systems (ANFIS) (44). Reportedly, compared to other AI models, the ANFIS model showed predictive power in pollutant removal ($R^2 = 0.99$). Thus, ANFIS with an optimal number of parameters was regarded as the best model to predict the efficiency of micropollutant removal.

On the other hand, some studies demonstrated weaker comparative performance of AI-based techniques. For example, ANN and ANFIS algorithms were compared with the traditional RSM for modelling and optimizing five key parameters—initial draw concentration, initial feed concentration, time, initial feed pH, and temperature—to maximize water flux and minimize RSF (45). ANN-based models demonstrated relatively poor predictive capabilities for both parameters: $R^2 = 0.7970$ for water flux and $R^2 = 0.7807$ for RSF. The ANFIS model was most accurate in predicting the RSF ($R^2 = 0.9427$), but it performed the worst for the water flux parameter ($R^2 = 0.7773$). RSM was actually the best model in this regard ($R^2 = 0.8529$).

Furthermore, a study demonstrated that an ANN model enhanced with RSM outperformed a stand-alone ANN model in accurately predicting membrane flux in the FO process (46).

Comparative studies for AI-based models have been carried out in OMBR systems as well. The performance of ANN and ANFIS was compared for predicting water flux, using inputs such as dissolved oxygen, mixed liquor suspended solids, electrical conductivity, and membrane type (47). While both models demonstrated high predictive power, ANFIS had better results for both types of membranes, TFC ($R^2$ = 0.9755) and CTA (R2 = 0.9861). The models had one key input in common, which was electric conductivity. A comparison of seven AI models was conducted to evaluate their effectiveness in predicting water flux and salinity in a laboratory-scale OMBR (48). The median prediction score for all models was 0.95 ($R^2$) for flux and slightly above 0.90 ($R^2$) for total dissolved solids. No single best model was identified; however, the RF model showed the best results for predicting the flux ($R^2$ = 0.987), while XGB performed best for total dissolved solids modelling ($R^2$ = 0.97; RMSE = 0.234). Finally, the SHapley Additive exPlanations (SHAP) model allowed us to identify the critical input (phosphorus concentration) in the FO process. Table 6.1 summarizes comparative studies and ranking methods.

**TABLE 6.1**
**Comparative studies of AI models in forward osmosis systems for water treatment**

| Author (Year) | Predictive Output | Models | $R^2$ | MRSE |
|---|---|---|---|---|
| Ibrar et al. (2022) (43) | Internal concentration polarization | CatBoost | 0.9900 | 0.0360 |
| | | ANN | 0.9700 | 0.0504 |
| | | RF | 0.8900 | 0.1220 |
| | | XGBoost | 0.8800 | 0.1430 |
| | | Linear Regression | 0.4800 | - |
| Aghilesh et al. (2021) (45) | Water flux | RSM | 0.8529 | 1.0280 |
| | | ANN | 0.7970 | 1.2200 |
| | | ANFIS | 0.7773 | 1.3060 |
| | Reverse solute flux | ANFIS | 0.9427 | 0.6350 |
| | | RSM | 0.9194 | 0.7430 |
| | | ANN | 0.7807 | 1.2340 |
| Hosseinzadeh et al. (2020) (47) | Water flux (TFC membrane) | ANFIS | 0.9755 | 0.2527 |
| | | ANN | 0.9404 | 0.4049 |
| | Water flux (CTA membrane) | ANFIS | 0.9861 | 0.1230 |
| | | ANN | 0.9817 | 0.1449 |
| Chang et al. (2022) (48) | Water flux | RF | 0.9870 | 0.0440 |
| | Total dissolved solids | XGBoost | 0.9700 | 0.2340 |

RSM: response surface methodology; ANFIS: adaptive neuro-fuzzy interference system; CatBoost: categorical boosting; ANN: artificial neural networks; RF: random forest; XGBoost: extreme gradient boosting; TFC: thin-film composite; CTA: cellulose triacetate

#### 6.1.4.2 AI Techniques for Membrane Fouling Prediction and Control

Membrane fouling remains a significant challenge in FO processes, leading to reduced performance and increased operational costs. AI techniques offer promising solutions for fouling prediction and control.

##### *6.1.4.2.1 Fouling Prediction Using ANNs*

ANNs have been employed to predict fouling behaviour by analysing operational data and membrane characteristics. These models can forecast the onset and progression of fouling, enabling proactive maintenance strategies. For example, ANNs have been used to predict membrane fouling in FO systems, demonstrating high prediction accuracy and supporting improved resource recovery from wastewater (49).

##### *6.1.4.2.2 Deep Learning and Imaging Techniques*

Deep learning models, particularly convolutional neural networks (CNNs), have been applied to analyse imaging data for fouling detection. OCT images provide detailed insights into membrane surface conditions, and CNNs can process these images to identify fouling patterns. A hybrid model combining deep learning with OCT imaging was developed to predict fouling in UF membranes, offering potential for adaptation to FO systems (50).

#### 6.1.4.3 Real-Time Monitoring and Control

In FO systems, early control is essential for membrane decision-making. Issues such as membrane fouling may have serious adverse consequences in terms of cost, performance, and sustainability of FO systems (51). Such a phenomenon as membrane fouling is dynamic in nature, thereby requiring continuous monitoring of the process. The traditional models are often based on simplified assumptions related to the fouling mechanism and fail to accurately predict it in both the short and long terms (52). Arguably, AI models are better equipped to handle this problem due to their data-driven nature and flexibility. Therefore, several studies of FO systems explored the process monitoring and control topic as well.

Real-time monitoring of membrane fouling was investigated using ANN models to enhance predictive accuracy and process control (53). In both cases, porosity, thickness, roughness, and density of an FO membrane were the key fouling characteristics examined. In the first study, OCT was used to have real-time image-based monitoring of the system. The applied ANN model demonstrated good predictive values across all parameters (including porosity $R^2 = 0.86$, thickness $R^2 = 0.90$, roughness $R^2 = 0.92$, and density $R^2 = 0.90$). In the second study, membrane fouling monitoring was supplemented with prediction and control of water flux and pollutant removal. With an improved ANN protocol, stronger prediction results were obtained for the fouling characteristics (porosity $R^2 = 0.85$, thickness $R^2 = 0.97$, roughness $R^2 = 0.97$, and density $R^2 = 0.98$). In addition, the model produced an excellent prediction for water flux ($R^2 = 0.92$) and a good prediction for contaminant removal ($R^2 = 0.87$). In both cases, RMSE was lower than 3%.

Early control and monitoring of the FO process have also been investigated using principal component analysis, which was applied to characterize and control membrane fouling based on seven identified fouling component inputs (54). The model was deemed reliable in identifying the underlying fouling factors and, therefore, controlling the fouling process. The study also discusses using an ANN model for membrane fouling in an OMBR system. The model showed a good predictive value for membrane fouling ($R^2 = 0.92$), water flux ($R^2 = 0.93$), and contaminant removal ($R^2 = 0.98$), thereby rendering the technique suitable for early control prediction in such systems.

## 6.2 BRIDGING INNOVATION AND ADOPTION IN FORWARD OSMOSIS: MARKET TRENDS, VIABILITY, AND SAFETY

### 6.2.1 Potential Applications

The future of FO technology in the food processing industry is poised for growth, driven by the expanding need for sustainable solutions and energy-efficient processes. Below are some of the key potential applications of FO:

1 Fermented Beverages and Plant-Based Beverages: The growing demand for plant-based beverages (like soy milk, almond milk, and oat milk) and fermented beverages (such as kombucha and kefir) creates an opportunity for FO in concentrating these products without exposing them to high temperatures. The low-temperature operation of FO preserved the sensory and nutritional qualities of these beverages, which are often sensitive to heat.

In terms of benefits, FO minimizes thermal degradation, retaining vitamins, enzymes, and probiotics essential for human health. The gentle processing maintains the natural flavours and aromas crucial for consumer acceptance. Furthermore, by reducing water activity, FO can enhance the microbial stability of the product.

2 Recovery of Valuable Nutrients from Dairy Side Streams: In the context of dairy processing, FO is applied to concentrate whey proteins, lactose, and minerals from dairy by-products. For instance, studies have demonstrated the effectiveness of FO in concentrating whey protein solutions, achieving higher concentrations with lower energy consumption compared to traditional methods. By concentrating valuable components from dairy by-products, FO transforms waste streams into valuable ingredients for food and animal feed applications. This not only reduces waste but also creates new revenue streams for dairy processors. Thereby, by minimizing waste, FO supports sustainable dairy processing practices. Furthermore, the recovery of proteins, lactose, and minerals through FO can lead to the development of high-value products such as WPCs, lactose powders, and mineral supplements. These products have significant market demand, offering economic benefits to dairy processors.

The application of FO in the recovery of valuable nutrients from

3 Low-alcohol and alcohol-free beverage processing
4 Cold brew and speciality coffee concentrates.

### 6.2.2 Market Trends and Economic Viability

The market for FO technology is expanding rapidly, driven by its advantages in energy savings, water recovery, and product quality preservation. As food processors continue to embrace sustainability practices, the demand for FO-based solutions is expected to rise.

As technological advancements continue to improve FO system efficiency, the cost of implementation will decrease. Reduced membrane fouling and improved DS recovery technologies are expected to make FO even more economically viable.

### 6.2.3 Life Cycle Assessments

LCA is a method for the environmental assessment of products and services, covering their life cycle from raw material extraction to waste management (including disposal and recycling). LCA is a comprehensive assessment that takes a product life cycle perspective and covers a range of environmental impacts. These unique features of LCA facilitate avoiding problem shifting from one life cycle stage to another stage or from one environmental impact to another impact. LCA is a method of quantification normalized by two agencies, namely the Society of Environmental Toxicology and Chemistry (SETAC) and the United Nations Environment Programme (UNEP), under the standard ISO 14044. It is typically broken into the following stages (Figure 6.3):

1 Goal and scope definition aim to define how big a part of the product life cycle will be taken in assessment and to what end the assessment will be serving. This criterion serving for system comparison and specific times is described in this step.
2 Life cycle inventory (LCI)—In this step, inventory analysis provides a description of material and energy flows within the product system and especially its interaction with the environment, consumed raw materials,

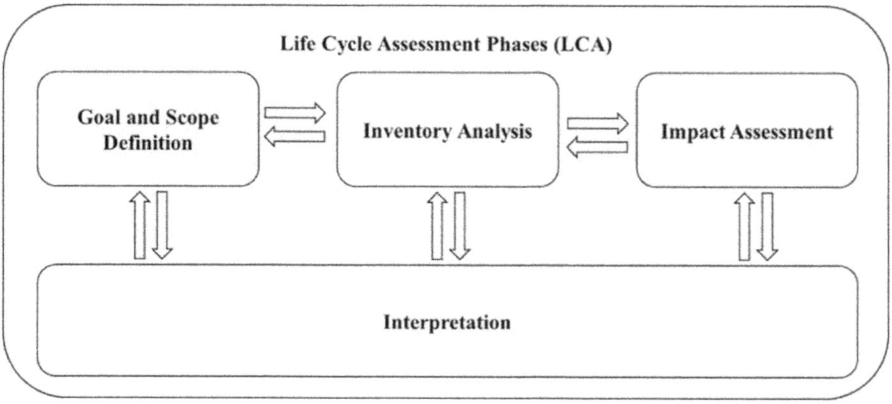


**FIGURE 6.3** Life cycle assessment (LCA) phases (ISO 14044)

and emission to the environment, followed by all important processes, subsidiary energy, and material flows.
3. Life cycle impact assessment (LCIA)—Details from inventory analysis serve for impact assessment. The indicator results of all impact categories are detailed in this step; the importance of every impact category is assessed by normalization and eventually also by weighting.
4. Interpretation—Interpretation of a life cycle involves critical review, determination of data sensitivity, and result presentation.

LCA methodology can be categorized into three levels based on technological details (55):

1. Conceptual LCA—the first level of LCA is based on limited environmental aspects of a few life cycle stages, where there is still some improvement potential existing for the manufacturers. These results might be useful for qualitative reporting of assessment results but are not suitable for corporate marketing or explicit publications of the LCA study.
2. Simplified LCA—This is the type of comprehensive assessment using generic datasets covering the whole life cycle of a product/system of processes. This time required and expenditures as well have reduced significantly here, which is a significant difference from the detailed LCA. This consists of a screening of life cycle stages, simplification of LCA results for future recommendations, and assuring the reliability of analysis results. This is often termed as "Streamlined LCA".
3. Detailed LCA—This type of LCA is a comprehensive one with full consideration of each life cycle stage with specific datasets and analysed in detail for further process improvement.

Furthermore, to assess the environmental impact of FO in food processing industries, it is necessary to define the life cycle stages of a typical FO system. The stages include the following:

1. Raw material acquisition—The production of polymeric materials like PSf, PA, and CA used for the fabrication of FO membranes consumes significant energy and raw materials. The process involved in membrane fabrication, such as IP, requires chemicals, energy, and water. The environmental impact during this stage is influenced by the type of materials used such as solvents and monomers. Furthermore, some FO membranes are modified with nanomaterials (such as carbon nanotubes and GO) to enhance membrane performance. The extraction and manufacturing of these materials also add to the environmental load.
2. FO system operation—The FO systems are energy-efficient, requiring lower pressures than technologies like RO or evaporation. The operational requirements in FO include the energy required to maintain osmotic pressure and regenerate the DSs. Typically, FO requires one-third to half the energy required for RO or a similar process. The FO system recovers water

from FSs (such as liquid food and beverages). This process can reduce the overall water footprint in food processing, but the water recovery rate and waste generation still need to be considered in LCA. Additionally, the regeneration and recovery of DSs play a significant role in FO system performance and environmental impact. Regenerable DSs have a lower environmental impact due to less frequent replacements.

3 End-of-life disposal or recycling—Similar to any membrane, FO membranes also have a finite lifespan, and their disposal or recycling is an important consideration. Polymeric membranes can be difficult to recycle, leading to landfill waste if not managed properly. The proper disposal of used DSs is critical, especially in food processing applications where safety and non-toxicity are the priorities. Some DSs can be regenerated and reused, reducing the waste and environmental impact.

Furthermore, once the life cycle stages are identified, the next step in an LCA is to access the environmental impacts associated with each phase. The following are the key categories that need to be considered when analysing FO technology.

1 Energy consumption and global warming potential—The energy use in the FO system is generally lower than that of RO, due to FO's low-pressure operation. Conventionally, a lower energy requirement means reduced global warming potential (GWP), which translates into a smaller carbon footprint in food processing. The LCA studies of FO revealed that it can be up to 50% more energy-efficient than traditional methods, especially for the concentration of heat-sensitive products.
2 Water usage—the FO system recovers a high volume of water, which can be reused in food processing operations, lowering the overall water footprint of the industry. This is particularly important for water-intensive industries such as beverage and dairy processing industries.
3 Waste generation and treatment—Waste minimization is a crucial aspect of sustainability. Unlike the traditional thermal evaporation process, this generates a significant amount of wastewater and solid waste. FO generates less waste, and when integrated with other processes (such as MD), it offers an opportunity for ZLD. Proper waste treatment and DS recovery/regeneration can reduce the overall environmental impact of the FO process.
4 Fouling and membrane lifetime—Membrane fouling reduces the overall operational efficiency of FO systems and contributes to the need for frequent maintenance, cleaning, and eventual membrane replacement. Innovations in anti-fouling membranes and membrane regeneration can help to mitigate the environmental impact of frequent membrane disposal and reduce the need for chemical cleaning agents.
5 Carbon footprint—The carbon footprint of FO systems is a direct result of their energy consumption. The use of renewable energy sources and the ongoing development of more efficient FO systems will further reduce the overall carbon footprint of the food processing operations.

Overall, LCA provides a valuable insight into the environmental performance of the FO process in the food and beverage processing industries. This analysis confirms that FO offers substantial energy saving, water recovery, and reduced waste generation compared to traditional methods. As the food processing industries continue to focus on sustainability, energy efficiency, and resource conservation, FO has the potential to play a leading role in transforming food manufacturing processes into more environmentally friendly and resource-efficient operations.

### 6.2.4 Regulatory and Safety Considerations

FO is already being applied in food processing for concentrating fruit juices, milk, whey, and coffee extracts while preserving sensitive flavours and nutrients. To enable its transition from promising demonstrations to widespread industrial adoption, strict compliance with food-grade regulatory and safety standards is essential. The food industry operates under stringent regulations to safeguard consumer health, preserve product quality, and minimize environmental impact. Similarly, the technology deployed in food processing must adhere to safety standards that govern food-grade materials, processes, and waste management.

This section explores the key regulatory requirements and safety considerations that govern the use of FO technology in food processing, focusing primarily on food safety regulations, DS safety, membrane material regulations, and the environmental impact of FO-based processes. Additionally, we will highlight the importance of compliance with international standards, including those set by regulatory bodies such as the US Food and Drug Administration (USFDA) and European Food Safety Authorities (EFSA), and other global health and safety agencies.

#### 6.2.4.1 Food Safety Regulations in FO Systems

1 Membrane material—The membrane materials used in FO systems must meet strict food-grade standards to ensure that no harmful substances leach into food products. This involves careful selection of raw materials, solvents, and monomers that do not contain harmful additives or impurities.

Regulatory bodies (such as the FDA and EFSA) assess the toxicity and safety of materials for food contact applications. For instance, in the United States, the Food and Drug Administration (FDA) regulates food contact substances under Title 21 CFR Part 177, which outlines guidelines for materials and additives used in direct food contact. Similarly, the EFSA oversees food contact materials under regulation (EC) No 1935/2004, ensuring that materials do not transfer constituents to food in quantities that could endanger human health.

The membrane materials should be inert, non-toxic, and resistant to degradation under processing conditions. Commonly used materials include CA (biocompatible and hydrophilic in nature), PA (offers high selectivity and chemical resistance), and PS (provides mechanical strength and thermal stability). After selection of membrane materials, the manufacturers must conduct migration testing to ensure that any potential leachables from the membrane materials remain within the limits set by regulatory bodies. This step involves simulating processing conditions and analysing the extent of substance transfer to food simulants.

2 DS—The DS plays a critical role in driving the osmosis process by generating the osmotic pressure needed to move water through the FO membrane. However, the DS must meet safety standards to ensure it does not contaminate the food product during the separation process. The application of DS such as sodium chloride, sucrose, citric acid, and food preservatives (including potassium sorbate and sodium benzoate) is commonly considered food-safe due to their non-toxic nature.

In food processing, it is essential to use food-safe DSs that do not contain any toxic substances. For instance, sodium chloride (NaCl), sucrose, and citric acid are commonly used in food-grade applications due to their non-toxic nature. The use of non-food-safe chemicals (e.g. ammonium-based compounds or synthetic polymers) could lead to contamination and pose a risk to consumer health. Furthermore, efficient regeneration of DS ensures improved FO performance. However, regulatory bodies are required to ensure that the regeneration process is safe and energy-efficient, while the regenerated DS is non-toxic and safe for food contact.

(iii) Compliance with Good Manufacturing Practices

The integration of FO technology into food production must adhere to Good Manufacturing Practices (GMP), which are guidelines set by regulatory bodies to ensure food safety and quality. GMP standards ensure that the equipment used in FO systems must be sanitary and maintained to prevent contamination. This includes the following:

- Cleaned in place (CIP) systems: automated cleaning processes that ensure thorough sanitation without disassembly.
- Material compatibility—Equipment materials should withstand cleaning agents and not corrode or degrade.

Operators should receive comprehensive training on hygiene practices, equipment operation, and emergency procedures to maintain a contamination-free environment. Maintaining detailed records of processing parameters, cleaning schedules, and maintenance activities is essential for traceability and regulatory compliance. This documentation supports audits and facilitates quick responses to any food safety incidents. In the United States, adherence to 21 CFR Part 117 (current good manufacturing practice, hazard analysis, and risk-based preventive controls for human food) is mandatory. The FDA's Food Safety Modernization Act (FSMA) further emphasizes preventive controls and risk-based approaches to food safety. In the EU, compliance with Regulation (EC) No. 852/2004 on the hygiene of foodstuffs is required.

Overall, implementing FO technology in juice and milk concentration processes offers numerous benefits but must be approached with a comprehensive understanding of food safety regulations. Ensuring membrane material safety, using approved DSs, and adhering to GMP are critical components in safeguarding consumer health and maintaining product quality. Continuous monitoring, validation, and adherence to regulatory standards are essential for the successful integration of FO systems in the food industry.

#### 6.2.4.2 Regulatory Considerations for Environmental Impact

In addition to food safety, environmental considerations are a key aspect of the regulatory landscape for FO technology in food processing. Environmental regulations focus on waste management, water recovery, and ZLD systems to minimize the ecological impact of food manufacturing processes.

##### *6.2.4.2.1 Zero-Liquid Discharge and Waste Minimization*

One of the benefits of FO technology in food processing is its ability to achieve ZLD, where all the water recovered from the FS is reused, and no liquid waste is produced. Regulations governing ZLD systems require food processors to optimize wastewater treatment and water reuse systems to reduce environmental pollution. For instance, in the United States, the Environmental Protection Agency (EPA) has stringent discharge standards for industrial wastewater, particularly in the food industry, where high organic content in wastewater can lead to severe environmental pollution. The EPA's effluent guidelines set national standards for wastewater discharge to surface waters and municipal sewage treatment plants. Similarly, the European Union's Water Framework Directive (WFD) requires food manufacturers to meet water quality standards and reduce pollutants in wastewater. The WFD focuses on reducing and removing pollution and ensuring sufficient water to support both wildlife and human needs.

FO-based systems that minimize water waste and help with water recovery are becoming a critical part of food processing facilities' efforts to meet these standards.

##### *6.2.4.2.2 Regulatory Compliance for Membrane Disposal and Recycling*

The disposal of membranes and DSs at the end of their life cycle is another critical aspect of environmental regulations in FO systems. Membrane waste from FO systems typically contains materials that may not be easily biodegradable, raising concerns about landfill waste.

The regulatory frameworks require that the membrane disposal adhere to waste management protocols that minimize environmental impact. This includes efforts to recycle and repurpose used FO membranes or incorporate biodegradable materials in membrane production. LCA has been proposed to monitor and control the environmental impact by selecting green operation strategies in different processes of membrane production and utilization.

Also, DSs must be disposed of or regenerated in a way that minimizes environmental contamination. Regulations mandate the safe disposal of used DSs in a manner that adheres to local environmental protection laws. For instance, the EPA emphasizes the importance of effective treatment technologies and strategies such as zero-effluent discharge approaches to reduce the impact of waste streams.

#### 6.2.4.3 Safety Standards in Food Processing with FO Technology

The safety of FO systems in food processing extends beyond just food and environmental safety; it also involves system design, operation, and maintenance. Food safety protocols must be strictly followed to ensure that FO systems operate within safe and sanitary conditions. This section delves into the critical components of safety standards in FO systems within the food processing industries.

1 Sanitation and hygiene protocols—The design of FO systems must incorporate sanitation standards to avoid contamination risks. This includes system cleaning and material selection. The regular cleaning protocols should be established to ensure that the system remains sanitary and that contaminants do not accumulate within the membranes or piping. Furthermore, the materials used in the system construction, such as membranes, pumps, and piping, should be made from food-grade, non-reactive materials that do not leach harmful substances. The equipment should be designed to facilitate easy cleaning and prevent microbial harbourage. Smooth surfaces with minimal crevices and appropriate drainage are some essential features.

Implementing a CIP system allows for efficient and effective cleaning of equipment without disassembly, ensuring consistent sanitation. Regular sanitization using approved agents helps in controlling microbial growth. It's crucial to validate the effectiveness of cleaning procedures through microbial testing.

Proper ventilation and air filtration systems are necessary to prevent airborne contaminants from compromising product safety. Effective waste disposal systems prevent cross-contamination and maintain a hygienic processing environment.

2 Operator training and safety—The proper training is critical to maintaining safe and efficient FO operations. Food processors need to ensure that workers handling the FO system must be trained on membrane handling/installation (to avoid damage and ensure the integrity of the system), DS management (to ensure safe handling, regeneration, and integrity of the system), and compliance with food safety protocols (to minimize contamination risks during food production).

Operators must be trained in the correct use of FO systems, understanding operational parameters and emergency procedures. The training should emphasize personal hygiene, proper use of protective clothing, and protocols to prevent contamination. Educating staff on Hazard Analysis and Critical Control Point (HACCP) principles ensures they can identify and control potential hazards in the processing line.

The safety protocols including personal protective equipment (PPE), which provides and enforces the use of appropriate PPE minimizes the risk of injury and contamination. Also, training should include emergency response procedures for incidents such as chemical spills or equipment malfunctions. Furthermore, continuous education for the handlers through ongoing training and regular assessment of the operator's performance ensures staff remain updated on best practices and regulatory changes.

3 Membrane fouling control—Membrane fouling is one of the key operational challenges in FO systems. Membrane fouling is one of the key operational challenges in FO systems, and recognizing the type of fouling is essential for effective control. Fouling reduces membrane efficiency, increases energy consumption, and compromises product quality. After understanding the type of membrane fouling, it is essential to identify the preventive measures. For instance, removing particulates and adjusting pH levels before processing can reduce fouling potential. While maintaining optimal flow rates and pressure minimizes the accumulation of foulants on membrane surfaces, establishing routine cleaning schedules prevents the build-up of contaminants. Selecting appropriate cleaning agents based on the type of fouling ensures effective membrane maintenance. Furthermore, tracking parameters such as flux rates and pressure drops helps in early detection of fouling.

The integration of FO technology in food processing industries offers significant benefits, including energy efficiency, water recovery, and improved product quality. However, its successful adoption hinges on compliance with food safety regulations, environmental standards, and safety protocols. By ensuring that membrane materials, DSs, and system designs meet stringent safety standards, food processors can safely utilize FO technology while achieving sustainability goals and maintaining product quality. As the industry moves towards more sustainable and resource-efficient food production practices, FO technology is well-positioned to play a pivotal role. However, careful attention to regulatory and safety considerations will be crucial in ensuring that the food industry fully realizes the benefits of this promising technology.

## REFERENCES

1. Yu F, Shi H, Shi J, Teng K, Xu Z, Qian X. High-performance forward osmosis membrane with ultra-fast water transport channel and ultra-thin polyamide layer. J Memb Sci. 2020 Dec;616:118611.
2. Wei Y, Yang Z, Wang L, Yu Y, Yang H, Jin H, et al. Facile ZIF–8 nanocrystals interlayered solvent–resistant thin–film nanocomposite membranes for enhanced solvent permeance and rejection. J Memb Sci. 2021 Oct;636:119586.
3. Yan M, Xi Y, Jiang N, Li Q, Zheng S, Hu Y, et al. High-performance thin film composite forward osmosis membrane for efficient rejection of antimony and phenol from wastewater: Characterization, performance, and MD-DFT simulation. J Memb Sci. 2024 Jun;703:122847.
4. Kumar P, Nandha N, Mondal M, Singh PS. Natural anthocyanin dye concentration through ultrathin polyester-amide hollow fiber forward osmosis membrane. Chem Eng J. 2025 Jul;516:163951.
5. Kim T, Park S, Lee Y, Kim J, Kim K. Thin film biocomposite membrane for forward osmosis supported by eggshell membrane. Membranes (Basel). 2022 Jan 30;12(2):166.

6. Xu D, Sun X, Wu S. A comprehensive review of the forward osmosis membrane modification strategies for membrane fouling mitigation in wastewater treatment. J Environ Chem Eng. 2025 Jun;13(3):116992.
7. Pejman M, Firouzjaei MD, Aktij SA, Das P, Zolghadr E, Jafarian H, et al. Improved antifouling and antibacterial properties of forward osmosis membranes through surface modification with zwitterions and silver-based metal organic frameworks. J Memb Sci. 2020 Oct;611:118352.
8. Chiao YH, Chen ST, Patra T, Hsu CH, Sengupta A, Hung WS, et al. Zwitterionic forward osmosis membrane modified by fast second interfacial polymerization with enhanced antifouling and antimicrobial properties for produced water pretreatment. Desalination. 2019 Nov;469:114090.
9. Eghbalazar T, Shakeri A. High-performance thin-film nanocomposite forward osmosis membranes modified with Poly(dopamine) coated UiO66-(COOH)2. Sep Purif Technol. 2021 Dec;277:119438.
10. Ghorbani F, Shakeri A, Ali Vafaei M, Salehi H. Polyoxometalate-cored supramolecular star polymers as a novel crosslinker for graphene oxide-based forward osmosis membranes: Anti-fouling, super hydrophilic and high water permeable. Sep Purif Technol. 2021 Jul;267:118578.
11. Sabry R, Ali HM, Ali SS, Gadallah H. Superior forward osmosis cellulosic membrane for water desalination and brine concentration. Sci Rep. 2025 May 14;15(1):16829.
12. Liu S, Tong X, Liu S, An D, Yan J, Chen Y, et al. Multi-functional tannic acid (TA)-Ferric complex coating for forward osmosis membrane with enhanced micropollutant removal and antifouling property. J Memb Sci. 2021 May;626:119171.
13. Akther N, Ali SM, Phuntsho S, Shon H. Surface modification of thin-film composite forward osmosis membranes with polyvinyl alcohol–graphene oxide composite hydrogels for antifouling properties. Desalination. 2020 Oct;491:114591.
14. Huang W, Zhao L, Zhang J, Wen H, Zhu Z, Wang P, et al. Template-etched sodium alginate hydrogel as the sublayer to improve the FO performance with double barriers for high metal ion rejection. Chem Eng J. 2021 Jun;413:127425.
15. Wen H, Xu N, Soyekwo F, Dou P, Liu C. Towards enhanced performance of fertilizer-drawn forward osmosis process coupled with sludge thickening using a thin-film nanocomposite membrane interlayered with Mxene scaffolded alginate hydrogel. J Memb Sci. 2023 Nov;685:121899.
16. Wen H, Liu C. Effect of the interlayer construction on the performances of the TFC-FO membranes: A review from materials perspective. Desalination. 2022 Nov;541:116033.
17. Giagnorio M, Tanis B, Hélix-Nielsen C, Aschmoneit FJ. Influence of porous support structure and the possible presence of active layer defects on FO membrane behaviour. Sep Purif Technol. 2024 Feb;330:125182.
18. Shen L, Wang F, Tian L, Zhang X, Ding C, Wang Y. High-performance thin-film composite membranes with surface functionalization by organic phosphonic acids. J Memb Sci. 2018 Oct;563:284–97.
19. Ghorbani F, Shakeri A, Ali Vafaei M, Salehi H. Polyoxometalate-cored supramolecular star polymers as a novel crosslinker for graphene oxide-based forward osmosis membranes: Anti-fouling, super hydrophilic and high water permeable. Sep Purif Technol. 2021 Jul;267:118578.
20. Zou WS, Wu P, Xiao B, Chen X, Kong W, Li Y, et al. Integration of nitrogen-doped carbon dots onto active layer of forward osmosis membrane for highly efficient antibacteria and enhanced membrane performances. J Environ Chem Eng. 2023 Apr;11(2):109468.
21. Seyedpour SF, Rahimpour A, Najafpour G. Facile in-situ assembly of silver-based MOFs to surface functionalization of TFC membrane: A novel approach toward long-lasting biofouling mitigation. J Memb Sci [Internet]. 2019;573:257–69. Available from: https://www.sciencedirect.com/science/article/pii/S0376738818330199

22. Zhang X, Cui HM, Gao Y, Yan ZW, Yan X, Chen Y, et al. Surfactant-induced intervention in interfacial polymerization to develop highly-selective thin-film composite membrane for forward osmosis process. Desalination. 2023 Jul;558:116617.
23. Ang WL, Mohammad AW, Johnson D, Hilal N. Unlocking the application potential of forward osmosis through integrated/hybrid process. Sci Total Environ. 2020 Mar;706:136047.
24. An X, Hu Y, Wang N, Zhou Z, Liu Z. Continuous juice concentration by integrating forward osmosis with membrane distillation using potassium sorbate preservative as a draw solute. J Memb Sci [Internet]. 2019;573(December 2018):192–9. Available from: https://doi.org/10.1016/j.memsci.2018.12.010
25. Akhtar A, Singh M, Subbiah S, Mohanty K. Sugarcane juice concentration using a novel aquaporin hollow fiber forward osmosis membrane. Food and Bioproducts Processing. 2021 Mar;126:195–206.
26. Menchik P, Moraru CI. Nonthermal concentration of liquid foods by a combination of reverse osmosis and forward osmosis. Acid whey: A case study. J Food Eng. 2019 Jul;253:40–8.
27. Qasim M, Darwish NA, Sarp S, Hilal N. Water desalination by forward (direct) osmosis phenomenon: A comprehensive review. Desalination. 2015 Oct;374:47–69.
28. Lambrechts R, Sheldon MS. Performance and energy consumption evaluation of a fertiliser drawn forward osmosis (FDFO) system for water recovery from brackish water. Desalination. 2019 Apr;456:64–73.
29. Lee DJ, Hsieh MH. Forward osmosis membrane processes for wastewater bioremediation: Research needs. Bioresour Technol. 2019 Oct;290:121795.
30. Trapote A, Albaladejo A, Simón P. Energy consumption in an urban wastewater treatment plant: the case of Murcia Region (Spain). Civil Engineering and Environmental Systems. 2014 Oct 2;31(4):304–10.
31. Xiao K, Liang S, Wang X, Chen C, Huang X. Current state and challenges of full-scale membrane bioreactor applications: A critical review. Bioresour Technol. 2019 Jan;271:473–81.
32. Suwaileh W, Johnson D, Jones D, Hilal N. An integrated fertilizer driven forward osmosis-renewables powered membrane distillation system for brackish water desalination: A combined experimental and theoretical approach. Desalination. 2019 Dec;471: 114126.
33. Liang YY. Solar-powered forward osmosis as a sustainable water treatment solution: A review. J Environ Chem Eng. 2025 Jun;13(3):116332.
34. Zhang C, Lu Y. Study on artificial intelligence: The state of the art and future prospects. J Ind Inf Integr. 2021 Sep;23:100224.
35. Pardeshi PM, Mungray AA, Mungray AK. Determination of optimum conditions in forward osmosis using a combined Taguchi–neural approach. Chem Eng Res Des. 2016 May;109:215–25.
36. Nam K, Heo S, Loy-Benitez J, Ifaei P, Yoo C. An autonomous operational trajectory searching system for an economic and environmental membrane bioreactor plant using deep reinforcement learning. Water Sci Technol. 2020 Apr 15;81(8):1578–87.
37. Arjmandi M, Peyravi M, Altaee A, Arjmandi A, Pourafshari Chenar M, Jahanshahi M, et al. A state-of-the-art protocol to minimize the internal concentration polarization in forward osmosis membranes. Desalination. 2020 Apr;480:114355.
38. Jawad J, Hawari AH, Zaidi S. Modeling of forward osmosis process using artificial neural networks (ANN) to predict the permeate flux. Desalination. 2020 Jun;484: 114427.
39. Kim J, Jeon J, Kim N, Kim S. Predicting flux of forward osmosis membrane module using deep learning. J Korean Soc Water and Wastewater. 2021 Feb 28;35(1):93–100.

40. Lu YX, Yuan H, Shao Y, Chand H, Wu Y, Yang YL, et al. Shedding light on the transfer of tetracycline in forward osmosis through experimental investigation and machine learning modeling. Chemosphere. 2023 Apr;319:137959.
41. Han H, Zhang S, Qiao J, Wang X. An intelligent detecting system for permeability prediction of MBR. Water Sci Technol. 2018 Jan 29;77(2):467–78.
42. Viet ND, Im SJ, Kim CM, Jang A. An osmotic membrane bioreactor–clarifier system with a deep learning model for simultaneous reduction of salt accumulation and membrane fouling. Chemosphere. 2021 Jun;272:129872.
43. Ibrar I, Yadav S, Braytee A, Altaee A, HosseinZadeh A, Samal AK, et al. Evaluation of machine learning algorithms to predict internal concentration polarization in forward osmosis. J Memb Sci. 2022 Mar;646:120257.
44. Viet ND, Jang A. Machine learning-based real-time prediction of micropollutant behaviour in forward osmosis membrane (waste)water treatment. J Clean Prod. 2023 Feb;389:136023.
45. Aghilesh K, Mungray A, Agarwal S, Ali J, Chandra Garg M. Performance optimisation of forward-osmosis membrane system using machine learning for the treatment of textile industry wastewater. J Clean Prod. 2021 Mar;289:125690.
46. Jawad J, Hawari A, Zaidi S. Modeling and sensitivity analysis of the forward osmosis process to predict membrane flux using a novel combination of neural network and response surface methodology techniques. Membranes (Basel). 2021 Jan 19;11(1):70.
47. Hosseinzadeh A, Zhou JL, Altaee A, Baziar M, Li X. Modeling water flux in osmotic membrane bioreactor by adaptive network-based fuzzy inference system and artificial neural network. Bioresour Technol. 2020 Aug;310:123391.
48. Chang HM, Xu Y, Chen SS, He Z. Enhanced understanding of osmotic membrane bioreactors through machine learning modeling of water flux and salinity. Sci Total Environ. 2022 Sep;838:156009.
49. Kazim H, Sabri M, Al-Othman A, Tawalbeh M. Artificial intelligence application in membrane processes and prediction of fouling for better resource recovery. J Res Rec. 2023 Apr 1;1(2):e703895. https://doi.org/10.52547/jrr.2303.1008
50. Chew CM, Aroua MK, Hussain MA. A practical hybrid modelling approach for the prediction of potential fouling parameters in ultrafiltration membrane water treatment plant. J Ind Engin Chem. 2017 Jan;45:145–55.
51. Behnam P, Faegh M, Khiadani M. A review on state-of-the-art applications of data-driven methods in desalination systems. Desalination. 2022 Jun;532:115744.
52. Mittal S, Gupta A, Srivastava S, Jain M. Artificial neural network based modeling of the vacuum membrane distillation process: Effects of operating parameters on membrane fouling. Chem Eng Process—Process Intensif. 2021 Jul;164:108403.
53. Im SJ, Nguyen VD, Jang A. Prediction of forward osmosis membrane engineering factors using artificial intelligence approach. J Environ Manage. 2022 Sep;318:115544.
54. Duc Viet N, Im SJ, Jang A. Characterization and control of membrane fouling during dewatering of activated sludge using a thin film composite forward osmosis membrane. J Hazard Mater. 2020 Sep;396:122736.
55. Farjana SH, Mahmud MAP, Huda N. Introduction to life cycle assessment. In: Life cycle assessment for sustainable mining. Elsevier; 2021. p. 1–13.

# 7 Conclusions

Over the past few decades, the food and dairy processing industries are undergoing a profound transformation, driven by an urgent need to adopt sustainable, energy-efficient, and non-thermal technologies. With this evolving landscape, FO has emerged as a pivotal innovation. Leveraging osmotic gradient instead of external pressure or heat, FO enables gentle, selective water removal—making it especially suited for sensitive food products such as juices, milk, whey, and functional beverages.

This book systematically examined the science, engineering, economics, and industrial relevance of FO, providing a panoramic view of its potential as a next-generation separation technique. The insights presented across the previous chapters collectively highlight that FO is no longer a conceptual novelty but a maturing technology with real-world applications and proven benefits.

In terms of key technological advancements, a major driver of the FO process has been the evolution of membrane materials including TFC membranes (offering high water permeability and reasonable solute rejection), biomimetic membranes (mimicking nature's selective water channels and achieving exceptional flux and selectivity), and zwitterionic and MOF-based membranes (enhanced anti-fouling performance and hydrophilicity crucial for long-term application). These advancements reflect a paradigm shift from purely mechanical membranes to smart, functional, and selective interfaces. In recent times, along with membrane materials, a recent shift has been observed in the development of responsive draw solutes. These smart solutes enhance energy recovery and simplify regeneration, aligning with green chemistry principles. Furthermore, a comprehensive mathematical model and computational simulations have improved FO system design. The modelling enables scale-up feasibility, customized configurations for various food matrices, and predictive control, especially when combined with AI technology.

In the food and beverage industry, the scope of FO extends beyond juice concentration to include dairy products, functional beverages, and plant extracts. Due to their mild operating condition, FO has enabled the recovery of polyphenols, protein, probiotics, and natural pigments from food processing wastewater streams. These applications show not only technical feasibility but also market relevance in meeting consumer demand for clean-label, nutrient-rich, and minimally processed foods. Despite these advantages, there are a few challenges spanning technical, economic, and regulatory domains. Table 7.1 presents a consolidated overview of these issues and proposed solutions (1). This strategic road map highlights that FO innovation is multidimensional, and its advancement depends on interdisciplinary solutions, scalable manufacturing, and strong policy support.

The continued development and implementation of FO in the liquid food and beverage processing sector will require targeted efforts in the following areas, including

- Membrane Material Research: Development of membranes with higher flux, enhanced selectivity, and lower fouling tendencies, using green and scalable fabrication techniques.


DOI: 10.1201/9781003657873-7

**TABLE 7.1**
**Summary of key challenges and future directions in forward osmosis technology (1)**

| Category | Key Challenge | Future Directions |
|---|---|---|
| Technical challenges | • ICP reduces osmotic efficiency | • Advanced manufacturing (3D printing and sacrificial templates) to optimize pore structure<br>• Self-healing or photocatalytic materials for fouling control |
| | • Membrane fouling from organic/inorganic contaminants | • Stimuli-responsive "smart" membranes for adaptive self-cleaning |
| Material limitations | • Trade-offs among permeability, selectivity, and mechanical strength<br>• High costs and scalability issues with advanced nanomaterials (e.g. graphene). | • Hybrid materials (MOF–polymer composites and COF) to decouple performance metrics<br>• Green synthesis (bio-derived nanomaterials) |
| Economic barriers | • High capital/operational costs compared to RO | Modular, plug-and-play FO units for niche markets (e.g. rural water supply)<br>Hybrid systems (FO–RO and FO–MD) to reduce energy and complexity |
| Environmental concerns | • Nanoparticle leaching and non-biodegradable membranes | • Biodegradable materials (PLA and chitosan)<br>• Circular design principles and life cycle assessments (LCAs) |
| Standardization gaps | • Disparate testing protocols hinder benchmarking | • Universal metrics (e.g. specific energy consumption) and ASTM/ISO standards |
| Future innovations | | • Hybrid systems: Integration with renewable energy (solar–thermal) and complementary technologies (e.g. anaerobic digestion)<br>• Smart membranes: AI-driven optimization and real-time adaptability<br>• Sustainability: Policy incentives for ZLD compliance and global collaborations |

ICP: internal concentration polarization; MOF: metal–organic framework; COF: covalent organic framework; FO: forward osmosis; RO: reverse osmosis; MD: membrane distillation; PLA: polylactic acid; ASTM: American Society for Testing and Materials; ISO: International Organization for Standardization; AI: artificial intelligence; ZLD: zero-liquid discharge

- Draw Solute Optimization: Innovation in non-toxic, low-cost, and responsive solutes with facile regeneration mechanisms that require minimal energy input.
- System Design and Hybridization: Increased emphasis on integrating FO with existing thermal, membrane, and biological processes to enhance system-level efficiency and adaptability.

- Digitization and AI Integration: Incorporating AI and real-time data analytics to predict membrane performance, optimize operations, and support automated systems.
- Policy and Regulation: Establishing standardized testing protocols, regulatory frameworks, and performance benchmarks to facilitate industry adoption and cross-border technology transfer.

FO has the potential to become a cornerstone technology in future food and beverage manufacturing systems. Its ability to concentrate liquids under ambient conditions, preserve nutritional and sensory quality, and reduce energy consumption aligns well with current trends towards sustainable and value-added processing. With advancements in membrane design, DS engineering, and system integration, FO can serve not only as an energy-efficient concentration method but also as a platform for circular resource recovery and ZLD solutions.

Future progress will depend on collaborative efforts across disciplines—including membrane science, process engineering, materials chemistry, and industrial policy. Through such interdisciplinary collaboration, FO can transition from an emerging technology to a mainstream solution, advancing both the economic competitiveness and environmental sustainability of global food production systems.

## REFERENCE

1. Jamadi-Torab J, Malekzadeh A, Rahbari-Sisakht M, Ismail AF. State-of-the-Art in forward osmosis membrane process: materials, fabrication, and applications. Chem Eng Process—Process Intensif. 2025 May;110385.

# Index

Note: Page locators in *italic* indicate a figure, and page locators in **bold** indicate a table on the corresponding page.

## R



## S



## T



## U



## V



## W

## X



## Z

For Product Safety Concerns and Information please contact our EU representative GPSR@taylorandfrancis.com
Taylor & Francis Verlag GmbH, Kaufingerstraße 24, 80331 München, Germany

www.ingramcontent.com/pod-product-compliance
Ingram Content Group UK Ltd.
Pitfield, Milton Keynes, MK11 3LW, UK
UKHW021506240926
475702UK00035B/499

* 9 7 8 1 0 4 1 1 0 5 2 6 8 *